T0331122

VIBRATION OF AXIALLY LOADED STRUCTURES

This book concerns the vibration and the stability of slender structural components. The loss of stability of structures is an important aspect of structural mechanics and is presented here in terms of dynamic behavior. A variety of structural components are analyzed with a view to predict their response to various (primarily axial) loading conditions. A number of different techniques are presented, with experimental verification from the laboratory. Practical applications are widespread, ranging from cables to space structures. The book presents methods by which the combined effects of vibration and buckling on various structures can be assessed. Vibrations and buckling are usually treated separately, but in this book their influence on each other is examined together, with examples when a combined approach is necessary. The avoidance of instability is the primary goal of this material.

Dr. Lawrence N. Virgin completed his doctorate in structural mechanics in 1986 at University College London. Since 1988, he has been at Duke University, where he teaches and conducts research in engineering mechanics. His interests are centered on the instability behavior of non-linear dynamics systems in the context of experimental vibrations, with applications including aeroelasticity, systems with discontinuities (impact and friction), fluid–structure interaction, and buckling. He is currently Gardner Professor and Chair of the Department of Civil and Environmental Engineering and holds a secondary appointment in the Department of Mechanical Engineering and Materials Science.

Vibration of Axially Loaded Structures

LAWRENCE N. VIRGIN

Duke University

CAMBRIDGE
UNIVERSITY PRESS

Shaftesbury Road, Cambridge CB2 8EA, United Kingdom

One Liberty Plaza, 20th Floor, New York, NY 10006, USA

477 Williamstown Road, Port Melbourne, VIC 3207, Australia

314–321, 3rd Floor, Plot 3, Splendor Forum, Jasola District Centre, New Delhi – 110025, India

103 Penang Road, #05–06/07, Visioncrest Commercial, Singapore 238467

Cambridge University Press is part of Cambridge University Press & Assessment, a department of the University of Cambridge.

We share the University's mission to contribute to society through the pursuit of education, learning and research at the highest international levels of excellence.

www.cambridge.org
Information on this title: www.cambridge.org/9780521880428

First published 2007
First paperback edition 2011

A catalogue record for this publication is available from the British Library

Library of Congress Cataloging-in-Publication data
Virgin, Lawrence N., 1960-
Vibration of axially loaded structures / Lawrence N. Virgin.
    p.   cm.
Includes bibliographical references and index.
ISBN 978-0-521-88042-8 (hardback)
1. Vibration.   2. Structural dynamics.   3. Elastic analysis (Engineering)
I. Title.
TA355.V57   2007
624.1´71–dc22   2007007410

ISBN   978-0-521-88042-8   Hardback
ISBN   978-1-107-40604-9   Paperback

This book is dedicated to my wife Lianne, my children Elliot and Hayley, and my parents Margaret and Alan

Contents

Foreword

The concept of stability is intrinsically a dynamical one. This is recognized even by the simplistic classical definition, which ignores the random disturbances of the real world and just inquires what would happen if a system were displaced to an adjacent position in phase space. So we are lucky, indeed, to have this well-conceived book written by a leading researcher who has mastered both nonlinear dynamics and the static bifurcations of elastic stability theory. The latter theory works well for conservative systems, for which powerful energy theorems are available, but needs augmenting by dynamical methods in the presence of loading that is either nonconservative or time dependent.

Lawrence Virgin has of course just the right background, having chosen (in his usual thoughtful way) to work first at University College London, then with Earl Dowell at Duke University. He is currently the Chair of the Department of Civil and Environmental Engineering at Duke (which has an active interdisciplinary program in nonlinear dynamics) and has enjoyed productive collaborations with Raymond Plaut (Virginia Polytechnic Institute and State University). His previous book, *Introduction to Experimental Nonlinear Dynamics* (also published by Cambridge University Press), brought a welcome sense of realism into the often esoteric field of nonlinear dynamics by focusing on experimental investigations, and I am delighted to see a similar emphasis in this new book titled *Vibration of Axially Loaded Structures*.

Understanding the buckling and vibration of structures under axial compression is of very great importance to structural and aerospace engineers, to whom this book is primarily addressed. They, together with readers from many other areas of mechanics, will be well served by Lawrence's latest offering. The book covers a wide field, including buckling, dynamics (both linear and nonlinear), theory, and experiments, all explained in a clear and lucid style. Especially valuable are the comprehensive lists of references, which nicely complement the text.

I can heartily recommend this book to all who want to see a wide-ranging and scholarly treatment that brings new insights to an important long-standing but still emerging field.

Michael Thompson, FRS
Cambridge, England

Preface

General Comments

Rationale and Scope

- The material covered by this book spans the areas of *vibration* and *buckling*. Both of these areas can be considered as subsets of structural mechanics and play a central role in the disciplines of civil, mechanical, and aerospace engineering.
- Although vibration and buckling are key elements in the teaching of advanced engineering, they are typically taught separately. However, the *interplay* of dynamics and stability in structural mechanics and its coverage in a single text provide an opportunity to present material in an interesting way.
- The quest for stronger, stiffer, and more lightweight structural systems is making the material covered in this book *increasingly* important in practical applications.
- By using axially loaded structures as a consistent theme, the book covers a wide variety of types of structure, methods of analysis, and potential applications without trying to cover too much. Experimental verification appears throughout.
- The level of material is appropriate for upper-level, advanced undergraduate classes, and graduate students, but researchers and practicing engineers will find plenty of interest too.
- The text is liberally illustrated by figures, and close to 500 technical references are given.

Acknowledgments

The material presented in this book contains a synthesis of material from the general literature together with results from my own research program. In terms of the latter, this is by no means a solo endeavor, and there are a number of people I would like to thank.

First, and foremost, much of the work I have conducted in this area in the past 20 years or so has been done with Raymond Plaut from Virginia Polytechnic Institute and State University. I have learned a considerable amount from his deep understanding of theoretical and applied mechanics as well as his attention to detail and meticulous approach to research. He also proofread this book, making useful suggestions and providing invaluable guidance.

My path along this road goes back to Terry Roberts in Cardiff, Michael Thompson in London, and Earl Dowell here in North Carolina. I have benefited immeasurably from their influence as mentors during my formative years (and beyond). In addition to my family, of course, I'd like to thank my friends and colleagues at Duke who have contributed to a supportive environment: Tod Laursen, John Dolbow, Henri Gavin, Ken Hall, Josiah Knight, and Bob Kielb.

I have had the privilege of working with many talented graduate students over a period of almost 20 years, and those whose research contributed directly or indirectly to material in this book include Phil Bayly, Kevin Murphy, Mike Todd, Kara Slade, Hui Chen, David Holland, Mike Hunter, Ilinca Stanciulescu, Sophia Santillan, and Ben Davis (who also diligently proofread the manuscript). Thanks to them all.

Lawrie Virgin
Durham, North Carolina

Plate I. Examples of slender structures in an aerospace context. Top, a solar sail model test; bottom, Predator B unmanned air vehicle. Courtesy of NASA.

This plate section is also available for download in colour from www.cambridge.org/9781107406049

Plate II. Examples of slender structures in an aerospace context. Top, Pathfinder; bottom, Space Shuttle. Courtesy of NASA.

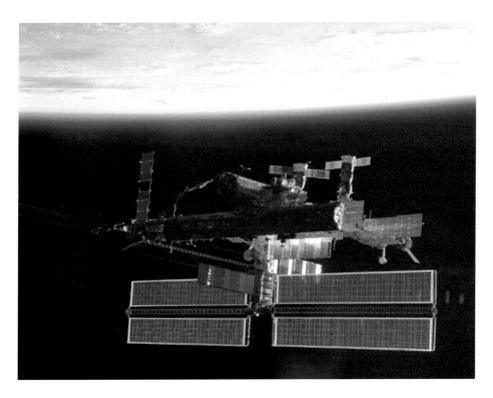

Plate III. Examples of slender structures in an aerospace context. Top, International Space Station; bottom, CH-46 Sea Knight helicopter. Courtesy of NASA.

Plate IV. Examples of slender structures in an aerospace context. Top, B2 Spirit stealth bomber; bottom, F-16C jet fighter. Courtesy of NASA.

Plate V. Examples of slender structures. Top, Golden Gate Bridge, San Francisco, courtesy of Naviose/age photostock; bottom, satellite radar dish, courtesy of NASA.

Plate VI. Examples of slender structures. Top, Trans Alaska Pipeline System above Tanana River, courtesy of Kearney/age photostock; bottom, offshore oil platform, courtesy of NOAA National Marine Sanctuaries.

Plate VII. Examples of slender structures. Top, USS *Dallas* submarine, courtesy of the U.S. Navy, photo by Paul Farley; bottom, Petronas Towers, Malaysia, courtesy of SuperStock/age photostock.

Plate VIII. Examples of slender structures. Top, U.S. Navy fleet oiler, courtesy of the U.S. Navy, Military Sealift Command; bottom, bridge on a guitar, courtesy of Mike Todd.

1 Context: The Point of Departure

In the engineered world (and in a good deal of the natural world), stable equilibrium, or some kind of stationary or steady-state behavior, is the order of the day. Systems are designed to operate in a predictable fashion to fulfill their intended functions despite disturbances and changing conditions. Control systems have been spectacularly successful in maintaining a desirable (stable[1]) response given inevitable uncertainty in modeling system physics. However, there are plenty of examples of systems becoming unstable – and often the consequences of instability are severe. This book looks at the interplay between vibrations and stability in elastic structures.

A brief view of an ecological system provides an effective analogy. The competition between certain species can be viewed as a coupled dynamic system in a slowly changing environment. External influences are provided by various factors including the climate, disease, and human influence. The delicate interaction is played out as conditions evolve and populations respond accordingly – usually in a correspondingly slow way also. However, an instability may occur leading to extinction on a relatively short time scale, perhaps when a disease (or massive meteorite) wipes out an entire population. This situation is not that dissimilar to the fluctuations of the stock markets (in which prediction of sudden changes is of concern to individuals and governments).

In an engineering context, we typically have considerable knowledge about the underlying physics and governing equations of our systems, are able to test a system both analytically and in the laboratory, and thus have a much better chance of assessing the robustness of a system, especially its propensity to failure. However, unforeseen circumstances do occur, and it would, of course, be remiss in a book concerning stability in engineering mechanics not to mention the Tacoma Narrows suspension bridge disaster. But many other bridges and buildings have collapsed, aircraft wings and rotorblades have a tendency to flutter, ships sometimes capsize, the tracks of a railroad will buckle from time to time, electric circuits sporadically exhibit unintended feedback, machine parts are prone to fatigue, and once in a while satellites disappear into deep space. What most of these systems have in common is that they were either subject to external influences with which they could not cope

---

[1] Some aeronautics control systems take advantage of a brief loss of stability for enhanced maneuverability.

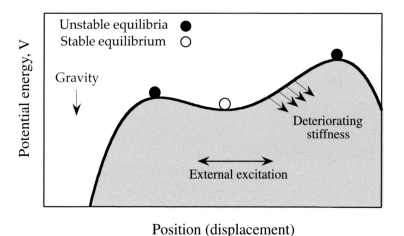

Position (displacement)

Figure 1.1. A deteriorating scenario.

or they changed. Perhaps an encounter with a rogue wave in the case of a ship, or collision with space debris in the case of the satellite. In this last instance an error in the units used in trajectory calculations may cause disaster but in the sense that the system was designed correctly but for the wrong conditions. Of course, there are always practical limits to how much safety or redundancy can be built into a system; the World Trade Center provided a sobering example. But it is also likely that a system is subject to slowly changing conditions, which may, of course, lead to catastrophe, but in a gradual deteriorating sense. It is with these systems that we have scope for monitoring and prediction, as their (dynamic) response may give clues about future performance.

Hence, given a (structural) system in some state of rest (equilibrium) or steady-state motion (an oscillation), we seek to understand those conditions that cause a change in the nominal response, and especially where such a change is *large* (and instability falls squarely into this category). The theoretical framework underlying this statement is of course based on Newton's laws and subsequent developments especially concerning concepts of energy. To crystallize this approach, consider the schematic diagram shown in Fig. 1.1.

Here we might consider the behavior of a small ball allowed to roll (under the influence of gravity) on a curved surface to represent a generic structural or mechanical system. The analogy is really brought into focus if we further assume that the curve is actually associated with the underlying potential energy of the system and that the surface causes a little energy dissipation as the ball rolls. Hence the bottom of the energy "well" is identified as a position of stable equilibrium, with linear theory based on a locally quadratic minimum. Linear stability theory will also tell us that the "hilltops" are points of *unstable* equilibrium. In both cases, the ball will remain at rest at these extremum values of the potential energy surface. However, the important behavior is observed if the system is subject to a *disturbance*. In the stable case, the ball might begin to oscillate but typically return to rest at the bottom of the well. In the unstable case, the ball picks up speed and

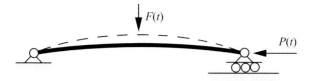

Figure 1.2. A slender axially loaded structure and its dynamic response.

departs the local neighborhood of the hilltop. These situations are well covered by linear stability theory providing the size of the perturbation is *small*.

Extending this concept further, it is natural to ask what happens

- if the morphology of the potential energy surface changes (typically slowly) such that the potential energy at a stable equilibrium position ceases to be a minimum,
- or if the ball is subject to a relatively *large* perturbation or disturbance that may push the ball well beyond the local neighborhood of the minimum.

These are the two situations depicted in Fig. 1.1. The former case is the basis of most studies in classical buckling. The application of an external axial load is assumed to take place quasi-statically, and buckling occurs (typically leading to large deflections) as the ball can no longer maintain its position. Many practical examples like this can be handled very effectively by use of statics. Most interest is naturally focused on the behavior of the system prior to buckling when the system is changing sufficiently slowly that kinetic energy can safely be ignored in the Lagrangian description (although it may still be useful to gain information based on dynamics). However, in the latter case, the application of a large (say, sudden or periodic) perturbation inevitably leads to a dynamic, perhaps unbounded, response. In fact, even in those cases in which a static approach works well, if we want to track the postcritical behavior, we may still need to use a dynamic approach, for example, one in which a system subject to a slowly increasing load results in a fast dynamic jump at buckling.

Figure 1.2 adds some specificity to the scope of the material covered in this book using the behavior exhibited by a vibrating thin beam:

- Figure 1.2 illustrates a beam undergoing small-amplitude free vibrations, that is, with $P(t) = F(t) = 0$. This is a thoroughly linear situation, with the straight configuration the only equilibrium and damping causing dynamic behavior to decay. Exact solutions are available; natural frequencies are constant and scale with the stiffness of the beam. For example, a longer beam is less stiff and thus natural frequencies are lower. Clamped boundary conditions lead to higher natural frequencies than simply supported, and so on.
- The presence of a constant axial load [but with $F(t) = 0$] tends to reduce the natural frequencies if the load is compressive and below its critical value. If the axial load is sufficiently large (i.e., greater than critical), postbuckled (nontrivial)

equilibria exist, and natural frequencies can be computed about these nontrivial equilibria.

- For laterally excited systems ($F(t) \not\equiv 0$ but with $P = 0$), we can have *resonance*. This may also occur about postbuckled equilibria when $P > P_{cr}$.
- If the axial load is a function of time (say, periodic), then the system may also lose stability (depending on the frequency of excitation) through *parametric resonance*.
- If the ends of the beam are both constrained against moving (in-plane) then membrane, or *stretching*, forces arise.
- In each of the preceding scenarios the vibration may have *large* amplitude.
- Many of these scenarios might occur simultaneously. For example, a postbuckled beam might *snap through* if excited laterally.

Thus this range of behavior encompasses both small-amplitude and large-amplitude motion about both trivial and nontrivial equilibria. Access to analytic solutions becomes restricted as the complexity (and nonlinearity) of the system increases. Damping oftens needs to be considered also. Although the example of the prismatic beam has been used here, extensions to other types of axially loaded structures, like plates and shells is easy to envision. Furthermore, some of these situations may lead to instability (both static and dynamic), which is of particular concern to engineers. It is worth mentioning that aerospace structures provide a natural context for much of this material; the continual quest for lighter vehicles naturally brings with it issues of vibration and stability.

Some practical examples of slender structures in aerospace engineering in which axial loads and dynamics may need to be considered are shown in Fig. 1.3. These images all portray aerospace systems. Spacecraft applications tend to be very lightweight: Thin-film solar sails designed for deep-space propulsion; high-altitude unmanned surveillence craft like the Predator; lightweight solar-powered high-endurance aircraft like the Pathfinder; the shuttle; international space station; rotorcraft; and military aircraft all possess slender structural components subject to a variety of loading conditions including vibration and axial-load effects.

Figure 1.4 shows some other examples of slender structures. They range from bridges to pipelines, telescopes to submarines, oil tankers to high-rise buildings. The vibrations of axially loaded structures also occur at very small scales, including the increasingly important range of applications in nanotechnology. The guitar string is an obvious case. The axial load in this case can only be tensile, but it is interesting to note the slightly angled bridge of the guitar – this accounts for the slight amount of bending stiffness in the thicker strings.

Hence this book is broadly divided into two main parts to cover these rather wide-ranging scenarios. A conventional division in the presentation of vibration problems is between free and forced vibration. That convention is somewhat followed here in the development of the material. However, there are occasions for which this division is not clear (e.g., an impulsive force can also be viewed as an initial velocity), but in terms of organizing the material, this seemed to be a natural

Figure 1.3. Examples of slender structures in an aerospace context. Courtesy of NASA. See color plates I–IV following page xvi.

Figure 1.4. More examples of slender structures. See color plates V–VIII following page xvi.

Figure 1.4. (continued) More examples of slender structures. See color plates V–VIII following page xvi.

choice. The next chapter will provide a brief overview of basic mechanics (which can be omitted by the more advanced reader), followed by a treatment of the interplay of dynamics and stability, without introducing too much in the way of mathematics, but still providing a flavor of the types of more practical structural systems considered later in the book.

2 Elements of Classical Mechanics

2.1 Introduction

This chapter develops the theoretical basis for the derivation of governing equations of motion. It starts with Newton's second law and then uses Hamilton's principle to derive Lagrange's equations. A number of conservation laws are introduced. The theory is developed initially for a single particle and extended to systems of particles where appropriate. The emphasis is placed on building the theory relevant to the types of physical system of interest in structural dynamics. Other than the usual limitations regarding relativistic and quantum effects, we also restrict ourselves to translational (rather than rotational) systems, which is largely a matter of coordinates. The majority of problems in this book involve systems in which the forces developed during elastic deformation play a crucial role. Certain standard problems in classical mechanics, for example the central force motion leading to the two-body problem or particle scattering, are not relevant here and are not considered. We shall see the important role played by energy methods in studying the dynamics of structures. Classical mechanics has a long history and in-depth treatment of the subject can be found in Goldstein [1], Whittaker [2], and Synge and Griffith [3] and, of course, going back to the early developments of Newton [4], Euler [5], and Lagrange [6].

2.2 Newton's Second Law

The natural starting point in any text covering an aspect of classical mechanics are Newton's laws of motion. They date back to 1686, with the second being the most important:

> *A body acted upon by a force moves in such a manner that the time rate of change of momentum equals the force.*

Mathematically we introduce the concept of a linear momentum vector $\mathbf{p}$ defined as the product of mass and velocity:

$$\mathbf{p} = m\mathbf{v}, \tag{2.1}$$

where m is the mass and $\mathbf{v}$ is the velocity vector. We can thus write Newton's second law as

$$\mathbf{F} = \frac{d\mathbf{p}}{dt} = \frac{d}{dt}(m\mathbf{v}), \qquad (2.2)$$

in which $\mathbf{F}$ is the force vector.

To apply this law we need to specify motion relative to a reference frame. If we define an absolute position vector, $\mathbf{r}$, in an inertial frame (i.e., a frame at rest or moving with a constant velocity relative to the "fixed" stars), then the corresponding absolute velocity vector is given by

$$\mathbf{v} = \frac{d\mathbf{r}}{dt} = \dot{\mathbf{r}}, \qquad (2.3)$$

where an overdot signifies a time derivative. Thus we can further express Newton's second law in its more familiar form as

$$\mathbf{F} = m\frac{d\mathbf{v}}{dt} = m\ddot{\mathbf{r}} = m\mathbf{a}, \qquad (2.4)$$

where $\mathbf{a}$ is an absolute acceleration vector and we have assumed m does not vary with time.

Equation (2.4) is a (set of) second-order ordinary differential equation fundamental to the study of mechanics. In general,

$$\mathbf{F} = \mathbf{F}(\mathbf{r}, \dot{\mathbf{r}}, t), \qquad (2.5)$$

and a solution $\mathbf{r}(t)$ that satisfies this equation can be obtained given appropriate initial conditions $\mathbf{r}(t_0)$ and $\dot{\mathbf{r}}(t_0)$. For the types of systems of relevance to the material covered in this book, these solutions are unique. The forces entering Eq. (2.5) arise from a number of different sources in structural dynamics: stiffness, inertia, excitation and damping being the most important. The SI units of force are newtons (N), where $1\,\text{N} = 1\,\text{kg m/s}^2$.

Clearly, if $\mathbf{F} = \mathbf{F}(t)$, then it would be a straightforward task to integrate Eq. (2.4) directly to obtain $\mathbf{v}(t)$ and then $\mathbf{r}(t)$. However, this will not typically be the case (as elastic forces tend to depend on the change in position), and a variety of techniques can be called on to solve differential equations. We observe at this point that solutions to equations of the type (2.4) will often involve *oscillations*, and also that there may not be analytic solutions available, especially in those situations in which *nonlinear* terms are present. Further discussion of nonlinearity and other aspects of differential equations are left to later chapters. However, the concept of *stability* (which will be developed continuously throughout this book) involves considering the manner in which closely adjacent solutions of Eq. (2.4) behave as a function of time, and specifically, when one of those solutions represents some kind of steady or equilibrium solution.

2.3 Energy and Work

Now suppose $\mathbf{F} = \mathbf{F}(\mathbf{r})$. We can obtain information about the solution to Eq. (2.4) by performing a path integral with respect to $\mathbf{r}$ along the trajectory:

$$\int_{\mathbf{r}(t_0)}^{\mathbf{r}(t)} \mathbf{F}(\mathbf{r}) \cdot d\mathbf{r} = \int_{t_0}^{t} \mathbf{F}(\mathbf{r}) \cdot \dot{\mathbf{r}} dt = m \int_{t_0}^{t} \frac{d^2\mathbf{r}}{dt^2} \cdot \frac{d\mathbf{r}}{dt} dt \tag{2.6}$$

$$= \frac{1}{2} m \int_{t_0}^{t} \frac{d}{dt}(\dot{r}^2) dt = \frac{1}{2} mv^2(t) - \frac{1}{2} mv^2(t_0), \tag{2.7}$$

which gives the magnitude of the velocity [rather than $\mathbf{r}(t)$] provided the integral on the left-hand side of Eq. (2.6) can be performed. This is not a straightforward matter because $\mathbf{r}(t)$ (which is unknown) appears in the upper limit and a path integral depends on the path of integration.

However, if we let the path of this integral [in Eq. (2.6)] be called C, then we can introduce the *work done* by the force $\mathbf{F}$ moving along this path as

$$W_C = \int_C \mathbf{F} \cdot d\mathbf{r}, \tag{2.8}$$

and, defining the kinetic energy as

$$T = \frac{1}{2} mv^2, \tag{2.9}$$

we can rewrite Eq. (2.7) as

$$W_C = T_2 - T_1, \tag{2.10}$$

which is a statement of the work – energy theorem. It turns out that there is a relatively large class of problems for which the work done for any admissible path between points 1 and 2 depends on only the end points of the path. In these cases forces are called *conservative*, and they play a dominant role in the static analysis of buckling, for example.

For a conservative force $\mathbf{F}(\mathbf{r})$, consider two paths C_1 and C_2 connecting two points $\mathbf{r}_1$ and $\mathbf{r}_2$. In this case we can write

$$\int_{C_1} \mathbf{F} \cdot d\mathbf{r} = \int_{C_2} \mathbf{F} \cdot d\mathbf{r}, \tag{2.11}$$

which implies that

$$\oint \mathbf{F} \cdot d\mathbf{r} = 0, \tag{2.12}$$

where the closed integral is performed from $\mathbf{r}_1$ to $\mathbf{r}_2$ and back again. We define the work done by a conservative force in moving a particle from a reference point, $\mathbf{r}_0$, to an arbitrary position $\mathbf{r}$ as the potential energy,

$$V(\mathbf{r}) = \int_{\mathbf{r}}^{\mathbf{r}_0} \mathbf{F}_c \cdot d\mathbf{r}, \tag{2.13}$$

and the work done in terms of the potential energy of end points we can write as

$$\int_{\mathbf{r}_1}^{\mathbf{r}_2} \mathbf{F}_c \cdot d\mathbf{r} = \int_{\mathbf{r}_1}^{\mathbf{r}_0} \mathbf{F}_c \cdot d\mathbf{r} - \int_{\mathbf{r}_2}^{\mathbf{r}_0} \mathbf{F}_c \cdot d\mathbf{r},$$

(2.14)

and therefore

$$\int_{\mathbf{r}_1}^{\mathbf{r}_2} \mathbf{F}_c \cdot d\mathbf{r} = V(\mathbf{r}_1) - V(\mathbf{r}_2) = V_1 - V_2.$$

(2.15)

The same result can be obtained if we write $\mathbf{F}$ as the gradient of the scalar function:

$$\mathbf{F} = -\nabla V,$$

(2.16)

where, for example, in Cartesian coordinates we have

$$\nabla \equiv \frac{\partial}{\partial x}\mathbf{i} + \frac{\partial}{\partial y}\mathbf{j} + \frac{\partial}{\partial z}\mathbf{k}.$$

(2.17)

The potential energy is defined to within an additive constant, but because the important behavior depends on the *change* in potential energy, this constant is usually chosen to facilitate the solution procedure (and often zero is a convenient choice).

Conservation of Energy. In the absence of external forcing or damping, the concept of conservation of total mechanical energy provides a useful framework for analyzing a dynamic system. Equating Eqs. (2.10) and (2.15) we have

$$T_2 - T_1 = V_1 - V_2,$$

(2.18)

and because we can assign $\mathbf{r}_2$ as any point on the path, then we obtain the conservation of energy

$$T + V = E,$$

(2.19)

where E is a constant and represents the total (mechanical) energy of the system. We can thus make this statement:

> *If the forces acting on a particle are conservative, then the total energy of the particle $(T + V)$ is conserved.*

These concepts are easily extended to include systems of particles, and a number of other conservation theorems can be developed. For example, if a particle is free from the effects of any force, then the linear momentum $\dot{\mathbf{p}} = 0$ and thus $\mathbf{p}$ is a constant. A similar expression can be developed in terms of angular momentum. Clearly these conserved quantities can play a significant role in facilitating a solution $\mathbf{r}(t)$ to a physical problem.

2.4 Virtual Work and D'Alembert's Principle

In practical situations it may be quite difficult to describe all the forces acting on a system in a vectorial context. We will see that this is one of the reasons that conducting an energy approach is often easier than using Newton's laws directly. However, it is possible to make use of a variational principle in mechanics to facilitate the solution procedure, and this involves the concept of virtual displacements.

Suppose we have a particle in equilibrium (described by a position vector **r**) under a set of forces **F**. If the position of the particle is subject to infinitesimal changes (i.e., small variations in the systems coordinates, compatible with any system constraints) then the total virtual work done is

$$\delta W = \left(\sum_{i=1}^{n} \mathbf{F_i}\right)\delta \mathbf{r} = 0 \tag{2.20}$$

for a system in equilibrium, where the symbol δ is given to instantaneous, virtual variations. This can be generalized for a number of particles and, indeed, for elastic bodies, which comprise the largest interest in this book.

We can thus state the principle of virtual work:

For a system of forces acting on a particle, the particle is in statical equilibrium if, when it is given any virtual displacement, the net work done by the forces is zero.

There are a number of ways in which this statement [and Eq. (2.20)] can be put to practical use. We can divide the forces into two categories: applied forces and constraint forces. It can be shown that the virtual work that is due to constraint forces acting through small virtual (termed *reversible*) displacements is zero, and the principle of virtual work is adjusted accordingly.

In applications to structural mechanics, it is convenient to also divide the work into two parts: that due to external loads and that due to internal forces, and thus

$$\delta W_e + \delta W_i = 0. \tag{2.21}$$

We can incorporate dynamics into the framework of virtual work by using D'Alembert's principle. We achieve this by writing Newton's second law as

$$\mathbf{F} - m\ddot{\mathbf{r}} = 0, \tag{2.22}$$

in which $m\ddot{\mathbf{r}}$ is called the inertia force. Therefore we can view this as a statement of dynamic equilibrium, and in simple structural dynamics problems this is often the easiest means of obtaining the equations of motion. The statement of virtual work can thus be written in a more general form for a system of N particles of mass m_i acted on by forces $\mathbf{F}_i$ as

$$\sum_{i=1}^{N}(\mathbf{F}_i - m_i\delta\ddot{\mathbf{r}}_i)\cdot\delta\mathbf{r}_i = 0, \tag{2.23}$$

and D'Alembert's principle may be stated thus:

The virtual work performed by the effective forces through infinitesimal virtual displacements compatible with the system constraints is zero.

Here, "effective forces" refers to the combination of regular and inertia forces.

However, in contrast to the energy approaches of the next section, Eq. (2.23) still describes motion in terms of physical, vectorial coordinates. A number of issues surround the independence of coordinates, for example, in systems possessing m equations of constraint the number of *degrees of freedom* (DOFs) is three fewer than the number of rectangular coordinates needed to describe the positions of all the particles. Suppose we have already eliminated the forces of constraint (because they do no work) and rewrite Eq. (2.23) as

$$\sum_{i=1}^{N} \mathbf{F}_i \cdot \delta \mathbf{r}_i = 0, \tag{2.24}$$

where $\mathbf{F}_i$ is now a combination of the applied and inertia forces. To satisfy equilibrium, however, we need independent coordinates for $\mathbf{F}_i = \mathbf{0}$ ($i = 1, 2, \ldots, N$). It can be shown that transforming from the $\mathbf{r}_i$ coordinates to *generalized* coordinates q_j and then taking infinitesimal virtual displacements leads to the virtual work being written in the form

$$\delta W = \left(\sum_{j=1}^{n} Q_j \right) \delta q_j, \tag{2.25}$$

where

$$Q_j = \sum_{i=1}^{N} \mathbf{F}_i \cdot \frac{\partial \mathbf{r}_i}{\partial q_j}, \quad j = 1, 2, \ldots n, \tag{2.26}$$

and the variations in $\mathbf{r}$ are in the q directions. The Q_j are called the generalized forces. Equilibrium is thus given by

$$Q_j = 0, \quad j = 1, 2, \ldots, n. \tag{2.27}$$

2.5 Hamilton's Principle and Lagrange's Equations

Although Newton's laws are remarkably useful, there are a number of limitations. These concern systems comprising particles at very small distances and also systems in which *very* high velocities are involved. These types of systems are of no concern in this book, but there are many circumstances in the macromechanical world for which determining all the forces present in a system is a challenging or even impossible task. An alternative approach is based on Hamilton's principle, which can be used to derive equations of motion via Lagrange's equations. Although they can be shown to be equivalent to Newton's second law, they provide a more powerful and global approach to solving problems in mechanics. A particular advantage is the flexibility in choosing coordinate systems. Attention is focused primarily on conservative systems, with a more thorough discussion of nonconservative forces left until later.

Hamilton's principle, for conservative systems, states

Of all possible paths along which a dynamical system may move from one point to another within a specified time interval (consistent with any constraints), the actual path followed is that which minimizes the time integral of the difference between the kinetic and potential energies.

The integral referred to in this statement is often called the *action I* and can be written as

$$I = \int_{t_1}^{t_2} (T - V)dt = \int_{t_1}^{t_2} \mathcal{L}dt, \tag{2.28}$$

where $\mathcal{L}$ is the *Lagrangian*.

Thus the issue is to find the minimum of this integral, a classic problem in the calculus of variations. If $\mathcal{L}$ depends on a single coordinate, say q [together with its time derivative $\dot{q}(t)$ and time t], then we need to find the trajectory $q(t)$ that minimizes

$$I = \int_{t_1}^{t_2} \mathcal{L}\left[q(t), \dot{q}(t), t\right] dt. \tag{2.29}$$

To do this we need to consider what other permissible trajectories do in comparison with $q(t)$: Any neighboring trajectory must make I increase relative to the minimum. We consider the behavior of a close-by trajectory given by

$$q(t) + \epsilon\phi(t), \tag{2.30}$$

where we suppose $q(t)$ is the path corresponding to the minimum, ϵ is a small value, and the function $\phi(t)$ is zero at the end points t_1 and t_2 but is otherwise any function of time. Equation (2.29) is thus transformed to

$$I = \int_{t_1}^{t_2} \mathcal{L}\left[q(t) + \epsilon\phi(t), \dot{q}(t) + \epsilon\dot{\phi}(t), t\right] dt. \tag{2.31}$$

Mathematically we express the condition for a minimum as

$$\frac{dI}{d\epsilon} = 0, \tag{2.32}$$

which then leads to

$$\int_{t_1}^{t_2} \left(\frac{\partial \mathcal{L}}{\partial q}\phi + \frac{\partial \mathcal{L}}{\partial \dot{q}}\dot{\phi}\right) dt = 0. \tag{2.33}$$

The second term in the integrand can be integrated by parts, leaving

$$\int_{t_1}^{t_2} \left(\frac{\partial \mathcal{L}}{\partial q} - \frac{d}{dt}\frac{\partial \mathcal{L}}{\partial \dot{q}}\right) \phi(t)dt = 0. \tag{2.34}$$

For this to be a minimum for arbitrary $\phi(t)$, the term in the large parentheses must be zero, which gives us Lagrange's equation:

$$\frac{\partial \mathcal{L}}{\partial q} - \frac{d}{dt}\frac{\partial \mathcal{L}}{\partial \dot{q}} = 0. \tag{2.35}$$

This can be extended to incorporate situations in which the trajectory depends on a number of independent coordinates $q_j(t)$:

$$\frac{\partial \mathcal{L}}{\partial q_j} - \frac{d}{dt} \frac{\partial \mathcal{L}}{\partial \dot{q}_j} = 0, \quad j = 1, 2, \ldots, n. \tag{2.36}$$

There is considerable advantage in the Lagrangian approach in that the coordinates need not be physically meaningful. Generalized coordinates consist of any set of quantities that fully describes the state of a system, and we may view our dynamic system as evolving in this *configuration space*. The generalized coordinates are often referred to as q_j, they are not unique, and their time derivatives are the generalized velocities $\dot{q}_j$.

2.5.1 Constraints

In many physical situations the motion of a system is subject to constraints. That is, there is some kind of kinematic restriction on the motion, usually involving a relation between coordinates, their rates of change, or time. Forces arise because of the constraints, but because they depend on the motion itself, they are not known *a priori*. However, if the constraint can be expressed as position coordinate relations (or just involve time explicitly) then it can be expressed in a differential form, is termed *holonomic*, and can be incorporated into the Lagrangian description without much difficulty. That the motion is restricted leads naturally to a reduced number of DOFs; that is, we seek to select independent generalized coordinates that do not violate the constraints, and, because the constraint forces do no virtual work, they do not appear in the equations of motion.

However, another class of problem involves constraints that influence the rates of change of generalized coordinates. They may be expressed as inequalities or as nonintegrable differential relations and are termed *nonholonomic*. They cannot be reduced to independent generalized coordinates, and appropriate equations of motion must include the constraints. In practice, the method of Lagrange multipliers is used [7], in which the generalized coordinates and constraint forces are obtained simultaneously. Holonomic constraint forces can also be handled in this way, although they are not of direct interest to the material covered in this book.

2.5.2 Conservation Laws

We have seen how certain quantities (e.g., the total mechanical energy) may be conserved. This was developed from basic definitions of work and energy and their relation to Newton's second law. We can show that similar relations can be developed by using the Lagrangian description.

There may often occur instances in which a symmetry property enables a considerable simplification to be made, which leads to the absence of a particular coordinate in the Lagrangian. Suppose the missing (or ignorable) coordinate is q_j; then its Lagrange's equations will be

$$\frac{d}{dt}\frac{\partial \mathcal{L}}{\partial \dot{q}_j} = 0,\tag{2.37}$$

which implies

$$\frac{\partial \mathcal{L}}{\partial \dot{q}_i} = \text{constant},\tag{2.38}$$

and because $\partial \mathcal{L}/\partial \dot{q}_j = m\dot{q}_j = p_j$, this means the generalized momentum is conserved, that is, we effectively have a constant of the motion.

Another class of problem involves those in which time does not appear explicitly in the Lagrangian, and thus

$$\frac{\partial \mathcal{L}}{\partial t} = 0.\tag{2.39}$$

In this case, we write the total derivative ($\mathcal{L}$ can change in time only through its dependence on the coordinates and velocities) as

$$\frac{d\mathcal{L}}{dt} = \frac{\partial \mathcal{L}}{\partial q}\frac{dq}{dt} + \frac{\partial \mathcal{L}}{\partial \dot{q}}\frac{d\dot{q}}{dt} = 0.\tag{2.40}$$

From Lagrange's equations, we have

$$\frac{\partial \mathcal{L}}{\partial q} = \frac{d}{dt}\frac{\partial \mathcal{L}}{\partial \dot{q}},\tag{2.41}$$

and therefore

$$\frac{d\mathcal{L}}{dt} = \left(\frac{d}{dt}\frac{\partial \mathcal{L}}{\partial \dot{q}}\right)\dot{q} + \frac{\partial \mathcal{L}}{\partial \dot{q}}\frac{d\dot{q}}{dt}.\tag{2.42}$$

We recognize that this is the derivative of a product, that is,

$$\frac{d\mathcal{L}}{dt} = \frac{d}{dt}\left(\frac{\partial \mathcal{L}}{\partial \dot{q}}\dot{q}\right),\tag{2.43}$$

which can be written as

$$\frac{d}{dt}\left(\frac{\partial \mathcal{L}}{\partial \dot{q}}\dot{q} - \mathcal{L}\right) = 0.\tag{2.44}$$

This can now be integrated, and the term in parentheses is therefore constant in time:

$$\dot{q}\frac{\partial \mathcal{L}}{\partial \dot{q}} - \mathcal{L} = H = \text{constant}.\tag{2.45}$$

If the potential energy does not depend explicitly on the velocities or time, we have $V = V(q)$, and, using $\mathcal{L} = T - V$, we obtain

$$\frac{\partial \mathcal{L}}{\partial \dot{q}} = \frac{\partial(T - V)}{\partial \dot{q}} = \frac{\partial T}{\partial \dot{q}},\tag{2.46}$$

and Eq. (2.45) becomes

$$\dot{q}\frac{\partial T}{\partial \dot{q}} - (T - V) = H.\tag{2.47}$$

Assuming the kinetic energy depends on only the generalized velocities and that this relation is the standard quadratic form, then we obtain

$$\dot{q}\frac{\partial T}{\partial \dot{q}} = 2T, \tag{2.48}$$

and Eq. (2.47) gives us the total energy of the system:

$$H = T + V. \tag{2.49}$$

Therefore, we have the result that H (termed the *Hamiltonian*) is equal to the total energy of the system if the Lagrangian does not depend explicitly on time, and the potential energy does not depend on velocity. These concepts can easily be extended to include systems of particles, provided the equations of transformation relating regular and generalized coordinates is also independent of time.

It is possible to express the Lagrangian (and specifically the velocities associated with the generalized coordinates) in terms of generalized momenta. An advantage of doing this includes the fact that it is often the momentum that is a conserved quantity, and the Hamiltonian also has more physical meaning (through its relation with energy for conserved systems) than the Lagrangian. It also results in a set of $2n$ first-order equations rather than the n second-order Lagrange's equations, and this may assist the development of numerical solutions. It turns out that the resulting *Hamilton's equations* have certain symmetric features that render them unchanged under transformation of coordinates and momenta, and they are often referred to as the *canonical* equations of motion.

2.6 Nonconservative Forces and Energy Dissipation

Not all forces are derivable from a potential, that is, there may not be a potential energy function V that satisfies Eq. (2.16) for a particular system. We can write the external forces acting on the system in the form

$$\mathbf{F}_i = \mathbf{F}_{Pi} + \mathbf{F}_{Di}, \tag{2.50}$$

where $\mathbf{F}_{Pi}$ is derivable from a potential $V = V(q_i)$ and $\mathbf{F}_{Di}$ is not. Thus we can also divide the virtual work into conservative and nonconservative parts,

$$\delta W = \delta W_P + \delta W_D. \tag{2.51}$$

The first term on the right-hand side of Eq. (2.51) is defined by Eq. (2.16), and, by virtue of Eq. (2.26), we can write Lagrange's equation as

$$\frac{\partial \mathcal{L}}{\partial q_j} - \frac{d}{dt}\frac{\partial \mathcal{L}}{\partial \dot{q}_j} = Q_{Dj}. \tag{2.52}$$

We can also rearrange Eq. (2.51) and write

$$\delta W_D = \delta W - \delta W_P = \mathbf{F} \cdot d\mathbf{r} - \mathbf{F}_P \cdot d\mathbf{r} = dT - (-dV) = d(T + V) = dE. \tag{2.53}$$

Integrating over the path, we get

$$\int_{\mathbf{r}_1}^{\mathbf{r}_2} \mathbf{F}_D \cdot d\mathbf{r} = \int_{E_1}^{E_2} dE = E_2 - E_1. \tag{2.54}$$

Thus the change in total energy is equal to the work done by the nonconservative forces.

In structural mechanics it is often found that there are two general classes of forces that do not arise from a potential function. In these cases, the total mechanical energy is not conserved. We will occasionally consider a load that is nonconservative in nature because of its direction changing (for example, following the slope at the end of a beam), but in general there are two classes of nonconservative forces encountered in mechanical systems. In cases in which the energy decreases we use the term *dissipative* forces. The main example is the loss of energy through damping, which in many cases relates to a force proportional to velocity. The other main type of nonconservative force is time dependent and often associated with the external *driving* of a system.

2.6.1 Damping

In mechanical systems, energy dissipation is inevitable. If we assume that a certain class of nonconservative force acting on a single particle is a function of velocity only, then

$$\mathbf{F}_{D_i} = -g(\mathbf{v}_i)\mathbf{v}_i, \tag{2.55}$$

where we also assume that the force is directed opposite to the velocity. Very often this relation will describe linear-viscous damping in unidirectional motion:

$$\delta W = \mathbf{F}_{i_x} = -c_{x_i}\dot{x}_i, \tag{2.56}$$

where c is the damping coefficient. The virtual work done by this dissipative force is

$$\delta W = \sum_i \mathbf{F} \cdot \delta \mathbf{r}$$

$$= -\sum_{i=1}^{n} c_{x_i}\dot{x}_i \delta x_i$$

$$= -\sum_{i=1}^{n} \left[\sum_{j=1}^{n} c_{x_i}\dot{x}_i \frac{\partial x_i}{\partial q_j} \delta q_j \right]$$

$$= -\sum_{j=1}^{n} \left[\frac{1}{2} \sum_{i=1}^{n} \frac{\partial}{\partial \dot{q}_j} c_{x_i}\dot{x}_i^2 \delta q_j \right]. \tag{2.57}$$

Thus the corresponding generalized force is given by

$$Q_{D_j} = -\frac{1}{2} \sum_{i=1}^{n} \frac{\partial}{\partial \dot{q}_j} c_{x_i}\dot{x}_i = -\frac{1}{2} \frac{\partial}{\partial \dot{q}_j} \sum_{i=1}^{n} c_{x_i}\dot{x}_i^2. \tag{2.58}$$

Introducing *Rayleigh's dissipation function D*,

$$D = \frac{1}{2} \sum_{i=1}^{n} c_{x_i} \dot{x}_i^2, \tag{2.59}$$

we can obtain the dissipative generalized forces from

$$\delta W = \sum_{j=1}^{n} Q_{D_j} \delta q_j = -\sum_{j=1}^{n} \frac{\partial D}{\partial \dot{q}_j} \delta q_j, \tag{2.60}$$

and thus a more general form of Lagrange's equation:

$$\frac{d}{dt} \frac{\partial \mathcal{L}}{\partial \dot{q}_j} - \frac{\partial \mathcal{L}}{\partial q_j} + \frac{\partial D}{\partial \dot{q}_j} = 0, \quad j = 1, 2, \ldots, n. \tag{2.61}$$

2.6.2 Time-Dependent Forces

We now consider another group of nonconservative forces, namely, time-dependent forces, as they crop up quite naturally in structural systems subject to periodic excitation or impulses, for example. Suppose a system is subject to forces $F(t)$ that depend on time (but are independent of the generalized coordinates), then a time-dependent Lagrangian will have a term in the form $qF(t)$. Clearly, this will then lead to the appearance of $F(t)$ in the resulting equation of motion.

Furthermore, the principle of impulse and momentum can be used via the Lagrangian approach. Consider an impulsive force

$$\hat{\mathbf{F}} = \int_{t_0}^{t_0 + \Delta t} \sum \mathbf{F}(t) dt. \tag{2.62}$$

Intgerating Lagrange's equation from $t_1 = t_0$ to $t_2 = t_0 + \Delta t$ and allowing $\Delta t \to 0$ leads to the impulsive form of Lagrange's equation:

$$\left. \frac{\partial T}{\partial \dot{q}_j} \right|_2 - \left. \frac{\partial T}{\partial \dot{q}_j} \right|_1 = \hat{Q}_j, \quad j = 1, 2, \ldots, n, \tag{2.63}$$

where $\hat{Q}_j$ is a generalized impulse.

It is also possible to derive a velocity-dependent potential for some problems, although this will not be encountered in the types of applications in this book. Finally, we note that extension of Lagrangian mechanics to continuous systems (with an infinite number of DOFs) will be dealt with in a later chapter.

In summary, we will often be in a position to write potential and kinetic energies and make use of Lagrange's equation:

$$\frac{d}{dt} \frac{\partial \mathcal{L}}{\partial \dot{q}_j} - \frac{\partial \mathcal{L}}{\partial q_j} + \frac{\partial D}{\partial \dot{q}_j} = q_j F_j(t), \quad j = 1, 2, \ldots, n. \tag{2.64}$$

Given the types of axially loaded elastic structures of primary interest in this book, we inevitably focus quite heavily on equations of motion of the type

$$M \ddot{q}_j + C \dot{q}_j + f(q_j, \lambda) = F(t), \quad j = 1, 2, \ldots, n, \tag{2.65}$$

in which we have a mass matrix M, damping matrix C, the stiffness matrix f, which depends (often nonlinearly) on external axial loads (λ), and both free [$F(t) = 0$] and forced vibrations are examined for a wide variety of slender structures, and where we often begin by assuming $C = 0$.

2.7 Strain Energy

We now introduce a brief description of strain energy, as this is a fundamental aspect of structural mechanics that we will repeatedly encounter throughout this book. In Section 2.3 the concepts of energy and work were introduced. Because many of the specific physical systems to be considered later in this book involve deformable bodies, it is important to understand the manner in which (elastic) energy is stored as strain energy, especially in bending.

We start with a basic definition of strain energy per unit volume, or strain-energy density U_0, for a uniaxially loaded system given by

$$U_0 = \int_0^{\epsilon_1} \sigma_x d\epsilon_x, \tag{2.66}$$

in which σ_x is stress and ϵ_x is strain. We see that the strain-energy density is equal to the area under the stress–strain from $\epsilon_x = 0$ to $\epsilon_x = \epsilon_1$. A thorough background to elasticity and the general description of fundamental issues in solid mechanics can be found in Langhaar [8]. The total strain energy stored in the solid is

$$U = \int_V U_0 \, dV. \tag{2.67}$$

For a linear isotropic elastic material, we have $\sigma_x = E\epsilon_x$, where E is Young's modulus, and, in this case, expression (2.66) becomes

$$U_0 = \frac{\sigma_x^2}{2E}. \tag{2.68}$$

The strain energy U_0 is always positive-definite, and the conservation of energy introduced in Subsection 2.5.2 can also be stated in terms of strain energy:

If an elastic body is in equilibrium under an external force system, then the internal strain energy that is due to deformation is equal to the work of the externally applied force system.

Application to Beams. We finish this chapter by briefly focusing on the strain energy associated with slender beams because they represent an important element in this book, but these concepts can be easily extended to strings, plates, and so on.

A prismatic bar of cross-sectional area A and length L is subjected to an axial load P (which passes through the centroid of the cross section). The stress $\sigma = P/A$ and thus the total strain energy in the bar is

$$U = \frac{P^2 L}{2AE}. \tag{2.69}$$

Suppose the beam is subjected to an applied lateral load rather than axial and has a second moment of area I. In this case, a bending moment causes a stress $\sigma_x = My/I$ (where y is the distance from the centroid) and thus a total strain energy of

$$U = \int_0^L \frac{M^2}{2EI} dx. \tag{2.70}$$

When standard beam theory [9] for a prismatic beam is used, this simplifies to the well-known expression

$$U = \frac{1}{2} EI \int_0^L \left(\frac{\partial^2 w}{\partial x^2} \right)^2 dx. \tag{2.71}$$

The interaction between axial and bending effects will be a central theme.

Throughout this book extensive use will be made of energy concepts [10]. Kinetic energy and the work done by external loads are added to the consideration of strain energy for a variety of structural systems. In simpler cases, we will make direct use of Newton's laws, but for complex systems, energy will provide a powerful (equilibrium and stability) framework in which to study the dynamics of axially loaded structures.

References

[1] H. Goldstein. *Classical Mechanics*. Addison-Wesley, 1980.
[2] E.T. Whittaker. *A Treatise on the Analytical Dynamics of Particles and Rigid Bodies*. Cambridge Mathematical Library, 1988.
[3] J.L. Synge and B. A. Griffith. *Principles of Mechanics*. McGraw-Hill, 1959.
[4] I. Newton. *Philosophiæ Naturalis Principia Mathematica*. London, 1687.
[5] L. Euler. *Methodus Inveniendi Lineas Curvas Maximi Minimive Proprietate Gaudentes*. Marcum Michaelem Bousquet, 1744.
[6] J.L. Lagrange. *Mecanique Analytique*. Courier, 1788.
[7] L. Meirovitch. *Principles and Techniques of Vibrations*. Prentice Hall, 1997.
[8] H.L. Langhaar. *Energy Methods in Applied Mechanics*. Wiley, 1962.
[9] S.P. Timoshenko and J.M. Gere. *Theory of Elastic Stability*, 2nd ed. McGraw-Hill, 1961.
[10] J.H. Argyris and S. Kelsey. *Energy Theorems and Structural Analysis*. Butterworth, 1960.

3 Dynamics in the Vicinity of Equilibrium

We are primarily interested in the concept of a slow evolution toward, and through, instability. The quasi-static change of a system parameter (typically the axial load) allows any instability mechanisms to unfold, especially if transient dynamics is present. Before moving on to consider a range of structural components, we investigate the response of a linear oscillator under various conditions, including a continuously deteriorating stiffness. This is followed by consideration of some aspects of the Lagrangian approach and the interaction of dynamics and stability. The final part of this chapter then uses a phenomenological model to describe the dynamics and stability of a simple physical system. Again, there is a large literature concerning the linear oscillations of mechanical systems. The interested reader can find good coverage of basic material in Refs. [1–7].

3.1 The Linear Oscillator

Consider the continuous-time evolution of a dynamical system governed by

$$\dot{\mathbf{x}} = \mathbf{F}(\mathbf{x}, t), \quad \mathbf{x} \in \mathbb{R}^n, \quad t \in \mathbb{R}, \tag{3.1}$$

where $\mathbf{x}$ is a state vector that describes the evolution of the system under the vector field $\mathbf{F}$. Given an initial condition, typically the values of the state vector prescribed at $t = 0$, that is, $\mathbf{x}(0)$, we can seek to solve system (3.1) to obtain a trajectory $\mathbf{x}(t)$, or orbit, along which the solution evolves with time. We then seek to ascertain the stability of the system, generally as a function of a (control) parameter μ, and thus consideration of

$$\dot{\mathbf{x}} = \mathbf{F}(\mathbf{x}, \mu, t) \quad \mathbf{x} \in \mathbb{R}^n, \quad t \in \mathbb{R}, \tag{3.2}$$

will play a central role in the material contained in this book [8, 9].

Application of Newton's second law relates acceleration and force (and hence position), and thus often results in a second-order ordinary differential equation of the form

$$\frac{d^2x}{dt^2} = -\omega_n^2 x, \tag{3.3}$$

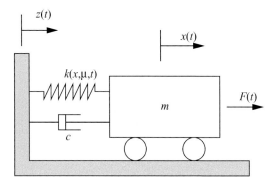

Figure 3.1. A spring–mass–damper system.

where ω_n is a constant (the natural frequency), and with $\dot{x} \equiv dx/dt$, we obtain the nondimensional governing equation of motion,

$$\ddot{x} + \omega_n^2 x = 0, \tag{3.4}$$

subject to the two initial conditions, $x(0)$, and $\dot{x}(0)$. This is the equation of motion governing the dynamic response of the spring–mass system shown in Fig. 3.1 with $\omega_n = \sqrt{k/m}$ (k and m constant) and all other parameters set equal to zero, that is, $c = F(t) = z(t) = 0$. Because Eq. (3.4) is a linear, homogeneous, ordinary differential equation with constant coefficients, we can write the solution as

$$x(t) = Ae^{st}. \tag{3.5}$$

Placing Eq. (3.5) into Eq. (3.4), we find that $s = \pm i\omega_n$, and thus the general form of the solution is given by

$$x(t) = Ae^{i\omega_n t} + Be^{-i\omega_n t}. \tag{3.6}$$

Alternatively, using Euler's identities, we can write

$$x(t) = C\cos(\omega_n t) + D\sin(\omega_n t). \tag{3.7}$$

To determine A and B (or C and D), we make use of the initial conditions to get

$$x(t) = x(0)\cos(\omega_n t) + \frac{\dot{x}(0)}{\omega_n}\sin(\omega_n t). \tag{3.8}$$

We can convert this system into a pair of coupled, first-order ordinary differential equations (in state-variable format) by introducing a new variable

$$y = \dot{x}, \tag{3.9}$$

and substituting in Eq. (3.4) gives

$$\dot{x} = y, \qquad \dot{y} = -\omega_n^2 x. \tag{3.10}$$

In matrix notation,

$$\begin{bmatrix} \dot{x} \\ \dot{y} \end{bmatrix} = \begin{bmatrix} 0 & 1 \\ -\omega_n^2 & 0 \end{bmatrix} \begin{bmatrix} x \\ y \end{bmatrix}. \tag{3.11}$$

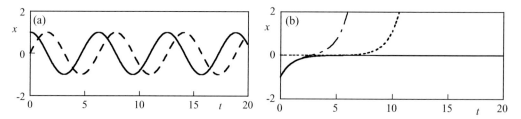

Figure 3.2. (a) Two trajectories exhibiting simple harmonic motion: $x(0) = 1, \dot{x}(0) = 0$, and $x(0) = 0, \dot{x}(0) = 1$; (b) solutions to Eq. (3.16). (i) $x(0) = -1, \dot{x}(0) = 1$ (solid curve); (ii) $x(0) = 0.0001, \dot{x}(0) = 0$ (dotted curve); and (iii) $x(0) = -0.99, \dot{x}(0) = 1$ (dot–dashed curve).

A plot of Eq. (3.8) (with $\omega_n = 1$) is shown in Fig. 3.2(a) for two typical sets of initial conditions.

At this point, we simply note that, from Eq. (3.8) and its derivative (to obtain velocity), we can envision the trajectory evolving with time in a repeating manner. Plotting position versus velocity (the phase plane) will be a useful way of displaying dynamic behavior later in this book, and in this (undamped) case it is apparent that the motion is described by ellipses. This is, of course, the periodic behavior we would expect for a simple spring–mass system with ω_n (assumed real, i.e., $\omega_n^2 > 0$) identified as the natural frequency of the oscillation. In terms of a heuristic concept of stability, we might consider this behavior to be neither stable or unstable, as any motion we might initiate does not decay or grow, but simply persists.

Solution (3.7) can also be written as

$$x(t) = \bar{A} \cos(\omega_n t + \phi), \tag{3.12}$$

where $\bar{A} = \sqrt{C^2 + D^2}$ is the amplitude and $\phi = \arctan(C/D)$ is the phase. Thus we see that the larger the initial conditions, the larger the area enclosed by the ellipses, that is,

$$x^2(t) + [\dot{x}(t)/\omega_n]^2 = \bar{A}^2. \tag{3.13}$$

The two trajectories shown in Fig. 3.2(a) differ by a phase $\phi = \pi/2$, and thus the dashed curve can be viewed as the corresponding velocity time series. Later, we will see how this relates to energy. However, the *form* of the resulting motion is independent of the initial conditions.

Suppose we have $\omega_n^2 < 0$. This is a situation that is difficult to envision, physically, but can occur, for example, in a nonlinear system if the spring stiffness becomes negative. Then the motion is governed by

$$\ddot{x} - \omega_n^2 x = 0. \tag{3.14}$$

Now adopting the solution $x(t) = Ae^{st}$ leads to $s = \pm \omega_n$, and thus

$$x(t) = Ae^{\omega_n t} + Be^{-\omega_n t}. \tag{3.15}$$

Using the definition of hyperbolic functions and the initial conditions, we also have

$$x(t) = x(0) \cosh \omega_n t + [\dot{x}(0)/\omega_n] \sinh \omega_n t. \tag{3.16}$$

In this case, we do *not* have a periodic solution: The positive exponent indicates that typically $x \to \infty$ as $t \to \infty$. Hence, our heuristic concept of stability indicates that this behavior is unstable. However, we also observe that we can choose very specific initial conditions (unlikely but nevertheless important cases) in which the trajectory will end up at the origin, that is, where the positive exponential term is completely suppressed, as well as the case in which the negative exponential term in Eq. (3.15) dominates for a short time before the trajectory is swept away. These cases are illustrated in Fig. 3.2(b). For all practical purposes, that is, arbitrary initial conditions, the motion is clearly unstable. The meaning of the special trajectory will be discussed at length later.

Damping. The preceding examples are somewhat unrealistic in terms of practical mechanics because they do not include energy dissipation [10]. With the inevitable presence of damping the question of stability becomes less ambiguous. Typical motion will then consist of a transient followed by some kind of recurrent long-term behavior. This brings us to the fundamentally important concept of *attractors*. These are the special solutions alluded to earlier, and they play a key role in organizing dynamic behavior in phase space (the space of the state variables). We shall also see that for nonlinear systems *unstable* solutions have an important influence on the general nature of solutions.

Suppose we now allow for some energy dissipation in the form of linear viscous damping, that is, $c \neq 0$ in Fig. 3.1. The equation of motion is now

$$\ddot{x} + 2\zeta \omega_n \dot{x} + \omega_n^2 x = 0, \tag{3.17}$$

into which a nondimensional damping *ratio*, $\zeta \equiv c/(2m\omega_n)$, has been introduced. Solutions to this equation now depend on the value of ζ. For underdamped systems, we have $\zeta < 1$ and solutions of the form

$$x(t) = e^{-\zeta \omega_n t} \left[\frac{\dot{x}(0) + \zeta \omega_n x(0)}{\omega_d} \sin \omega_d t + x(0) \cos \omega_d t \right], \tag{3.18}$$

where the damped natural frequency ω_d is given by

$$\omega_d = \omega_n \sqrt{1 - \zeta^2}. \tag{3.19}$$

A typical underdamped response ($\zeta = 0.1$) is shown in Figs. 3.3(a) and 3.3(b) as a time series and phase portrait, respectively. The origin in Fig. 3.3(b) indicates a position of asymptotically stable equilibrium; that is, any disturbance leads to a dynamic response that moves smoothly back to equilibrium. The trajectory gradually spirals down to this rest state: We can imagine a family of trajectories forming a *flow* as time evolves. Because this equilibrium is unique, the whole of the phase space is the attracting set for all initial conditions and disturbances [11]. Damping in this range, e.g., $\zeta \approx 0.1$, is quite typical for mechanical and structural systems.

For a heavily (or overdamped) system $\zeta > 1$, and in this case the form of the solution is

$$x(t) = A e^{(-\zeta + \sqrt{\zeta^2 - 1})\omega_n t} + B e^{(-\zeta - \sqrt{\zeta^2 - 1})\omega_n t}, \tag{3.20}$$

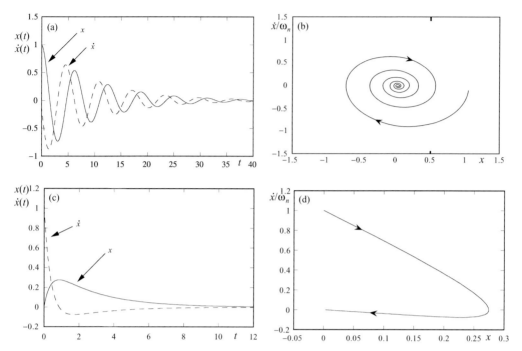

Figure 3.3. (a) Time series and (b) phase portraits for underdamped (oscillatory) motion, $x(0) = 1.0$, $\dot{x}(0) = 0.0$, $\zeta = 0.1$; (c) and (d), overdamped (nonoscillitory) motion, $x(0) = 0.0$, $\dot{x}(0) = 1.0$, $\zeta = 1.5$.

where

$$A = \frac{\dot{x}(0) + (\zeta + \sqrt{\zeta^2 - 1})\omega_n x(0)}{2\omega_n\sqrt{\zeta^2 - 1}}, \tag{3.21}$$

and

$$B = \frac{-\dot{x}(0) - (\zeta - \sqrt{\zeta^2 - 1})\omega_n x(0)}{2\omega_n\sqrt{\zeta^2 - 1}}. \tag{3.22}$$

The motion is a generally monotonically decreasing function of time and may take a relatively long time to overcome rather heavy damping forces on the way to equilibrium. A typical case is also shown in Figs. 3.3(c) and 3.3(d).

The boundary between these two cases is the *critically* damped case, i.e., $\zeta = 1$. In the context of this book, we will regularly encounter the situation in which the stiffness of a system degrades, and given the definition of ζ we expect not only a reduction in the natural frequency but also an increase in the damping ratio.

Returning to the state-variable-matrix format of the linear oscillator, we therefore have

$$\begin{bmatrix} \dot{x} \\ \dot{y} \end{bmatrix} = \begin{bmatrix} 0 & 1 \\ -\omega_n^2 & -2\zeta\omega_n \end{bmatrix} \begin{bmatrix} x \\ y \end{bmatrix}. \tag{3.23}$$

We can also write the solution in terms of the eigenvalues of the state matrix, that is, the roots of the characteristic equation

$$\lambda^2 + 2\zeta\omega_n\lambda + \omega_n^2 = 0. \tag{3.24}$$

Critical damping thus relates to the discriminant's being equal to zero.

Given the scenario of a system losing stability, we can usefully view all the response possibilities of this type of linear system according to the location of the roots in the complex plane. For example, having two complex roots with negative real parts corresponds to an exponentially decaying oscillation. Summarizing these outcomes leads to Fig. 3.4 [12]. In general, we will have a system with positive stiffness and damping and thus a root structure corresponding to the upper-right-hand quadrant. Critical damping corresponds to the parabola, and phase portraits and eigenvalues are indicated for various combinations of the natural frequency and damping. The system eigenvectors organize the transient behavior in the phase portrait. Some useful terminology here includes the *spiral* or *focus* for decaying oscillatory motion (also called a *sink*), the *node* for overdamped motion, the *inflected node* for equal eigenvalues (and thus including the critically damped case), and the *saddle* for the motion characterized by having both a stable and unstable direction

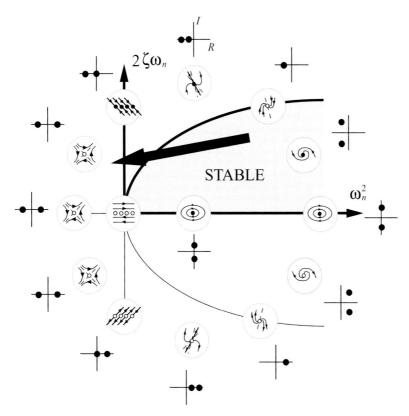

Figure 3.4. Phase portraits and root structure of a linear oscillator (after Thompson [12]).

(eigenvector) with instability becoming dominant. We can also view the undamped case as a *center*.

We shall focus more on higher-order dynamical systems in the next chapter, but more formally we state (see Jordan and Smith [13]) that if we have a dynamical system $\dot{x} = Ax$, where A is constant with eigenvalues $\lambda_i, i = 1, 2, ..., n$, then

(1) if the system is stable, $\text{Re}\{\lambda_i\} \leq 0, i = 1, 2, \ldots, n$;
(2) if either $\text{Re}\{\lambda_i\} < 0, i = 1, 2, \ldots, n$, or if $\text{Re}\{\lambda_i\} \leq 0, i = 1, 2, \ldots, n$ and there is no repeated eigenvalue, the system is uniformly stable;
(3) the system is asymptotically stable if and only if $\text{Re}\{\lambda_i\} < 0, i = 1, 2, \ldots, n$ [and then it is also uniformly stable, by (2)];
(4) if $\text{Re}\{\lambda_i\} > 0$ for any i, the solution is unstable.

We thus observe what will typically happen when the stiffness of the system degrades (e.g., because of an axial load acting on a slender structure), and this is indicated by the large arrow in Fig. 3.4. For a small amount of damping, the eigenvalues start off as a complex-conjugate pair with negative real parts. As the stiffness (and hence natural frequency) reduces, the eigenvalues merge on the negative real axis, and then their magnitudes diverge, with one entering the positive half-plane. Thus instability occurs, and solutions grow without bound.

Although the preceding description relates to a single-degree-of-freedom (SDOF) linear oscillator, this type of scenario is encountered to a large extent within a variety of high-order and nonlinear systems. The geometric view afforded by a consideration of the root structure and phase portraits of *families* of solutions about equilibrium points is very useful. We will make extensive use of linearization to utilize this view *locally* to equilibrium within a nonlinear context [14].

3.2 Oscillator with a Slow Sweep of Frequency

Consider again the spring–mass system shown in Fig. 3.1. Again assume that there is no damping or external forcing ($c = F(t) = z(t) = 0$), and that the spring stiffness decays linearly (in time) from a base value $k = 1$ at $t = 0$. We assume this decay is very slow, and characterized by a small parameter ϵ. In this case we can write the governing equation of motion as

$$\ddot{x} + \mu^2(t)x = 0, \tag{3.25}$$

in which

$$\mu^2(t) = \frac{k}{m}(1 - \epsilon t), \tag{3.26}$$

that is, the system will lose stability when $t \to 1/\epsilon$. If we assume that the evolution of the stiffness change is very slow ($\epsilon \ll 1$), that is, much slower than the natural oscillatory response of the system, then we have a number of approaches available for obtaining a solution $x(t)$ [15]. A direct numerical solution of Eq. (3.25) is easily obtained. Alternatively, a perturbation approach can be applied, e.g., using the multiple-scales technique [16], we obtain (to the leading term in the expansion)

$$x(t, \epsilon) = x^+(t^+, \epsilon) = \sqrt{\frac{\mu(0)}{\mu(\tilde{t})}} (\cos t^+), \qquad (3.27)$$

where the initial conditions are taken as unity, the plus sign indicates a solution forward in time, and

$$t^+ = \frac{1}{\epsilon} \int_0^{\tilde{t}} \mu(\tilde{s})d\tilde{s}. \qquad (3.28)$$

Inserting the specific form for the linear sweep [Eq. (3.26)], we get a solution

$$x(t) = (1 - \epsilon t)^{-1/4} \cos\left[-\frac{2}{3\epsilon}(1 - \epsilon t)^{3/2} + \frac{2}{3\epsilon} \right], \qquad (3.29)$$

where $k/m = 1$ has been used.

Consider the specific case of $\epsilon = 0.01$. In this case, we would expect the system to lose stability near $t = 100$. The numerical solution of Eq. (3.25) is shown in Fig. 3.5(a). Part (b) shows the perturbation solution [Eq. (3.29)], and part (c) shows an exact analytic solution in terms of Airy functions [17]. Part (c) also illustrates that the frequency and stability characteristics for this type of linear system are not influenced by initial conditions (which are different in this final case). In each case, the system loses stability when the stiffness has dropped to zero. We can view this

(a)

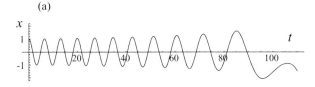

(b)

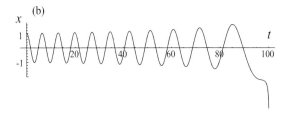

Decreasing frequency

(c)

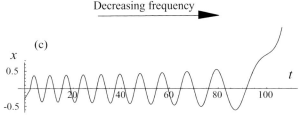

Figure 3.5. Some sweeps towards instability: (a) Numerical solution to Eq. (3.25), (b) a perturbation solution, and (c) an exact solution obtained with Airy functions.

scenario as buckling. However, in most practical systems, the decay in stiffness is not so simple, and the following chapters investigate a more comprehensive set of circumstances, including the effects of multiple degrees of freedom (MDOF), damping, and various geometrical effects.

3.3 Dynamics and Stability

In this section, we introduce some important concepts relevant to the equilibrium and stability of conservative systems, together with extensions to dissipative systems of the type in which we are mainly interested.

It is a simple matter to write the potential and kinetic energy for the spring–mass system:

$$V = \frac{1}{2}kx^2, \quad T = \frac{1}{2}m\dot{x}^2. \tag{3.30}$$

Applying Lagrange's equation yields Eq. (3.4). In Eq. (3.4), we can use $\ddot{x} = \dot{x}d\dot{x}/dx$, separate variables, and confirm that energy is conserved. In this simple case, both the potential and kinetic energies were positive and quadratic. Throughout this book we will encounter systems characterized by a potential-energy function that is not necessarily quadratic and whose form changes with a (load) parameter.

We also note the relation between energy and the natural frequency,

$$\omega_n^2 = \frac{d^2V}{dx^2} \Big/ \frac{d^2T}{d\dot{x}^2} \equiv \frac{V_{11}}{T_{11}}, \tag{3.31}$$

in which the subscripts refer to differentiation with respect to generalized position (for the potential energy) and generalized velocity (for the kinetic energy). This will form the basis of a number of approximate techniques, including Rayleigh's method [18], to be described later.

3.3.1 Stability Concepts

We have seen how, for conservative, nongyroscopic forces, we can write the potential energy as

$$V = V(Q_i), \tag{3.32}$$

and Lagrange's equation tells us that the condition for equilibrium (i.e., a stationary state, or no motion) is given by

$$V_i \equiv \frac{\partial V}{\partial q_i} = 0 \tag{3.33}$$

(for all i), which can be thus stated in words:

A stationary value of the total potential energy with respect to the generalized coordinates is necessary and sufficient for the equilibrium of the system.

Because we are primarily interested in systems that have a smooth potential-energy function [19], we can develop a Taylor series expansion about equilibrium,

$$V = V^E + \sum_{i=1}^{i=n} \frac{\partial V}{\partial Q_i}\bigg|^E q_i + \frac{1}{2}\sum_{i=1}^{i=n}\sum_{j=1}^{j=n} \frac{\partial^2 V}{\partial Q_i \partial Q_j}\bigg|^E q_i q_j + \cdots +, \qquad (3.34)$$

where incremental coordinates $q_i \equiv Q_i - Q_i^E$ have been introduced. Now, if we make use of the tensor summation convention and define

$$\frac{\partial^2 V}{\partial Q_i \partial Q_j}\bigg|^E \equiv V_{ij}^E, \qquad (3.35)$$

we obtain the dominant quadratic form

$$V = \frac{1}{2} V_{ij}^E q_i q_j + \cdots + \qquad (3.36)$$

because $V^E \equiv V(Q_i^E)$ is an arbitrary constant (which we generally choose equal to zero), and the linear term automatically drops out by virtue of Eq. (3.33).

So far, we still have not fully considered stability in terms of energy. Although the notion of equilibrium enables considerable progress to be made in linear structural analysis, the presence of compressive axial loads demands further scrutiny. A theorem, which goes back to Lagrange, states [20]

A complete relative minimum of the total potential energy with respect to the generalized coordinates is necessary and sufficient for the stability of an equilibrium state of the system.

We must therefore examine the local form of the potential energy in the vicinity of equilibrium; that is, we need to determine the conditions for which Eq. (3.36) is a minimum, and for this we need V_{ij} to be positive-definite.

For the types of axially loaded structures considered in the next few chapters, we can write the dominant (quadratic) form of the potential energy as

$$V = \frac{1}{2} V_{ij}^E q_i q_j + \cdots + \qquad (3.37)$$

$$= \frac{1}{2}\left(U_{ij} q_i q_j - P^k \epsilon_{ij}^k q_i q_j \right), \qquad (3.38)$$

where U_{ij} is the strain energy and P^k is a set of loads with corresponding movement ϵ_{ij}^k. In matrix notation, we can also write

$$V = \frac{1}{2} V_{ij}^E q_i q_j = \frac{1}{2} \mathbf{q}^T K \mathbf{q}, \qquad (3.39)$$

in which K is the effective stiffness matrix. For conservative systems, this matrix is symmetric.

Similarly, most systems of interest will have a quadratic kinetic energy of the form

$$T = \frac{1}{2}T_{ij}^E \dot{q}_i \dot{q}_j = \frac{1}{2}\dot{\mathbf{q}}^T M \dot{\mathbf{q}}, \tag{3.40}$$

in which M is the mass matrix. Placing the general expressions for the energy in Lagrange's equation then yields

$$T_{ij}^E \ddot{q}_i + V_{ij}^E q_i = 0 \tag{3.41}$$

in terms of the generalized coordinates q_i. It can be shown (and will be in a later chapter) that the coordinates can be transformed into principal coordinates u_i such that the equations of motion become decoupled:

$$T_{ii}^E \ddot{u}_i + V_{ii}^E u_i = 0. \tag{3.42}$$

In this case, we can write all the natural frequencies:

$$\omega_i^2 = \frac{V_{ii}^E}{T_{ii}^E}. \tag{3.43}$$

Of course, there are a number of important issues underlying these last few expressions. Further aspects of these concepts for MDOF and continuous systems are considered in the next chapter.

3.4 Bifurcations

We have already seen how the loss of stiffness in a linear oscillator leads to instability. In a practical situation, the stiffness may not degrade in a linear fashion, and instability may not lead to solutions that lose stability completely. The behavior of the linear oscillator provides an informative *local* view of behavior, but in a practical situation, we might expect nonlinear effects to limit the response in some way. Bifurcation theory can be used to classify instability phenomena based on generic behavior [5, 21–25]. In other words, as a control parameter is varied (e.g., the axial load on a structure), what happens to a system as a critical condition is reached? In this section, we find some generic forms of bifurcation, assess the role played by initial imperfections, and relate these back to the form of underlying potential energy, thus paving the way for the analysis of the dynamics and buckling of structures to be considered throughout this book [26].

Clearly the passage of an eigenvalue through to the positive real half-plane leads to a qualitative change in the phase portrait, that is, the behavior of trajectories in the local vicinity of an equilibrium point. As a parameter is (slowly) varied, the response of a system changes (often gradually), but it is the *qualitative* change in the dynamics that is classified as a bifurcation [5]. Although an elementary classification of bifurcations is based on a one-dimensional (1D) description, we will focus attention on two dimensions, which are fundamentally 1D (based on center manifold theory [27]), as we are primarily interested in oscillations that result from application of Newton's second law to mechanical systems.

3.4.1 The Saddle-Node Bifurcation

The *saddle-node bifurcation* is the fundamental instability mechanism of a system under the action of a single control parameter. The simplest form of a saddle-node bifurcation is given by

$$\dot{x} = \mu - x^2. \qquad (3.44)$$

The control parameter μ and coordinate x are linked quadratically [28]. However, to maintain a meaningful relationship with vibration, we incorporate this relation into the context of a lightly damped oscillator:

$$\ddot{x} + 0.1\dot{x} + x^2 - \mu = 0. \qquad (3.45)$$

Equilibrium corresponds to the rest state, and thus

$$x_e = \pm\sqrt{\mu}. \qquad (3.46)$$

The stability of these equilibria can be determined in a number of ways, and we start by considering the oscillations resulting from a small perturbation. Let $x = x_e + \delta$, where δ is a small deviation from equilibrium. Placing this in Eq. (3.45), we get

$$\ddot{\delta} + 0.1\dot{\delta} + x_e^2 + 2x_e\delta + \delta^2 - \mu = 0. \qquad (3.47)$$

By definition $x_e^2 - \mu = 0$, and neglecting δ^2 (because δ is small) we obtain

$$\ddot{\delta} + 0.1\dot{\delta} + 2x_e\delta = 0. \qquad (3.48)$$

This describes the dynamic response of small perturbations about equilibrium. Substituting in the expression for equilibrium, Eq. (3.46), results in

$$\ddot{\delta} + 0.1\dot{\delta} \pm 2\sqrt{\mu}\delta = 0. \qquad (3.49)$$

Taking the positive sign we have a response that oscillates with a frequency a little less than $\omega_n^2 = 2\sqrt{\mu}$. The damping causes the motion to decay back to equilibrium.

Taking the negative sign we have negative stiffness and a solution that grows with time. These are, of course, familiar from the start of this chapter. They can also be located in the root structure of Fig. 3.4. We can also arrive at the same conclusion by using energy considerations. The potential energy associated with the saddle-node can be written as

$$V = \frac{x^3}{3} - \mu x \qquad (3.50)$$

and equilibrium from

$$V_1 \equiv \frac{dV}{dx} = x^2 - \mu. \qquad (3.51)$$

We have already seen how the sign of the curvature of the potential energy governs stability:

$$V_{11} = 2x, \qquad (3.52)$$

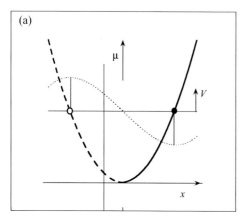 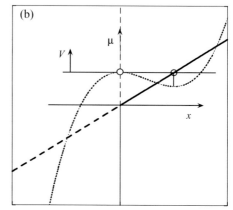

Figure 3.6. (a) A saddle-node bifurcation and (b) a transcritical bifurcation.

which can be evaluated about equilibrium. When $x_e = \sqrt{\mu}$, the second derivative of the potential-energy function is positive, indicating that this is a minimum and hence is stable. The opposite conclusion can be drawn from the other equilibrium branch, thus confirming the results of the stability properties based on the decay or growth of local perturbations.

A typical graphical representation of this situation is shown in Fig. 3.6(a). Suppose we have a system with a relatively large positive μ. In this case, there is a stable and an unstable equilibrium, characterized by a local minimum and a local maximum of the underlying potential energy, respectively. We can imagine the oscillations of a small ball rolling on this potential-energy *surface* (shown dotted). As the value of μ is reduced the two equilibria come together (the frequency of small oscillations will decrease and effective damping increases) as the potential surface flattens out. Just prior to coalescence the stable equilibrium can be thought of as a node, and the unstable equilibrium remains a saddle. Hence their approach (at the critical point) is called a saddle-node bifurcation. No equilibria exist for negative μ and trajectories would simply be swept away. This instability is also sometimes referred to as a fold or limit point.

3.4.2 Bifurcations from a Trivial Equilibrium

Although the saddle-node is the key stability transition in a system under the action of a single control, there are many systems in mechanics in which some kind of initial symmetry is present. An example is the *transcritical bifurcation*. In the context of a second-order ordinary differential equation, we can write

$$\ddot{x} + 0.1\dot{x} + x^2 - \mu x = 0. \tag{3.53}$$

Following the same approach as for the saddle-node we obtain the situation illustrated in Fig. 3.6(b). Here there is a fundamental (trivial) equilibrium for negative μ that loses stability as μ passes the through the origin (from negative to positive). The

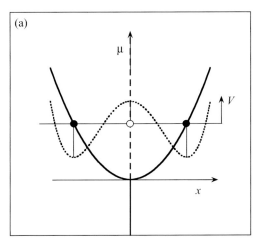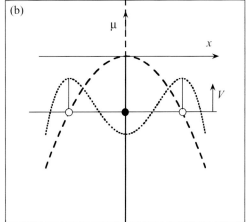

Figure 3.7. (a) A supercritical pitchfork bifurcation and (b) a subcritical pitchfork bifurcation.

other equilibrium becomes stable at this point, and deflection occurs in the positive
x direction.

The final pair of bifurcations is associated with the loss of stability of the trivial
solution and has global symmetry. It represents an important class of instability in
structural mechanics: supercritical and subcritical pitchfork bifurcations [19]. For
the *supercritical pitchfork* bifurcation we can consider the oscillator

$$\ddot{x} + 0.1\dot{x} + x^3 - \mu x = 0. \tag{3.54}$$

Again, we observe the $x_e = 0$ solution, which is stable for $\mu < 0$. At $\mu = 0$ a secondary equilibrium intersects the fundamental, and it can be shown that the two
(symmetric) nontrivial solutions are stable [see Fig. 3.7(a)]. This situation corresponds to the classic "double-well" potential that is also shown superimposed for a
specific (positive) value for μ. The corresponding *subcritical pitchfork* bifurcation,

$$\ddot{x} + 0.1\dot{x} + x^3 + \mu x = 0, \tag{3.55}$$

is shown in Fig. 3.7(b). In this case, suppose we start from a negative value of μ.
The trivial equilibrium is again stable but now, when the critical point is reached,
the system becomes completely unstable. Furthermore, as the critical point is approached, the potential-energy maxima associated with the adjacent saddles start to
erode the size of allowable perturbations. This is an important consequence of the
nonlinearity in the system. Although these last two bifurcations have the same stable trivial equilibrium and critical points, they have quite different consequences if
encountered in practice. Hence they are sometimes characterised as *safe* or *unsafe*
according to whether a local or adjacent postcritical stable equilibrium is available.

3.4.3 Initial Imperfections

It has already been mentioned that initial geometric imperfections or load eccen-
tricities may have a relatively profound effect on stability. We consider this type of

effect and its influence on the supercritical pitchfork. Incorporating a small offset causes Eq. (3.54) to be altered to

$$\ddot{x} + 0.1\dot{x} + x^3 - \mu x + \epsilon = 0, \tag{3.56}$$

where ϵ is a small parameter, which breaks the symmetry. Figure 3.8(a) shows how the instability transition is changed. Now, for large negative μ we have a primary equilibrium slighlty offset from $x = 0$, and this simply grows as μ approaches and then passes beyond the critical value for the perfect geometry (the origin). There is also a *complementary* solution for negative x, but this would not ordinarily be obtained under a natural loading history, that is, as μ is monotonically increased. However, the complementary solution does possess a critical point, and this is actu-ally a saddle-node bifurcation (which would be encountered if μ was initially large and x negative, and then μ were reduced). We also note the small tilt in the po-tential-energy function. Furthermore, the complementary solution has an effect on very large-amplitude motion and strong disturbances, and this will be revisited in a later chapter.

Initial imperfections have little effect on the saddle-node bifurcation, but have a very important effect on the subcritical pitchfork bifurcation, which is termed *im-perfection sensitive*; that is, the magnitude of the critical load is considerably reduced in the presence of imperfections. For the transcritical bifurcation, the reduction of the maximum critical load occurs for some imperfections but not all.

Finally, we relate the natural frequency to governing potential energy (which depends on the control parameter) and these aforementioned generic instabilities. We have already seen that a typical static instability is associated with the loss of stiffness and hence a natural frequency that drops to zero. We have also seen how Rayleigh's method relates the square of the natural frequency to the second deriva-tive of potential energy evaluated at equilibrium (and assuming a quadratic form for the kinetic energy, and thus a constant second derivative). Thus we can extract

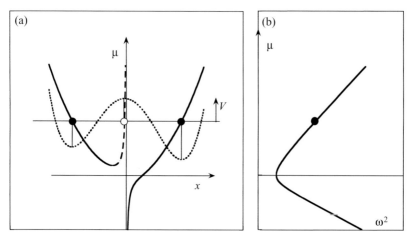

Figure 3.8. (a) A perturbed supercritical pitchfork bifurcation and (b) corresponding natural fre-quency (for the primary branch).

a relation between the natural frequency and the control parameter in the presence of an imperfection. This is shown schematically in Fig. 3.8(b). We see that the natural frequency of small oscillations reduces to a minimum in the vicinity of $\mu = 0$ but then starts to increase as the system moves along the post-critical equilibrium path. This characteristic curve will be repeatedly encountered throughout this book. Often we will see that the control is a loading parameter and the natural frequency relates to small lateral oscillations.

3.4.4 Bifurcations of Maps

There is an analogous set of generic instabilities for discrete maps, that is, dynamical systems governed by equations of the form

$$\mathbf{x}_{i+1} = F(\mathbf{x}_i, \mu). \tag{3.57}$$

For example, in the linear 1D system we would have $x_{i+1} = ax_i$, where a is a constant, and we can immediately write the solution $x_i = x_0 a^i$. From this we see that as a parameter μ is changed we can use similar stability arguments to continuous systems but now with the magnitude of the (mapping) eigenvalues $|a|$ required to be less than one, that is, inside the unit circle. Discrete systems are considered later in this book, when we look at periodically excited systems (e.g., a column with a pulsating end load), and Poincaré sampling. It will be seen that the loss of stability of periodic motion may occur quite routinely in forced nonlinear vibration problems.

3.5 A Simple Demonstration Model

Now we bring a number of the preceding issues together by way of a demonstration model. Consider the slender, flexible system shown in Fig. 3.9. This cable model was attributed to Brooke Benjamin by Iooss and Joseph [21]. This simple system exhibits an unstable symmetric (subcritical pitchfork) bifurcation that subsequently stabilizes for large deflections. We conduct a qualitative analysis of this system by associating a *load* parameter μ with the length of the cable (measured from the critical value), and a displacement q associated with a general out-of-plane deflection [29]. The flexible cable (sometimes used as curtain wire) is somewhat unusual because it possesses a moment–curvature relation that exhibits a softening spring effect, that is, the subcritical bifurcation manifests itself as a sudden motion from in-plane to a drooped out-of-plane position [30].

A *qualitative* form of the underlying potential function [31, 32] for this system can be written as

$$V = \frac{1}{720}q^6 - \frac{1}{24}q^4 - \frac{1}{2}\mu q^2. \tag{3.58}$$

That is, a function reflecting the global symmetry of the system and anticipated equilibria, which are given by the solutions of

$$V_1 = \frac{1}{120}q^5 - \frac{1}{6}q^3 - \mu q = 0. \tag{3.59}$$

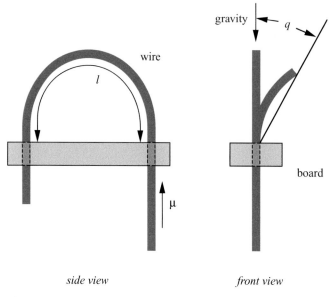

Figure 3.9. Schematic of an example bifurcation problem.

Thus equilibrium curves are given by

$$q = 0,$$

$$\mu = -\frac{1}{6}q^2 + \frac{1}{120}q^4, \tag{3.60}$$

and are shown in Fig. 3.10(a). The second derivative of potential energy is

$$V_{11} = \frac{1}{24}q^4 - \frac{1}{2}q^2 - \mu = 0, \tag{3.61}$$

and evaluating this expression along the equilibrium paths of Eqs. (3.60) we get

$$V_{11}^f = -\mu,$$

$$V_{11}^p = \frac{1}{24}(10 \pm \sqrt{100 + 120\lambda})^2$$

$$-\frac{1}{2}(10 \pm \sqrt{100 + 120\lambda}) - \mu. \tag{3.62}$$

The signs of these expressions thus indicate stability. Assuming a simple quadratic form for the kinetic energy, we can view Eqs. (3.62) as representing the natural frequencies. These are plotted as functions of the control (the length of the cable) in Fig. 3.10(b). We see a linear decay as the critical value is reached, followed by a finite jump to a higher frequency associated with the heavily drooped equilibrium. An initial imperfection can again be incorporated into the analysis starting from

$$V = \frac{1}{720}q^6 - \frac{1}{24}q^4 - \frac{1}{2}\mu q^2 + \epsilon q. \tag{3.63}$$

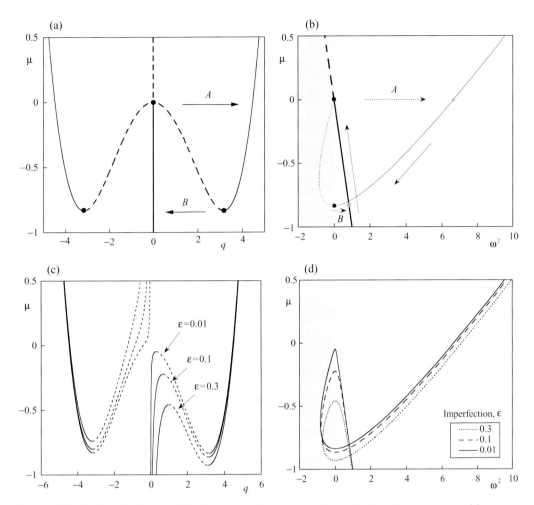

Figure 3.10. (a) Equilibrium and (b) dynamics of the wire for the initially perfect geometry; (c) and (d) include the effect of a small initial imperfection.

Plots of these relations are shown in Figs. 3.10(c) and 3.10(d) for some representative initial imperfections.

A Note on Linearization. Consider the specific case of $\mu = 0.0$ and $\epsilon = 0.0$. The governing equation of motion becomes

$$\ddot{q} + \frac{1}{120}q^5 - \frac{1}{6}q^3 = 0. \tag{3.64}$$

Expanding as a series about $q = q_e + \delta$ leads to

$$\ddot{\delta} + \delta \left[\frac{q_e^4}{24} - \frac{q_e^2}{2} \right] = 0, \tag{3.65}$$

and at this value the *remote* equilibrium position is at $q_e = \pm 4.48$ and thus $\omega^2 = 6.67$ (in addition to zero frequency of the upright configuration).

A Note on Damping. Incorporating a little viscous damping in our model gives

$$\ddot{q} + \beta \dot{q} + \frac{1}{120}q^5 - \frac{1}{6}q^3 - \mu q = 0. \tag{3.66}$$

Considering dynamics along the fundamental equilibrium path, that is, about $q_e = 0$, we have

$$\ddot{q} + \beta \dot{q} - \mu q = 0. \tag{3.67}$$

Assuming the solution is of the form $q = Ae^{\lambda t}$ we obtain the characteristic equation

$$\lambda^2 + \beta \lambda - \mu = 0, \tag{3.68}$$

with roots

$$\lambda = -\frac{\beta}{2} \pm \frac{1}{2}\sqrt{\beta^2 + 4\mu}. \tag{3.69}$$

We recall the familiar expression for damped natural frequencies,

$$\omega_d = \omega_n \sqrt{1 - \zeta^2}, \tag{3.70}$$

where $\zeta \equiv \beta/2\sqrt{-\mu}$, that is, because of the decaying stiffness we observe that the response becomes *critically damped* just prior to buckling when $\beta = 2\sqrt{-\mu}$. For example, with $\beta = 0.1$, oscillations would cease when $\mu = -0.0025$.

The length of the cable can be made to evolve as a linear function of time, and some numerical simulations are shown in Fig. 3.11 in which the rate of change of the length has been set such that the critical condition is reached after 300 time units. A small overshoot is encountered before the cable rapidly moves to one of its remote

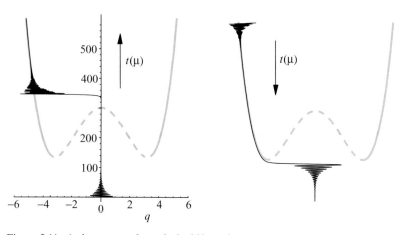

Figure 3.11. A slow sweep through the bifurcations.

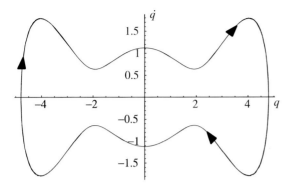

Figure 3.12. An undamped phase trajectory, $\mu = -0.5$, $q(0) = 4.8$, $\dot{q}(0) = 0$.

equilibria. Also shown in this figure is a reverse sweep when the length of the cable is gradually shortened. A region of hysteresis is readily observed.

So far, we have been mainly interested in the dynamic behavior in the vicinity of equilibrium. Figure 3.12 shows a phase trajectory based on a numerical simulation of the equation of motion when the length of the cable is such that the cable is stable in both the upright position or one of two highly drooped configurations. In this case the damping has been removed and initial conditions set such that the motion traverses across all the equilibrium positions. There are five equilibria at this length: stable at -4.04, 4.04, 0.0 and unstable at -1.92, 1.92. Of course, the presence of damping would cause any oscillations to decay such that the trajectory would end up at one of the (three) available attractors. This type of large-amplitude oscillation will be considered in the final chapter.

3.6 Experiments

To give the preceding behavior some practical context, some *qualitative* experimental work is now described for verification purpose [33]. A series of tests was conducted with the softening elastic cable. Four sets of hole separation were used: 10, 15, 20, and 25 cm. For each case, the length of the cable was increased and the lateral deflection was measured a short distance up from the base. The experimental system is shown in three stages of deformation in Fig. 3.13. In a uniform gravitational field the loop will reach a critical value, at which point it flops suddenly to one side, characteristic of a subcritical pitchfork bifurcation. Figure 3.14(a) shows the measured (equilibrium) results. In each case, a sudden jump to a severely drooped configuration is apparent, as well as the hysteresis on reduction of the cable length. In this case, the cable length is scaled by the hole separation. Frequencies were also measured (with a laser velocity vibrometer), and these are shown for a hole separation of 20 cm in Fig. 3.14(b).

Frequencies became increasingly difficult to measure very close to the critical length because of the shrinking nature of the attractive domain of initial condi-

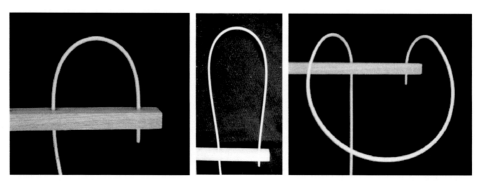

Figure 3.13. Front views of the loop: (a) short arc length, i.e., $\mu \ll 0$, prebuckled; (b) longer arc length, i.e., $\mu < 0$, prebuckled; (c) long length, i.e., $\mu > 0$ (or displaced from the trivial equilibrium if in region of hysteresis), postbuckled [33].

tions surrounding the equilibrium state. Also, because the stiffness is decreasing, the damping *ratio* is increasing and motion becomes increasingly sluggish. Frequencies associated with out-of-plane motion about both the prebuckled (upright) and postbuckled (drooped) configurations are plotted. The frequency reduces in a linear fashion as the critical length is approached, as suggested by the qualitative analysis summarized in Fig. 3.10(d), with $\mu = 0$ corresponding to a dimensional (arc) length of approximately 1 m. Frequencies corresponding to in-plane and twisting motion were also measured, although these were not included in the simple theoretical model of course. Taking measurements about the highly drooped configuration was difficult for the same reasons as previously mentioned, and these data do not correspond well to the simplistic theoretical model. A detailed analysis of this system can be found in Plaut and Virgin [33].

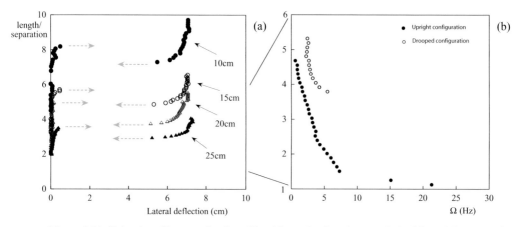

Figure 3.14. Behavior of loop made of a cable with a softening characteristic: (a) equilibrium paths for four different loops, (b) lowest frequency for the loop with a 20-cm hole separation (with an expanded y axis) [33].

References

[1]   L. Meirovitch. *Principles and Techniques of Vibrations*. Prentice Hall, 1997.

[2]   W.T. Thomson. *Theory of Vibration with Applications*. Prentice Hall, 1981.

[3]   B.H. Tongue. *Principles of Vibration*. Oxford University Press, 1996.

[4]   J.P. den Hartog. *Mechanical Vibrations*. Dover, 1984.

[5]   R. Seydel. *Practical Bifurcation and Stability Analysis*. Springer, 1994.

[6]   D.J. Inman. *Engineering Vibration*. Prentice Hall, 2000.

[7]   S.G. Kelly. *Fundamentals of Mechanical Vibrations*. McGraw-Hill, 1993.

[8]   J. Guckenheimer and P.J. Holmes. *Nonlinear Oscillations, Dynamical Systems, and Bifurcations of Vector Fields*. Springer-Verlag, 1983.

[9]   V.I. Arnold. *Geometrical Methods in the Theory of Ordinary Differential Equations*. Springer-Verlag, 1988.

[10] S.H. Crandall. The role of damping in vibration theory. *Journal of Sound and Vibration*, 11:3–18, 1970.

[11] R.H. Abraham and C.D. Shaw. *Dynamics: The Geometry of Behavior*. Aerial Press, 1982.

[12] J.M.T. Thompson. *Instabilities and Catastrophes in Science and Engineering*. Wiley, 1982.

[13] D.W. Jordan and P. Smith. *Nonlinear Ordinary Differential Equations*. Oxford University Press, 1999.

[14] A.H. Nayfeh and B. Balachandran. *Applied Nonlinear Dynamics*. Wiley, 1995.

[15] Y.A. Mitropolskii. *Problems of the Asymptotic Theory of Nonstationary Vibrations*. Daniel Davey, 1965.

[16] J. Kevorkian and J.D. Cole. *Perturbation Methods in Applied Mathematics*. Springer-Verlag, 1981.

[17] O. Vallee and M. Soares. *Airy Functions and Applications to Physics*. World Scientific, 2004.

[18] G. Temple and W.G. Bickley. *Rayleigh's Principle and Its Applications to Engineering*. Oxford University Press, 1933.

[19] J.M.T. Thompson and G.W. Hunt. *Elastic Instability Phenomena*. Wiley, 1984.

[20] H.L. Langhaar. *Energy Methods in Applied Mechanics*. Wiley, 1962.

[21] G. Iooss and D.D. Joseph. *Elementary Stability and Bifurcation Theory*. Springer-Verlag, 1980.

[22] M. Kubicek and M. Marek. *Computational Methods in Bifurcation Theory and Dissipative Structures*. Springer-Verlag, 1983.

[23] M. Golubitsky and D.G. Schaeffer. *Singularities and Groups in Bifurcation Theory*. Springer-Verlag, 1985.

[24] K. Huseyin. *Multiple Parameter Stability Theory and Its Applications*. Oxford University Press, 1986.

[25] H. Troger and A. Steindl. *Nonlinear Stability and Bifurcation Theory: An Introduction for Engineers and Applied Scientists*. Springer-Verlag, 1991.

[26] F. Bleich. *Buckling Strength of Metal Structures*. McGraw-Hill, 1952.

[27] R.D. Cook. *Concepts and Applications of Finite Element Analysis*. Wiley, 1981.

[28] L.N. Virgin. Parametric studies of the dynamic evolution through a fold. *Journal of Sound and Vibration*, 110:99–109, 1986.

[29] D. Hui and J.S. Hansen. The swallowtail and butterfly cuspoids and their application in the initial post-buckling of single-mode structural systems. *Quarterly of Applied Mathematics*, 38:17–36, 1980.

[30] L.N. Virgin and R.H. Plaut. Postbuckling and vibrations of linearly elastic and softening columns under self-weight. *International Journal of Solids and Structures*, 41:4989–5001, 2004.

[31] R. Thom. *Structural Stability and Morphogenesis*. Benjamin, 1975.

[32] T. Poston and I. Stewart. *Catastrophe Theory and Its Applications*. Pitman, 1978.

[33] R.H. Plaut and L.N. Virgin. Three-dimensional postbuckling and vibration of vertical half-loop under self-weight. *International Journal of Solids and Structures*, 41:4975–88, 2004.

4 Higher-Order Systems

4.1 Introduction

In the previous two chapters we focused on issues of modeling, equilibrium, and stability largely associated with SDOF systems. As such, the dynamic and stability behavior of a mass was largely characterized, for example, by a single frequency. However, there are many examples of systems with more than a SDOF, in which dynamics and stability issues are more involved. Certain characteristics of a physical system can be lumped at discrete locations, and this leads to *sets* of coupled ordinary differential equations. Linear algebra plays a key role in their analysis. Of course, most real structures are continuous and have an infinite number of DOFs. Governing equations of motion are typically partial differential equations (depending on both space and time), with boundary, as well as initial, conditions needing to be satisfied for a complete solution. Unlike typical (static) bending problems, which lead to inhomogeneous differential equations, the systems of primary focus in this book concern nontrivial homogeneous differential equations and often, in the analysis process, we will need to formulate and solve an eigenvalue problem: *algebraic*, for finite DOFs, and *differential*, for infinite DOFs. This will also typically involve the use of various approximation techniques, discretizations, and computational methods [1, 2]. This chapter serves the purpose of expanding the theoretical basis of dynamics and stability to this wider class of problems.

4.2 Multiple-Degree-of-Freedom Systems

Returning to the Lagrangian description (Section 2.5), we again focus attention on conservative systems and develop Lagrange's equation in matrix form. Restating Lagrange's equation, we have

$$\frac{\partial \mathcal{L}}{\partial q_j} - \frac{d}{dt}\frac{\partial \mathcal{L}}{\partial \dot{q}_j} = 0, \quad j = 1, 2, \dots, n, \tag{4.1}$$

in which again $\mathcal{L} \equiv T - V$. If we introduce the generalized coordinate and velocity vectors $\mathbf{q} = [q_1, q_2, \dots, q_n]^T$ and $\dot{\mathbf{q}} = [\dot{q}_1, \dot{q}_2, \dots, \dot{q}_n]^T$, respectively, we can rewrite Lagrange's equation in matrix form:

$$\frac{\partial \mathcal{L}}{\partial \mathbf{q}} - \frac{d}{dt}\frac{\partial \mathcal{L}}{\partial \dot{\mathbf{q}}} = 0. \tag{4.2}$$

The kinetic energy,

$$T = \frac{1}{2}\sum_{i=1}^{n}\sum_{j=1}^{n} m_{ij}\dot{q}_i\dot{q}_j = \frac{1}{2}\dot{\mathbf{q}}^T M\dot{\mathbf{q}}, \tag{4.3}$$

is typically a (quadratic) function of the generalized velocities (explicitly), $m_{ij} = m_{ij}(q_1, q_2, \ldots, q_n)$ and $M = [m_{ij}] = M^T$ is a symmetric $n \times n$ mass matrix of coefficients.

The potential energy is commonly a function of the generalized coordinates only; that is, $V = V(\mathbf{q})$. For linear conservative systems, we have again a quadratic form in which

$$V = \frac{1}{2}\sum_{i=1}^{n}\sum_{j=1}^{n} k_{ij} q_i q_j = \frac{1}{2}\mathbf{q}^T K\mathbf{q}, \tag{4.4}$$

and $K = [k_{ij}] = K^T$ is a symmetric $n \times n$ stiffness matrix. Throughout this book we will be interested in systems in which the work done by external loads (P) moving through a distance (e.g., end shortening, e) tends to diminish the strain energy stored in bending (U); that is, $V = U - Pe$.

4.2.1 The Algebraic Eigenvalue Problem

For *linear* conservative systems, we have a set of n simultaneous ordinary differential equations

$$M\ddot{\mathbf{q}}(t) + K\mathbf{q}(t) = 0, \tag{4.5}$$

where M and K are real symmetric matrices. Just as for the SDOF system we can assume an exponential form for the solution

$$\mathbf{q}(t) = e^{st}\mathbf{u}(t), \tag{4.6}$$

in which s is a constant and $\mathbf{u}$ is a constant n-dimensional vector. Plugging this solution into Eq. (4.5) leads to the algebraic eigenvalue problem

$$K\mathbf{u} = \lambda M\mathbf{u}, \tag{4.7}$$

where $\lambda = -s^2$, which will later be identified with frequency. We are often interested in the values of the parameter λ (and thus the natural frequencies) that lead to nontrivial solutions.

This is a standard problem in linear algebra, and we can facilitate the solution by taking advantage of certain properties of the matrices K and M. They are both real and symmetric, and M is also positive-definite (i.e., all its principal minor determinants are positive—Sylvester's criterion [2]). Equation (4.7) can be transformed into the standard form

$$A\mathbf{v} = \lambda\mathbf{v}, \tag{4.8}$$

in which $M = Q^T Q$, $A = (Q^T)^{-1} K Q^{-1}$, and the linear transformation $Q\mathbf{u} = \mathbf{v}$ has been used. Equation (4.8) can be written in the form

$$(A - \lambda I)\mathbf{v} = 0, \tag{4.9}$$

which admits nontrivial solutions if and only if the determinant of coefficients is zero, that is,

$$\det(A - \lambda I) = 0. \tag{4.10}$$

This is the *characteristic* equation, a generalization of the simple expression first encountered in Eq. (3.24), the solution of which corresponds to the roots of an nth-order polynomial. The values of λ that make $\det(A - \lambda I)$ zero are the eigenvalues of the system, and each eigenvalue λ_r has a corresponding eigenvector $\mathbf{v}_r$ that can be obtained from

$$(A - \lambda_r I)\mathbf{v}_r = 0, \quad r = 1, 2, \ldots, n, \tag{4.11}$$

assuming $\mathbf{v}_r \neq 0$. Various results from linear algebra can be used to show that because the matrix A is a real symmetric matrix then the eigenvalues and eigenvectors are also real, and the eigenvectors are orthogonal with respect to A and have arbitrary absolute magnitude. Some of these properties can be used to facilitate the solution to certain problems [2].

4.2.2 Normal Modes

One of the cornerstones of vibration theory is modal analysis. A brief description is given here, and we will return to this subject when considering distributed systems at the end of this chapter. We look again at the system

$$M\ddot{\mathbf{q}}(t) + K\mathbf{q}(t) = 0 \tag{4.12}$$

and seek a solution in the form

$$\mathbf{q}(t) = \Phi \mathbf{r}(t), \tag{4.13}$$

where we have introduced the matrix of normalized mode shapes (eigenvectors):

$$\Phi = [\mathbf{v}_1, \mathbf{v}_2, \ldots, \mathbf{v}_n]. \tag{4.14}$$

Now we substitute Eq. (4.13) into Eq. (4.12) and, premultiplying the result by Φ^T, we get

$$\Phi^T M \Phi \ddot{\mathbf{r}}(t) + \Phi^T K \Phi \mathbf{r}(t) = 0, \tag{4.15}$$

which can be written as

$$M_D \ddot{\mathbf{r}}(t) + K_D \mathbf{r}(t) = 0, \tag{4.16}$$

where $M_D = \Phi^T M \Phi$ and $K_D = \Phi^T K \Phi$. Alternatively, we can write

$$\ddot{\mathbf{r}}(t) + \omega_D^2 \mathbf{r}(t) = 0, \tag{4.17}$$

where $\omega_D^2 = M_D^{-1} K_D$. These equations are now uncoupled in modal (principal) co-ordinates $\mathbf{r}(t)$, and ω_D is a diagonal matrix consisting of the natural frequencies of the system.

It also turns out that this decoupling can be achieved if the (linear-viscous) damping is *proportional*; that is, it takes the form

$$D = aM + bK, \tag{4.18}$$

where a and b are real scalars, and it acts on $\dot{\mathbf{r}}(t)$. We emphasize that, although there is limited physical justification for this assumption, it results in strong analytical advantages, and because damping is difficult to model (and measure) and we are concerned with lightly damped systems this assumption is not unreasonable. Care must also be taken to transform the initial conditions and any external (often periodic) driving in the system. Initially we are focused on unforced systems, but we will revisit forced systems later [3].

4.2.3 Equilibrium, Linearization, and Stability

When structures are subject to axial loads, they may experience a loss of stability that is manifest in the local growth of motion in the vicinity of equilibrium: buckling occurs. We have already seen that unstable motion is characterized by the stiffness becoming negative which leads to an imaginary frequency ($\omega^2 < 0$) for a SDOF system, and we will now see how this relates to the behavior of MDOF systems.

Just as in the SDOF case it is the sign of the eigenvalues that is important for stability. It can be shown that the sign of the eigenvalues depends only on the matrix K. Typically, we will have a system which (at least *initially*) has a positive definite stiffness matrix. In this case, all the eigenvalues are positive and we can relate the eigenvalues to the natural frequencies using

$$\lambda_j = \omega_j^2 \quad j = 1, 2, \ldots n. \tag{4.19}$$

Returning to the original equations of motion (4.5), employing Euler identities, and assuming light, proportional damping, we thus have exponentially decaying, harmonic solutions in the jth mode of the form

$$r_j(t) = A_j e^{-\zeta_j \omega_j t} \cos(\omega_{dj} t - \phi_{dj}), \tag{4.20}$$

where the A's and ϕ's depend on the initial conditions [see (3.12)]. In terms of the original physical coordinates, we have

$$\mathbf{q}(t) = \sum_{j=1}^{n} \mathbf{v}_j r_j(t), \tag{4.21}$$

and thus

$$\mathbf{q}(t) = \sum_{j=1}^{n} \mathbf{v}_j A_j e^{-\zeta_j \omega_j t} \cos(\omega_{dj} t - \phi_{dj}). \tag{4.22}$$

It can be shown that careful choice of initial conditions can result in decaying sinusoidal motion in an isolated mode.

Although Eq. (4.20) represents a specific solution to the problem, the *general* solution of Eq. (4.5) will consist of a linear combination of modes, and will have the form (ignoring damping for the moment)

$$\mathbf{q}(t) = \sum_{j=1}^{n} \mathbf{q}_j(t) = \sum_{j=1}^{n} A_j \cos(\omega_j t - \phi_j)\mathbf{u}_j, \tag{4.23}$$

where the constants A_j and ϕ_j again depend on the initial conditions.

In similarity with the SDOF response, we can have (and in fact we are especially interested in) a situation in which the stiffness *ceases* to be positive. In this case, when the stiffness matrix may be non-positive definite, the eigenvalues may be negative (implying imaginary frequencies) and the exponents in the solution take the form

$$\pm\sqrt{|-\lambda_j|}, \tag{4.24}$$

and we see that the root with the positive sign leads to exponentially growing (unstable) motion. Even with just a single negative eigenvalue, the system is unstable because motions grow and become large [as in Eq. (3.15)].

The local phase portraits depicted in Fig. 3.4 can be extended to higher-order systems by generalizing the evolution of a trajectory in phase space, where now the phase space is defined by the $2n$-dimensional vector $\mathbf{x} = [\mathbf{q}^T \dot{\mathbf{q}}^T]$. Of course, this is not so easy to visualize and sometimes a phase *projection* is used. We are often interested in the behavior associated with equilibrium points; that is, $\mathbf{q} = \mathbf{q}_e$ is a constant, which implies $\dot{\mathbf{q}} = \dot{\mathbf{q}}_0 = 0$. And when $\mathbf{q}_0 = 0$, we have the trivial equilibrium solution.

Equilibrium is thus associated with constant solutions of Lagrange's equation, and for the systems of most interest in this book, the kinetic energy is a function of the generalized velocities only (i.e., it does not influence equilibrium), we have the familiar result

$$\left.\frac{\partial V}{\partial \mathbf{q}}\right|_{\mathbf{q}=\mathbf{q}_0} = 0, \tag{4.25}$$

that is, the potential energy at an equilibrium point has a stationary value. The potential energy is locally a quadratic form and the stiffness matrix is symmetric. The stiffness matrix is positive-definite if the quadratic form is positive for all nonzero vectors. However, for an elastic structure, we will often encounter a stiffness matrix that depends (nonlinearly) on a set of initial loads

$$\mathbf{K}(\mu)\mathbf{u} = \mathbf{f}, \tag{4.26}$$

where $\mathbf{K}$ is a matrix of *incremental* stiffness coefficients (a local Hooke's law), $\mathbf{u}$ is a vector of small generalized displacements, and $\mathbf{f}$ is a vector of small load increments. There are various methods used to convert this into a standard form typically including incremental linearization of $\mathbf{K}(\mu)$.

Returning to the question of stability, we are thus primarily concerned with continuous dynamical systems of the type

$$\dot{\mathbf{x}} = \mathbf{F}(\mathbf{x}, t), \tag{4.27}$$

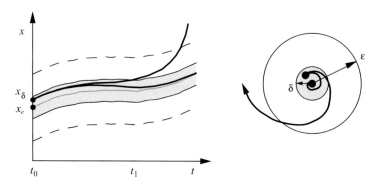

Figure 4.1. Lyapunov stability. For stability, close-by solutions remain within a small distance of equilibrium.

where $\mathbf{x}$ is an n-dimensional state vector and $\mathbf{F}$ is the corresponding (smooth) vector field. Given a set of initial conditions $\mathbf{x}_0$ we then have a unique solution at any later time. We are especially interested in the behavior of certain special solutions, for example, an equilibrium point $\mathbf{x}_e$ defined by

$$\mathbf{F}(\mathbf{x}_e, t) = 0. \tag{4.28}$$

From a stability point of view the important question to ask is: *what happens to a (homogeneous) dynamical system, that is, its state, if it is slightly perturbed from equilibrium,* $\mathbf{x}_e$ (which will typically be located at the origin). We will also later be interested in what happens to the stability of a system as its parameters are changed. Consider the trajectory $\mathbf{x}(t)$ in phase space coordinates, perturbed slightly at time $t = 0$ from equilibrium. We wish to monitor the distance (separation) between the perturbed trajectory and the equilibrium point as a function of time, and in order to do this, we introduce the concept of distance based on the Euclidian norm

$$\mathbf{r} = \|\mathbf{x} - \mathbf{x}_e\| = \sqrt{(\mathbf{x} - \mathbf{x}_e)^T (\mathbf{x} - \mathbf{x}_e)}. \tag{4.29}$$

We are now in a position to introduce the definitions of stability due to Lyapunov [4]:

(1) The equilibrium point $\mathbf{x}_e$ is Lyapunov stable if for any arbitrary positive ϵ and time t_0, there exists a $\delta(\epsilon) > 0$ such that if $\|\mathbf{r}(t_0)\| < \delta$ then $\|\mathbf{r}(t_0)\| < \epsilon$ for all $t \geq t_0$.
(2) The equilibrium point $\mathbf{x}_e$ is asymptotically stable if it is Lyapunov stable and $\lim_{t \to \infty} \|\mathbf{r}(\mathbf{t})\| = 0$.
(3) The equilibrium point $\mathbf{x}_e$ is unstable if there is an $\epsilon_a > 0$ such that for arbitrarily small $\delta > 0$ there is a motion $\mathbf{x}_a(t)$ for which $\|\mathbf{r}_a(t_0)\| < \delta$ and $\|\mathbf{r}_a(t_1)\| > \epsilon$ at some time $t_1 > t_0$.

Thus we see that with these definitions instability occurs when the perturbed trajectory reaches the sphere of radius ϵ in finite time. We can view this situation graphically as shown in Fig. 4.1. Clearly, for conservative systems we expect Lyapunov definition (1) to apply [5]. In typical structural systems, we would expect some damping to be present, and assuming it is of the form of linear viscous damping (derivable from Rayleigh's dissipation function) we might expect asymptotic stability to represent typical dynamical behavior for an autonomous structural system. Hence, after a small disturbance, a transient slowly decays back to the rest position. The presence of damping can considerably complicate the dynamic analysis of MDOF systems, although the adoption of proportional damping, as mentioned earlier, facilitates the modal analysis approach [6, 7].

We can return to some of the responses of the SDOF system to see how this more general definition encompasses the simple cases. In Figure 3.4, we showed how the response of a linear oscillator depended on the stiffness and damping. Thus we see the upper-right quadrant as corresponding to areas of aysmptotic stability. Since the primary focus of this book is axially loaded structures we will repeatedly encounter a decay in stiffness and thus a transition to instability (crossing the y axis). Although a decay in damping leads to instability as well, this type of *flutter* instability (usually initiated via a Hopf bifurcation [8]) is often associated with nonconservative forces (e.g., in aeroelasticity [9]) and is not a feature central to the dynamics of axially loaded structures of the type considered in this book, although one or two examples are given.

Considering disturbances about an equilibrium state, we write

$$\mathbf{x}(t) = \mathbf{x}_e + \eta(t), \tag{4.30}$$

which can be substituted back into Eq. (4.27) to give

$$\frac{d}{dt}[\mathbf{x}_e(t) + \eta(t)] = \mathbf{F}[\mathbf{x}_e(t) + \eta(t), t], \tag{4.31}$$

and because

$$\frac{d}{dt}\mathbf{x}_e(t) = \mathbf{F}[\mathbf{x}_e(t), t], \tag{4.32}$$

we can write

$$\frac{d}{dt}\eta(t) = \mathbf{F}[\mathbf{x}_e(t) + \eta(t), t] - \mathbf{F}[\mathbf{x}_e(t), t]. \tag{4.33}$$

This is the *variational equation*, and it plays a crucial role in determining the stability associated with certain solutions of dynamical systems.

We can expand the right-hand side of Eq. (4.33) by using

$$\mathbf{F}[\mathbf{x}_e(t) + \eta(t), t] = \mathbf{F}[\mathbf{x}_e(t), t] + \mathbf{DF}(\mathbf{x}_e, t)\eta(t) + \cdots +, \tag{4.34}$$

where $\mathbf{DF}(\mathbf{x}_e, t)$ is the matrix of first derivatives (the Jacobian) of $\mathbf{F}(\mathbf{x}_e, t)$ evaluated at $\mathbf{x}_e$:

$$\mathbf{DF}(\mathbf{x}_e, t) = \begin{bmatrix} \partial F_1/\partial x_1 & \partial F_1/\partial x_2 & \cdots & \partial F_1/\partial x_N \\ \partial F_2/\partial x_1 & \partial F_2/\partial x_2 & \cdots & \partial F_2/\partial x_N \\ \cdot & & & \cdot \\ \cdot & & & \cdot \\ \cdot & & & \cdot \\ \partial F_N/\partial x_1 & & \cdots & \partial F_N/\partial x_N \end{bmatrix}_{\mathbf{x}=\mathbf{x}_e}. \tag{4.35}$$

Now, assuming the perturbations are *small*, we can truncate the higher-order terms in Eq. (4.34) and thus we can write Eq. (4.33) as

$$\frac{d\eta}{dt} = \mathbf{DF}(\mathbf{x}_e, t)\eta(t). \tag{4.36}$$

This is called the linear variational equation for the system. In general, the coefficients of the Jacobian are time varying, and thus we have a *non-autonomous* system. We shall pay specific attention to systems in which the coefficients are periodic in Chapter 14, as this is a case commonly encountered in forced vibrations. For now, we focus on the stability of equilibrium, that is, $\mathbf{x}_e(t) = \mathbf{x}_e$ represents a fixed point in phase space, and the coefficients in the Jacobian matrix are constant, that is, an *autonomous system*.

Thus the focus on linear systems is justified on the grounds that nonlinear systems can often be linearized in the vicinity of equilibrium. By restricting ourselves to small neighborhoods about an equilibrium point, we can make use of much of the preceding linear theory. In terms of energy, and within a free-vibration context, we thus consider the behavior resulting from a small disturbance from an equilibrium point,

$$\mathbf{q}(t) = \mathbf{q}_0 + \delta(t), \quad \dot{\mathbf{q}}(t) = \dot{\delta}(t), \tag{4.37}$$

where δ and $\dot{\delta}$ are perturbation vectors. Because the nonlinearity will typically appear in the generalized positions (and assuming a smooth nonlinearity), we expand the potential energy as a Taylor series:

$$U(q_1, q_2, \dots, q_n) = U(\mathbf{q}_0) + \delta^T \frac{\partial U}{\partial \mathbf{q}}\bigg|_{\mathbf{q}=\mathbf{q}_0} + \frac{1}{2}\delta^T K\delta, \tag{4.38}$$

where higher-order terms have been neglected. K is the stiffness matrix (evaluated at equilibrium). The first term on the right-hand side is an arbitrary constant scalar (which we can choose to be zero), the second term is equal to zero by virtue of the equilibrium condition, and hence we can follow the earlier theory to arrive at the linearized equations of motion:

$$M\ddot{\delta} + K\delta = 0, \tag{4.39}$$

where K is related to the change in potential energy about an equilibrium position and is generally a function of a set of external (axial) loads, and M is a mass matrix.

Now suppose we revisit the stability of a set of ordinary differential equations in terms of the phase space. That is, we solve the algebraic eigenvalue problem arising from the equations of motion in first-order state-variable format. We also add a little damping (with a matrix of damping coefficients C) and introduce the perturbation state-vector as $\mathbf{x}(t) = [\mathbf{q}^T \dot{\mathbf{q}}(t)^T]$, so that the set of equations (4.5) can be written (following Meirovitch [2]) as

$$\dot{\mathbf{x}}(t) = A\mathbf{x}(t), \tag{4.40}$$

where

$$A = \begin{bmatrix} 0 & I \\ -M^{-1}K & -M^{-1}C \end{bmatrix}. \tag{4.41}$$

We see that this is the matrix generalization of Eq. (3.11). Assuming an exponential form for the solution

$$\mathbf{x}(t) = e^{\lambda t}\mathbf{x} \tag{4.42}$$

and placing in Eq. (4.40) (canceling $e^{\lambda t}$), we get

$$A\mathbf{x} = \lambda\mathbf{x}. \tag{4.43}$$

This is again the algebraic eigenvalue problem. A set of n second-order ordinary differential equations can always be converted to a set of $2n$ first-order ordinary differential equations. The general solution is given by

$$\mathbf{x}(t) = \sum_{j=1}^{2n} c_j e^{\lambda_j t} \mathbf{x}_j, \tag{4.44}$$

where the coefficients c_j are generally complex and depend on the initial conditions. Specifically, the eigenvalues, which can be written as

$$\lambda_j = \alpha_j + i\beta_j, \quad j = 1, 2, \ldots, 2n, \tag{4.45}$$

govern both the frequency and stability of the response by means of their imaginary and real parts, respectively. With positive stiffness, and in the absence of damping, we expect purely imaginary eigenvalues, that is, $\alpha_j = 0$ (which is a stable situation according to part 2 of the Lyapunov stability definition). We can think of eigenvalues distributed along the imaginary axis in the complex plane (akin to the positive x axis in Fig. 3.4). With the addition of damping, we expect to have $\alpha_j < 0$ and asymptotic stability (including both the focus and node illustrated in Fig. 3.4, which depend on the *level* of damping). Analyzing vibration problems by solving a set of second-order systems is equivalent to converting and then solving a set of (twice as many) first-order systems. The eigenvalues of the two approaches are linked by Eq. (4.19). Although the second-order approach allows a more physical feel (in terms of natural frequencies and damping ratios), the first-order approach is popular in control theory, is well suited to systems in which coupled modes are present, is an approach very well suited to software packages like MATLAB [10], and, as we have seen, allows a more direct assessment of stability.

We shall often encounter a situation in which axial loading reduces the lateral stiffness of a structure such that the potential energy can be written as

$$V = U - W = \frac{1}{2}\sum_{i=1}^{n}\sum_{j=1}^{n}(a_{ij} - Pb_{ij})q_iq_j, \tag{4.46}$$

where the strain energy U (and matrix a_{ij}) are positive definite and the work done by the axial loads (and b_{ij}) are either indefinite or positive definite (semi-definite).

Placing this in the context of a set of lightly damped oscillators (and assuming a single load for simplicity) we then have

$$M\ddot{\mathbf{q}} + C\dot{\mathbf{q}} + (K_L - PK_G)\mathbf{q} = 0, \tag{4.47}$$

where C is the damping matrix and K_G is a matrix relating to the (prestressing) effect of the axial load. We typically start with $P = 0$ and thus have initial stability, and then a major question is to determine the specific (or critical) values of P that cause instability, and how this is reflected in dynamic behavior.

4.2.4 Routh–Hurwitz Criterion

In the previous subsection, we saw how the stability of a system depended on the sign of the real parts of the eigenvalues [Eq. (4.45)]. To determine these signs, we need not solve the complete eigenvalue problem. Rather, a technique, referred to as the Routh–Hurwitz criterion, can be used [11]. Writing the characteristic equation in the form

$$\lambda^n + a_1\lambda^{n-1} + a_2\lambda^{n-2} + \cdots + a_n = 0, \tag{4.48}$$

it can be shown that the system is asymptotically stable (i.e., the real parts of the eigenvalues are negative) if and only if the principal minors Δ of the $n \times n$ matrix

$$\begin{bmatrix} a_1 & 1 & 0 & 0 & \cdots & & 0 \\ a_3 & a_2 & a_1 & 1 & \cdots & & 0 \\ a_5 & a_4 & a_3 & a_2 & a_1 & 1 & \cdots & 0 \\ \cdot & & & & & & \cdot \\ \cdot & & & & & & \cdot \\ \cdot & & & & & & \cdot \\ \cdot & & \cdot & \cdot & \cdot & & a_n \end{bmatrix} \tag{4.49}$$

are all positive, that is,

$$\Delta_1 = a_1 > 0,$$

$$\Delta_2 = \begin{vmatrix} a_1 & 1 \\ a_3 & a_2 \end{vmatrix} > 0, \tag{4.50}$$

$$\Delta_3 = \begin{vmatrix} a_1 & 1 & 0 \\ a_3 & a_2 & a_1 \\ a_5 & a_4 & a_3 \end{vmatrix} > 0,$$

and so on. A necessary condition for asymptotic stability is that all the a's be positive. Computationally, this criterion may be more convenient to evaluate than the full determinant.

4.2.5 Lyapunov Functions

As seen in the previous subsection, we can draw important conclusions about the stability of a dynamical system without obtaining a complete solution to the equations of motion; computing the Lyapunov stability of a system directly may not be the most straightforward approach. Also, much of the preceding stability theory was based on linear behavior (in the vicinity of equilibrium). Nonlinearity and global stability can be assessed by use of the concept of Lyapunov functions [7, 12, 13].

This is sometimes referred to as Lyapunov's direct, or second, method [7], and it has been found to be especially useful in applications involving nonconservative forces and in control theory [14]. The concept is again based on assessing stability without solving the equations of motion and is closely related to energy for conservative systems. We introduce the Lyapunov function $E(\mathbf{x})$, which we assume is a continuous and differentiable scalar function of the vector $\mathbf{x}(t)$ and satisfies the following conditions:

(1) $E(\mathbf{x}) > 0$, for all values of $\mathbf{x}(t) \neq 0$,
(2) $\dot{E}(\mathbf{x}) \leq 0$, for all values of $\mathbf{x}(t) \neq 0$.

Now, if it can be shown that a Lyapunov function exists for a system, then that system is stable and is aymptotically stable if the time derivative of E is less than zero. Finding a Lyapunov function is not necessarily easy, and even if one cannot be found it does not imply that the system is unstable.

The total energy of a system is often a useful place to start when searching for a Lyapunov function. With this in mind we take a brief look at our general damped MDOF system:

$$M\ddot{\mathbf{q}} + C\dot{\mathbf{q}} + K\mathbf{q} = 0, \tag{4.51}$$

and assuming the mass, damping, and stiffness matrices are all symmetric and positive-definite, we can confirm the asymptotic stability of the system. Let a Lyapunov function be based on the total energy for this system as

$$E[\mathbf{q}(t), \dot{\mathbf{q}}(t)] = \frac{1}{2}[\dot{\mathbf{q}}^T(t)M\dot{\mathbf{q}}(t) + \mathbf{q}^T(t)K\mathbf{q}(t)]. \tag{4.52}$$

Clearly, this satisfies the first condition of Lyapunov's direct method, that is, $E(\mathbf{q}) > 0$. For the second part, we have

$$\dot{E}[\mathbf{q}(t)\dot{\mathbf{q}}(t)] = \dot{\mathbf{q}}^T M\ddot{\mathbf{q}} + \dot{\mathbf{q}}^T K\mathbf{q}, \tag{4.53}$$

which, by premultiplying Eq. (4.51) by $\dot{\mathbf{q}}^T(t)$, we obtain

$$\dot{E}[\mathbf{q}(t)\dot{\mathbf{q}}(t)] = -\dot{\mathbf{q}}^T C\dot{\mathbf{q}} \leq 0, \tag{4.54}$$

where C is positive-definite. That a damped spring–mass system is stable is no surprise, but for highly nonlinear and nonconservative problems this is often a convenient approach.

4.2.6 Rayleigh's Quotient

The previous subsections showed how it may be possible to estimate the stability of a system without having to solve the equations of motion. In a similar vein, it is possible to estimate the natural frequencies of a system (and hence stability) based on an approach developed by Lord Rayleigh for eigenvalue problems [15]. We used this approach in Chapter 3 for a SDOF system, but extend it here to MDOF systems, where it finds most utility in estimating the lowest natural frequency.

Given the standard eigenvalue problem,

$$A\mathbf{v}_r = \lambda_r \mathbf{v}_r, \quad r = 1, 2, \ldots, n, \tag{4.55}$$

where A is a real symmetric positive-definite $n \times n$ matrix, we can express the eigenvalues in the form

$$\lambda_r = \frac{\mathbf{v}_r^T A \mathbf{v}_r}{\mathbf{v}_r^T \mathbf{v}_r}, \quad r = 1, 2, \ldots, n, \tag{4.56}$$

in which the eigenvalues are ordered $\lambda_1 \leq \lambda_2 \leq \cdots \lambda_n$. The Rayleigh quotient is

$$\lambda(\mathbf{v}) = R(\mathbf{v}) = \frac{\mathbf{v}^T A \mathbf{v}}{\mathbf{v}^T \mathbf{v}}. \tag{4.57}$$

Again for a MDOF system [Eq. (4.5)] we have the eigenvalue problem

$$K\mathbf{u} = \lambda M\mathbf{u}, \tag{4.58}$$

and thus Rayleigh's quotient in this case is

$$R(\lambda, \mathbf{u}) = \frac{\mathbf{u}^T K \mathbf{u}}{\mathbf{u}^T M \mathbf{u}}. \tag{4.59}$$

We note that the numerator is the maximum potential energy and the denominator is sometimes referred to as the *zero-frequency* kinetic energy during an oscillation [16]. What makes this approach especially useful in a practical context is that the assumed vector need not be terribly accurate to produce a reasonable eigenvalue estimate. This is partly due to the fact that it is *stationary* when perturbed around any of the actual system eigenvectors. It can be shown that this corresponds to an upper bound (and hence if an alternative vector is chosen that gives a eigenvalue lower than the previous one, this is necessarily a better estimate). This approach will be generalized in the next section, but suffice to say here that if the chosen vectors **u** satisfy certain boundary conditions and resemble the actual mode shapes, then estimates of the lowest natural frequency (usually the most important) are often reasonably accurate.

A few other issues to consider before moving on to continuous systems are worth mentioning at this point. The definitions of stability introduced here are largely based (or at least specialized) for conservative systems (with the addition of a little damping) and the question of stability of an equilibrium position. Toward the end of this book we will focus on forced vibration problems, in which the stability of periodic behavior will need to be considered. In this case we will need to extend the concepts of stability to include orbital stability. Another aspect of the general eigenvalue problem is how the eigenvalues change as a system parameter changes. Of course, in the types of problems encountered in this book we will often have the case that the stiffness matrix degrades in some sense, but there are a number of instances (including in control theory) for which a minor perturbation of the eigenvalue problem is a key issue. Finally, computational aspects of the algebraic eigenvalue problem are cornerstones of efficient algorithms in software applications [17]. This will also be touched on later.

4.3 Distributed Systems

We have gone from SDOF to MDOF systems and now this is extended further to continuous systems, which can be thought of as having an infinite number of DOFs. Much of the preceding theory can be extended to these distributed systems, and indeed many methods have been developed to reduce the number of DOFs [2].

We start this section by returning to Hamilton's principle in order to derive Lagrange's equations and hence the equations of motion. In contrast to finite-dimensional systems we will see that this spatially continuous analysis results in a boundary-value problem characterized by a partial differential equation with appropriate boundary conditions. We focus our attention on beamlike structures characterized by deflection $w(x, t)$ and signify the boundary conditions at 0 and L over the 1D domain x. Hamilton's principle for conservative systems can be written (following Meirovitch [2]) as

$$\int_{t_1}^{t_2} \delta \mathcal{L} dt = 0, \quad \delta w(t = t_1) = \delta w(t = t_2) = 0, \tag{4.60}$$

where $\mathcal{L} = T - V$, and

$$T = \int_0^L \hat{T}(\dot{w}) \, dx, \tag{4.61}$$

$$V = \int_0^L \hat{V}(w, w', w'') \, dx. \tag{4.62}$$

The hats on $\hat{T}$ and $\hat{V}$ imply an energy density, an overdot is a time derivative, a prime is a spatial derivative, and we assume that the potential energy in this case may be a function of the second derivative of the displacement w, which will be the case for the strain energy of systems in bending, for example.

We need to take the variation of the Lagrangian in Eq. (4.60) and to do this (in terms of a Lagrangian density) we have

$$\delta \hat{\mathcal{L}} = \frac{\partial \hat{\mathcal{L}}}{\partial w} \delta w + \frac{\partial \hat{\mathcal{L}}}{\partial w'} \delta w' + \frac{\partial \hat{\mathcal{L}}}{\partial w''} \delta w'' + \frac{\partial \hat{\mathcal{L}}}{\partial \dot{w}} \delta \dot{w}, \tag{4.63}$$

and thus, evaluating the integrals and placing in Eq. (4.60) we get

$$
\int_{t_1}^{t_2} \left\{ \int_0^L \left[\frac{\partial \hat{\mathcal{L}}}{\partial w} - \frac{\partial}{\partial x}\left(\frac{\partial \hat{\mathcal{L}}}{\partial w'}\right) + \frac{\partial^2}{\partial x^2}\left(\frac{\partial \hat{\mathcal{L}}}{\partial w''}\right) - \frac{\partial}{\partial t}\left(\frac{\partial \hat{\mathcal{L}}}{\partial \dot{w}}\right) \right] \delta w\, dx \right.
$$

$$
- \left[\frac{\partial \hat{\mathcal{L}}}{\partial w'} - \frac{\partial}{\partial x}\left(\frac{\partial \hat{\mathcal{L}}}{\partial w''}\right) \right]\Bigg|_{x=0} \delta w(0, t)
$$

$$
+ \left[\frac{\partial \hat{\mathcal{L}}}{\partial w'} - \frac{\partial}{\partial x}\left(\frac{\partial \hat{\mathcal{L}}}{\partial w''}\right) \right]\Bigg|_{x=L} \delta w(L, t)
$$

$$
\left. - \frac{\partial \hat{\mathcal{L}}}{\partial w''}\Bigg|_{x=0} \delta w'(0, t) + \frac{\partial \hat{\mathcal{L}}}{\partial w''}\Bigg|_{x=L} \delta w'(L, t) \right\} dt = 0.
$$

We next choose the virtual displacements such that the variations at $(0, t)$ and (L, t) are zero and then we are left with

$$
\frac{\partial \hat{\mathcal{L}}}{\partial w} - \frac{\partial}{\partial x}\left(\frac{\partial \hat{\mathcal{L}}}{\partial w'}\right) + \frac{\partial^2}{\partial x^2}\left(\frac{\partial \hat{\mathcal{L}}}{\partial w''}\right) - \frac{\partial}{\partial t}\left(\frac{\partial \hat{\mathcal{L}}}{\partial \dot{w}}\right) = 0. \tag{4.64}
$$

This is called the Lagrange differential equation of motion. It is a partial differential equation with possible boundary conditions at $x = 0$,

$$
- \left[\frac{\partial \hat{\mathcal{L}}}{\partial w'} - \frac{\partial}{\partial x}\left(\frac{\partial \hat{\mathcal{L}}}{\partial w''}\right) \right]\Bigg|_{x=0} = 0, \tag{4.65}
$$

$$
w(0, t) = 0, \tag{4.66}
$$

$$
- \frac{\partial \hat{\mathcal{L}}}{\partial w''}\Bigg|_{x=0} = 0, \tag{4.67}
$$

$$
w'(0, t) = 0, \tag{4.68}
$$

and at $x = L$,

$$
\left[\frac{\partial \hat{\mathcal{L}}}{\partial w'} - \frac{\partial}{\partial x}\left(\frac{\partial \hat{\mathcal{L}}}{\partial w''}\right) \right]\Bigg|_{x=L} = 0, \tag{4.69}
$$

$$
w(L, t) = 0, \tag{4.70}
$$

$$
\frac{\partial \hat{\mathcal{L}}}{\partial w''}\Bigg|_{x=L} = 0, \tag{4.71}
$$

$$
w'(L, t) = 0. \tag{4.72}
$$

The system must satisfy two boundary conditions at each end (assuming the Lagrangian depends on the second spatial derivative of the displacement). Given homogeneous boundary conditions we can simplify Eq. (4.64) to the form of Eq. (2.36). Specific examples will be given in later chapters in which this type of boundary-value problem is solved by use of a variety of techniques.

4.3.1 The Differential Eigenvalue Problem

It will often be the case that the partial differential of motion can be subject to the separation of variables (temporal and spatial), that is, by assuming the motion is a sum of terms of the form

$$w(\mathbf{x}, t) = F(t)W(\mathbf{x}). \tag{4.73}$$

This allows for the consideration of ordinary differential equations. The temporal part of the solution will typically consist of harmonic motion (certainly for stable conservative systems) satisfying the initial conditions (see Chapter 2). The spatial part of the solution, together with the appropriate boundary conditions, constitutes a differential eigenvalue problem.

The actual form of the partial differential equation depends of course on the specific physics of the problem (e.g., we have already developed the case typically encountered for Euler–Bernoulli beams) including the boundary conditions. However, there are a few general conclusions that we can draw by using the concept of differential operators [18]. The general form is given by

$$\mathcal{K}[w(\mathbf{x}, t)] + \mathcal{M}\left[\frac{\partial^2 w(\mathbf{x}, t)}{\partial t^2}\right] = 0, \tag{4.74}$$

where $\mathcal{K}$ and $\mathcal{M}$ are (linear) homogeneous differential operators defined (for example) by

$$\mathcal{K} = a_1 + a_2\frac{\partial}{\partial x} + a_3\frac{\partial}{\partial y} + a_4\frac{\partial^2}{\partial x^2} + a_5\frac{\partial^2}{\partial x \partial y} + \cdots + . \tag{4.75}$$

$\mathcal{M}$ is very often a simple constant, and the spatial vector $\mathbf{x}$ is defined over a domain D. The form of the operator (i.e., the values of the coefficients) depends on the specific physical problem, for example, the general preceding form is appropriate for two-dimensional (2D) problems, including plates. Assuming harmonic motion we obtain the eigenvalue problem

$$\mathcal{K}[W(\mathbf{x})] = \lambda\mathcal{M}[W(\mathbf{x})], \tag{4.76}$$

where we recognize $\lambda = \omega^2$, and the boundary conditions

$$\mathcal{B}_\mu[W(\mathbf{x})] = 0, \tag{4.77}$$

where $\mathcal{B}_\mu$ is also a linear differential operator. We expect the trivial solution $W(\mathbf{x}) = 0$ to be present, but it is the values of the parameter λ (associated with the natural frequencies of the system) that gives rise to nontrivial solutions $[W(\mathbf{x}) \neq 0$ and satisfying the boundary conditions] that are of central importance. There are infinitely many of these values, and they are the eigenvalues of the system and the corresponding $W(\mathbf{x})$ are eigenfunctions. We observe certain similarities between the algebraic eigenvalue problem and the differential eigenvalue problem with the operators $\mathcal{K}$ and $\mathcal{M}$ having their relation with the stiffness and mass matrices of MDOF systems.

The boundary conditions did not appear explicitly in the algebraic eigenvalue problem for MDOF (matrix) systems. Essentially, they were contained in the matrices themselves, but for continuous systems they are a very important part of the problem and influence the functions $W(\mathbf{x})$. Eigenfunctions satisfy both the differential equation and the boundary conditions exactly. Comparison (test) and admissible functions satisfy certain of the boundary conditions, that is, both geometric and natural, and geometric, respectively. They also form the basis of a number of solution methods to be considered later. Differential operators may also be self-adjoint (typically for linear conservative systems), which is analogous to symmetric matrices in the algebraic eigenvalue problem [18]. Similarly the eigenvalue problem is said to be positive-definite if all its eigenvalues are greater than zero and the eigenfunctions exhibit orthogonality (to be discussed shortly).

4.3.2 Solution Methods

Modal Analysis and Truncation
For some differential eigenvalue problems, a closed-form solution may be available, and this will often be the case for linear systems with relatively simple geometry. The expansion theorem [2] allows the solution of a boundary-value problem by transforming it into an infinite set of ordinary differential equations—modal equations. Again we focus attention on self-adjoint systems, as well as on systems in which the eigenvalues do not depend on the boundary conditions.

Earlier, we saw how it was possible to decouple the equations of motion for a MDOF system by transforming the problem into modal coordinates. This idea carries through to infinite-dimensional systems [19]. Using operator notation, we typically have the system

$$\mathcal{K}[w(\mathbf{x}, t)] + \mathcal{C}\left[\frac{\partial w(\mathbf{x}, t)}{\partial t}\right] + \mathcal{M}\left[\frac{\partial^2 w(\mathbf{x}, t)}{\partial t^2}\right] = 0. \tag{4.78}$$

Assuming harmonic motion and without damping, we get the eigenvalue problem

$$\mathcal{K}[W(\mathbf{x})] = \omega^2 \mathcal{M}[W(\mathbf{x})], \tag{4.79}$$

with the boundary conditions

$$\mathcal{B}_\mu[W(\mathbf{x})] = 0, \quad \mu = 1, 2, \ldots, p, \tag{4.80}$$

that is, assuming that the boundary conditions are homogeneous (every term involves W) and that the differential expression is of the order of $2p$ [2].

Let us consider two distinct solutions of the eigenvalue problem: (λ_r, w_r) and (λ_s, w_s). Because of the self-adjoint nature of the system,

$$\int_D w_r \mathcal{M} w_s dD = \int_D w_s \mathcal{M} w_r dD, \tag{4.81}$$

$$\int_D w_r \mathcal{K} w_s dD = \int_D w_s \mathcal{K} w_r dD \tag{4.82}$$

we have the conditions of orthogonality of the eigenfunctions:

$$\int_D w_r \mathcal{M} w_s dD = 0 \quad \text{for } \omega_r^2 \neq \omega_s^2, \tag{4.83}$$

$$\int_D w_r \mathcal{K} w_s dD = 0 \quad \text{for } \omega_r^2 \neq \omega_s^2. \tag{4.84}$$

The magnitude of the eigenvectors is arbitrary, and thus it makes sense to normalize them in a consistent way. This can be achieved with

$$\int_D w_r \mathcal{M} w_s dD = \delta_{rs}, \quad r, s = 1, 2, \ldots, \tag{4.85}$$

$$\int_D w_r \mathcal{K} w_s dD = \delta_{rs} \omega_r^2, \quad r, s = 1, 2, \ldots, \tag{4.86}$$

where δ_{rs} is the Kronecker delta.

The expansion theorem [20] tells us that the system response is given by a linear combination of the eigenfunctions:

$$w(\mathbf{x}, t) = \sum_{r=1}^{\infty} a_r(t) w_r(\mathbf{x}). \tag{4.87}$$

We furthermore assume that proportional damping is now included, that is,

$$\int_D w_r \mathcal{C} w_s dD = \delta_{rs} 2\zeta_r \omega_r, \quad r, s = 1, 2, \ldots, \tag{4.88}$$

where ζ_r contains the modal damping ratios.

Placing Eq. (4.87) into Eq. (4.78) leads to

$$\mathcal{K}\left[\sum_{r=1}^{\infty} w_r(\mathbf{x}) a_r(t)\right] + \mathcal{C}\left[\frac{\partial}{\partial t}\sum_{r=1}^{\infty} w_r(\mathbf{x}) a_r(t)\right] + \mathcal{M}\left[\frac{\partial^2}{\partial t^2}\sum_{r=1}^{\infty} w_r(\mathbf{x}) a_r(t)\right] = 0. \tag{4.89}$$

Multiplying by $w_r(\mathbf{x})$ and integrating over the domain results in

$$\sum_{r=1}^{\infty} a_r(t)\delta_{rs}\omega_r^2 + \sum_{r=1}^{\infty} \dot{a}_r(t)\delta_{rs}2\zeta_r\omega_r + \sum_{r=1}^{\infty} \ddot{a}_r(t)\delta_{rs} = 0. \tag{4.90}$$

But orthogonality tells us this equation holds only if $r = s$, and thus

$$\ddot{a}_r(t) + 2\zeta_r\omega_r\dot{a}_r(t) + \omega_r^2 a_r(t) = 0, \quad r = 1, 2, \ldots. \tag{4.91}$$

Thus we arrive at an infinite set of (uncoupled) ordinary differential equations in terms of normal modes, assuming the diagonal form exists. This form is somewhat familiar from the consideration of principal coordinates in MDOF systems. Because, in practical situations, it is only the lowest few modes that are important (the higher modes typically do not contribute very much to the solution) it may be possible to use a truncated modal model. The following sections introduce a couple of popular (approximate) approaches to solving boundary-value problems that are developed from the concept of basis functions.

Rayleigh–Ritz

For continuous systems we can write Rayleigh's quotient (see Subsection 4.2.6) as

$$\omega^2 = R[w(\mathbf{x})] = \frac{\int_D w(\mathbf{x})\mathcal{K}[w(\mathbf{x})]dD}{\int_D w(\mathbf{x})\mathcal{M}[w(\mathbf{x})]dD}, \tag{4.92}$$

which is stationary in the class of kinematically admissible functions, at the eigenvalues. This is a very useful property for estimating especially the lowest eigenvalue of a continuous system [21]. The Rayleigh–Ritz method consists of replacing the differential eigenvalue problem with a set of algebraic eigenvalue problems and is based on assuming a solution of the form

$$w(\mathbf{x}) = \sum_{n=1}^{N} a_n(t)\theta_n(\mathbf{x}). \tag{4.93}$$

If the functions $\theta_n(\mathbf{x})$ satisfy both the differential equation and the boundary conditions exactly, then we would have the eigenvalues, but in general we will be faced with choosing these as trial functions. Clearly, a trial function that satisfies all the boundary conditions (both geometric, involving displacements/slopes, and natural, involving forces/moments) is desirable. These are generally called *comparison* or *test* functions. However, in general it will be easier to come up with trial functions that satisfy only the geometric boundary conditions. These are typically called *admissible* functions and tend to result in quite accurate estimates for the eigenvalues, especially as more are taken in the assumed solution, and because the approach yields an upper bound, we know that the solution will tend to approach the exact answer from above. This is very similar to the case for MDOF systems whereas now we seek trial *functions* rather than trial vectors.

We also note the relation with Rayleigh's energy method used in the previous chapter for a SDOF system. Rayleigh's principle states that the Rayleigh quotient has a minimum value, which is the square of the lowest frequency of vibration, in the neighborhood of the fundamental mode:

$$\omega_1^2 = \min R(w) = \min \frac{V_{\max}}{T_{\text{ref}}}, \tag{4.94}$$

where T_{ref} is the reference kinetic energy (referred to as the *zero-frequency* kinetic energy in Chapter 3). The Rayleigh–Ritz method is also closely related to the method of assumed modes [1].

Weighted Residuals–Galerkin

An alternative approach to obtaining approximate solutions to distributed parameter systems is based on the method of weighted residuals, the best-known technique being the Galerkin method [22]. This approach is not restricted to self-adjoint systems and generally requires the use of comparison functions rather than admissible functions, but when used for self-adjoint systems is equivalent to the Rayleigh–Ritz method. The central idea is that the error between an approximate solution and the exact solution is minimized.

Given a linear differential operator $\mathcal{L}$, we assume the solution to the differential equation $\mathcal{L}[w(\mathbf{x})] = 0$ (with homogeneous boundary conditions) is based on

$$w^{(r)}(\mathbf{x}, t) = \sum_{r=1}^{N} a_r(t)\phi_r(\mathbf{x}), \qquad (4.95)$$

where the functions $\phi_r(\mathbf{x})$ are linearly independent, form a complete set, satisfy the boundary conditions, and have unknown coefficients $a_r(t)$. For an infinite set of functions, we get the exact solution, with the requirement that the function $\mathcal{L}[w^{(r)}(\mathbf{x}, t)]$ be orthogonal to all the functions $\phi_r(\mathbf{x})$. However, because we have N functions rather than an infinite set, we know that Eq. (4.95) will not typically satisfy the differential equation exactly, and we will get a remainder (or residual) that is orthogonal to the space of the chosen functions:

$$\int_D \mathcal{L}[w^{(r)}(\mathbf{x}, t)]\phi_s(\mathbf{x})dD = 0, \qquad r, s = 1, 2, \ldots, N. \qquad (4.96)$$

Thus, suppose we have again a typical (self-adjoint with homogeneous boundary conditions) system given by Eq. (4.78). Placing Eq. (4.95) in Eq. (4.78), multiplying by $\phi_s(\mathbf{x})$, and integrating over the whole domain, we get

$$M\ddot{a}(t) + C\dot{a}(t) + Ka(t) = 0, \qquad (4.97)$$

where M is an $N \times N$ mass matrix of coefficients m_{sr} given by

$$m_{sr} = \int_D \phi_s(\mathbf{x})\mathcal{M}[\phi_r(\mathbf{x})]dD, \qquad r, s = 1, 2, \ldots, N, \qquad (4.98)$$

and where the stiffness K and (proportionally damped) C can be described similarly. As mentioned previously, the Rayleigh–Ritz and Galerkin methods produce identical results for conservative systems.

Other techniques based on weighted residuals include collocation and least squares. Furthermore, a more sophisticated approach based on Lyapunov-Schmidt reduction can be used as an alternative, especially in certain situations [23]. Specific techniques have also been developed for bifurcation and nonlinear eigenvalue problems [24].

The Finite-Element Method

Perhaps the technique with the greatest utility for solving continuous structural systems is the finite-element (FE) method (FEM) [25–29]. This is a huge subject, and obviously only a brief introduction is given here. In fact, FE analysis (FEA) will be outlined in more detail for specific structural forms later in this book (e.g., plane frames and plates). It is especially powerful for complicated structures and forms the basis of a vast array of commercially available software. Here, a brief introduction is given within the context of general boundary-value problems.

This technique really comes into its own for geometrically complex structures, because (unlike the Rayleigh–Ritz method) trial functions (often relatively low-order polynomials) are defined over subdomains of a structure in a process of discretization. Certain continuity conditions are enforced between subdomains at nodes, and a process of assembly brings all the elements together to form stiffness and mass matrices in standard vibration problems. From this point, the lumped-parameter model of the structure can be handled as a MDOF system, albeit with typically very high-dimensional matrices.

4.3.3 Context Revisited

Before we leave this chapter and embark on the study of slender, axially loaded structures, a few comments are in order about more specific issues covered in the rest of this book. We are primarily interested in structural systems characterized according to the following scheme.

- *Strain energy*. This a quantity associated with structural deformation. In linear theory, this has a quadratic form (small elastic deformations) and leads to a symmetric stiffness matrix, or stiffness operator. For zero deformation we typically have the trivial equilibrium state. For large deflections, both stretching (membrane) and bending effects contribute to strain energy.
- *Work done by axial loads*. External loads are often imparted to a slender structure axially, as well as laterally, and may be due to dead (gravity) loading, for example. They generally lead to a geometric stiffness matrix that tends to diminish lateral (bending) stiffness as a function of the loading. If compressive, then they may lead to instability (the stiffness matrix becomes singular) and the appearance of nontrivial equilibria. If time-periodic axial loads are present, then parametric resonance is possible.
- *Kinetic energy*. Again this is typically a quadratic form, but does not affect equilibrium. It may be imparted to a system in the form of sudden loading. Vibration can be considered as an oscillatory exhange of kinetic and potential energy.
- *Damping*. Small levels of damping can be modeled by Rayleigh's dissipation function. Also, it does not affect equilibrium. Often it turns the stability of conservative systems into asymptotically stable systems. Damping is always present in mechanical systems, although its effect can sometimes be neglected.
- *External forcing*. Many structures are subject to excitation, which is often periodic. The resulting problem of resonance provides an important motivation for study. Loading is sometimes suddenly applied and may cause instability *in the large*, even though the local equilibrium is stable. External forcing increases the phase space because governing equations become inhomogenous (nonautonomous) and may lead to a wider spectrum of behavior, especially for nonlinear systems.

References

[1]   D.J. Inman. *Engineering Vibration*. Prentice Hall, 2000.
[2]   L. Meirovitch. *Principles and Techniques of Vibrations*. Prentice Hall, 1997.

[3]   Y.G. Panovko and I.I. Gubanova. *Stability and Oscillations of Elastic Systems – Paradoxes, Fallacies and New Concepts*. Consultants Bureau, New York, 1965.

[4]   A.M. Lyapunov. *The General Problem of the Stability of Motion*. Princeton University Press, 1947.

[5]   A.P. Seyranian and A.A. Mailybaev. *Multiparameter Stability Theory with Mechanical Applications*. World Scientific, 2003.

[6]   K. Huseyin. *Nonlinear Theory of Elastic Stability*. Noordhoff, 1975.

[7]   H.H.E. Leipholz. *Stability Theory*. Wiley, 1987.

[8]   S.H. Strogatz. *Nonlinear Dynamics and Chaos*. Addison-Wesley, 1994.

[9]   E.H. Dowell. *Aeroelasticity of Plates and Shells*. Noordhoff, 1975.

[10] MATLAB. User's guide. Technical report, The Math Works, 1989.

[11] C. Hayashi. *Nonlinear Oscillations in Physical Systems*. Princeton University Press, 1964.

[12] J. LaSalle and S. Lefschetz. *Stability by Liapunov's Direct Method with Applications*. Academic, 1961.

[13] A.M. Lyapunov. *Stability of Motion (Collected Papers)*. Academic, 1966.

[14] K. Ogata. *System Dynamics*. Prentice Hall, 1998.

[15] Lord Rayleigh (John William Strutt). *The Theory of Sound*. Dover, 1945.

[16] B.H. Tongue. *Principles of Vibration*. Oxford University Press, 1996.

[17] F.W. Williams and W.H. Wittrick. Exact buckling and frequency calculations surveyed. *Journal of Structural Engineering*, 109:169–87, 1983.

[18] R. Courant and D. Hilbert. *Methods of Mathematical Physics*. Wiley Classics Library, 1989.

[19] E.H. Dowell and D.M. Tang. *Dynamics of Very High Dimensional Systems*. World Scientific, 2003.

[20] J.H. Argyris and S. Kelsey. *Energy Theorems and Structural Analysis*. Butterworth, 1960.

[21] G. Temple and W.G. Bickley. *Rayleigh's Principle and Its Applications to Engineering*. Oxford University Press, 1933.

[22] W.J. Duncan. Galerkin's method in mechanics and differential equations. *Reports and Memoranda No. 1798*, Aeronautical Research Council London (England), 1937.

[23] H. Troger and A. Steindl. *Nonlinear Stability and Bifurcation Theory: An Introduction for Engineers and Applied Scientists*. Springer-Verlag, 1991.

[24] H.B. Keller. Numerical solution of bifurcation and nonlinear eigenvalue problems. In P. Rabinowitz, editor, *Applications of Bifurcation Theory*. Academic Press, 1977, pp. 359–89.

[25] T.J.R. Hughes. *Finite Element Method-Linear Static and Dynamic Finite Element Analysis*. Prentice Hall, 2000.

[26] O.C. Zienkiewicz and R.L. Taylor. *The Finite Element Method*. McGraw Hill, 1989.

[27] K.J. Bathe. *Finite Element Procedures*. Prentice Hall, 1995.

[28] M.A. Crisfield. *Nonlinear Finite Element Analysis of Solids and Structures, Vol. 1: Essentials*. Wiley, 1997.

[29] M.A. Crisfield. *Nonlinear Finite Element Analysis of Solids and Structures, Vol. 2: Advanced Topics*. Wiley, 1997.

5 Discrete-Link Models

5.1 Introduction

We first consider a number of discrete-link models in which system properties are concentrated at specific locations. The motivation for considering simple mechanical models is that most of the concepts of dynamics and stability issues encountered with continuous systems (e.g., beams, plates) can be observed with discrete systems but are somewhat easier to analyze [1–7]. In fact the governing equations will tend to be algebraic rather than differential (at least in space), and it is natural to start with a look at systems in which the behavior of the system is completely described by just a single degree of freedom.

5.2 An Inverted Pendulum

Consider the simple hinged cantilever illustrated in Fig. 5.1. This system consists of a concentrated mass supported by a massless but rigid bar of length L. A torsional spring supplies a linear restoring force that is proportional to the angle of rotation of the hinge (in either direction), with spring coefficient K. The angle of rotation θ thus describes the location of the mass at any given instant of time. Typically, the vertical force is simply $P = Mg$, but here we assume that an axial load of magnitude P acts independently of the constant mass M. This assumption will be reexamined later. For a simple system like this, we can easily use any of the fundamental approaches introduced in Chapter 2 for writing the governing equation of motion.

Approaching this problem by using Lagrange's equation, we can write the total potential energy of the system V as consisting of two parts: U, the strain energy stored in the spring as it deflects, and V_P, the potential energy associated with the work done by the axial load as the mass moves through a vertical distance. Thus

$$V = U + V_P = \frac{1}{2}K\theta^2 - PL(1 - \cos\theta). \tag{5.1}$$

Similarly, the kinetic energy T is given by

$$T = \frac{1}{2}ML^2\dot{\theta}^2, \tag{5.2}$$

and placing these into Lagrange's equations, we obtain

$$ML^2\ddot{\theta} + K\theta - PL\sin\theta = 0, \tag{5.3}$$

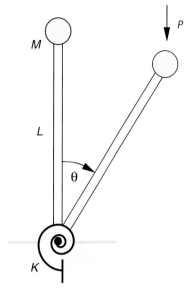

Figure 5.1. A simple hinged bar (the inverted pendulum model).

and using $\sin\theta = \theta$ for small θ we see an effective natural frequency $\omega^2 = K/(ML^2) - P/(ML)$. Equation (5.3) can be nondimensionalized by assuming

$$\omega_n^2 = K/ML^2, \qquad p = PL/K, \tag{5.4}$$

to give

$$\ddot{\theta} + \omega_n^2(\theta - p\sin\theta) = 0. \tag{5.5}$$

Equation (5.5) is a *nonlinear* second-order, homogeneous, ordinary differential equation with constant coefficients.

5.2.1 Static Behavior

Let's consider the underlying equilibrium behavior, which we can easily obtain by setting the time-dependent terms equal to zero:

$$\omega_n^2(\theta - p\sin\theta) = 0. \tag{5.6}$$

This is the first variation of the potential energy and could have been obtained from a direct application of the principle of stationary potential energy. Clearly, we have the trivial (or fundamental) equilibrium state for the perfectly upright position $\theta = 0$. However, we see that another (postbuckled) solution to Eq. (5.6) is (for $p > 1$)

$$p = \frac{\theta}{\sin\theta}. \tag{5.7}$$

Equation (5.7) is plotted together with the trivial solution in Fig. 5.2(a). These paths intersect at $p = 1$, that is, $P = K/L$. This is the critical load of the structure at which point the trivial equilibrium position gives way to an inclined position. We establish

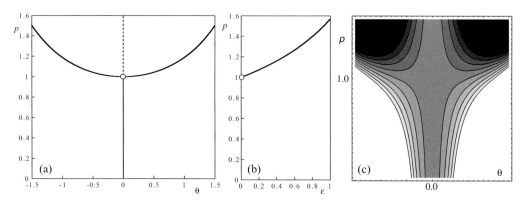

Figure 5.2. (a) The inverted pendulum and (b) equilibrium paths; (c) potential-energy contours (truncated for extreme levels).

that the trivial equilibrium path is unstable for loads greater than $p = 1$ by examining the curvature of the total potential energy in the vicinity of equilibrium. For stability it is sufficient that

$$\frac{d^2V}{d\theta^2} > 0, \tag{5.8}$$

and thus for $\theta = 0$ the preceding condition is satisfied only if $p < 1$.

To test the stability of the secondary (postbuckled) solution, we evaluate Eq. (5.8) for the system along the secondary path described by Eq. (5.7), which results in

$$\frac{d^2V}{d\theta^2} = 1 - p\cos\theta. \tag{5.9}$$

This is clearly positive (and indicative of stability) provided that

$$p < \frac{1}{\cos\theta}. \tag{5.10}$$

Therefore, for physically reasonable deflections, i.e., $\theta < \pi/2$, the postcritical equilibrium paths are stable because placing Eq. (5.7) into condition (5.10) leads to

$$1 - \theta\cot\theta < 0. \tag{5.11}$$

Supposing the end load were gradually increased, we would follow the y axis with no deflection until the critical value was reached, at which point (and under further increase in P), the system would rotate in either the positive or the negative direction. We recognize this type of stable-symmetric bifurcation as a supercritical pitchfork bifurcation. Figure 5.2(b) is a plot of the vertical deflection $\epsilon = L(1 - \cos\theta)$ as a function of axial load, and contours of total potential energy are shown in Fig. 5.2(c) also as functions of load p. In this last figure, the darkness of the gray shade reflects the depth of the energy well at a specific level of loading, and hence, although for $\theta = 0$ there is a constant shading, it actually turns from a relative minimum to a relative maximum when p passes through unity.

5.2.2 Geometric Imperfections

In a practical situation, we would expect the system to have a preferred direction of deformation: It is unreasonable to expect the system to be *perfectly* vertical at initial equilibrium [8]. Indeed, our earlier experience of symmetric branching behavior highlighted the importance of breaking the symmetry in order to generate generic results. Thus a small bias or geometric imperfection is built into the model by assuming the torsional spring is unstressed when $\theta = \theta_0$ (rather than when $\theta = 0$). This changes the total potential-energy expression to

$$V = \frac{1}{2}K(\theta - \theta_0)^2 - PL(\cos\theta_0 - \cos\theta), \qquad (5.12)$$

where we still use $\theta = 0$ as the origin and a corresponding equilibrium condition

$$\frac{dV}{d\theta} = (\theta - \theta_0) - p\sin\theta = 0, \qquad (5.13)$$

where V is now in nondimensional form, which can be rearranged to give

$$p = \frac{(\theta - \theta_0)}{\sin\theta}. \qquad (5.14)$$

We see that the trivial equilibrium state is lost, and now as p is increased the system deflects continuously. It is no longer a discrete bifurcational event. Equilibrium paths for typical small values of the imperfection parameter are shown in Fig. 5.3. It is interesting to note the appearance of a new equilibrium path for deflections in the opposite direction to the initial imperfection (and for $p > 1$). These are termed complementary and would not normally be observed in a natural loading history.

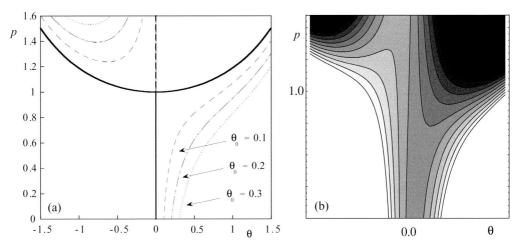

Figure 5.3. (a) The equilibrium paths and (b) potential energy of the imperfect inverted pendulum model (truncated for extreme levels).

The stability of the imperfect equilibrium paths is still obtained from the second derivative of the total potential energy:

$$\frac{d^2V}{d\theta^2} = 1 - p\cos\theta, \tag{5.15}$$

which, when evaluated on the primary (i.e., from zero-load) equilibrium curve, indicates stability [because $p < 1/(\cos\theta_e)$]. For the complementary path, part of the curve is stable and part unstable. In Fig. 5.3(a), they are identified by different dashed curves for different initial imperfections. Their stability of course can be determined from the potential energy, but stability considerations are deferred to the dynamic criterion discussed in the next subsection. Also plotted in this figure [part (b)] is a contour plot of the total potential energy for the specific case of $\theta_0 = 0.1$. The darker shades correspond to lower levels of potential energy for a given loading. Following the minimum as a function of load would thus lead to gradually increasing deflection θ in the positive direction, that is, the same direction as the initial deflection. We also see how the complementary path is represented by a (remote) minimum beyond a local maximum.

5.2.3 Dynamic Behavior

We now turn to the dynamic response of the inverted pendulum model. We know that the trivial equilibrium solution is stable provided the magnitude of the axial load is less than its critical value, and thus we would expect small oscillations about the origin given some initial disturbance from a stable equilibrium state. Expanding the sine term in Eq. (5.5) about zero and dropping higher-order terms leads to

$$\ddot{\theta} + \omega_n^2(1 - p)\theta = 0. \tag{5.16}$$

Here, we are assuming that although the mass provides the axial load the inertia is independent. We shall lift this restriction a little later. The linearized (effective) natural frequency drops toward zero as the critical load is approached,

$$\omega_n \to 0, \quad p \to 1, \tag{5.17}$$

and, with harmonic motion $\theta(t) = c\sin\omega t$, there is a linear relationship between the applied load and the square of the natural frequency $\omega^2 = \omega_n^2(1 - p)$, where $p = PL/K$ and $\omega_n^2 = K/ML^2$. We can thus observe this decay if we plot a time series in which the load is made a slowly increasing function of time in exactly the same way that was considered in Chapter 3.

We would expect to have oscillatory behavior about the stable postbuckled paths. We return to Eq. (5.5) and expand about a general equilibrium path

$$\theta = \theta_e + \delta. \tag{5.18}$$

Placing this in Eq. (5.5) we have

$$\ddot{\delta} + \omega_n^2[(\theta_e + \delta) - p\sin(\theta_e + \delta)] = 0. \tag{5.19}$$

Assuming δ is small such that $\cos \delta \approx 1$ and $\sin \delta \approx \delta$, and because $\theta_e - p \sin \theta_e = 0$, we are then left with the linearized (variational) equation of motion:

$$\ddot{\delta} + \omega_n^2[1 - p \cos \theta_e]\delta = 0. \tag{5.20}$$

For example, suppose $p = 1.1$; then we have $\theta_e = \pm 0.75$ rad, which corresponds to $\omega^2 = 0.35$. Hence we obtain the result that although the natural frequency drops to zero as the critical load is approached, it then starts to increase as the load is increased into the (stiffening) postbuckled range. We shall return to this type of behavior in the next section.

If we are not to be restricted to relatively small-amplitude oscillations about equilibrium then we must solve the nonlinear equation of motion, and this can be easily accomplished numerically [9]. We can also relate time and stiffness linearly— this provides a useful way to visualize the dynamics of the system in the vicinity of equilibrium while the load is slowly increased (as done in the introduction). Such a scheme can be incorporated numerically by

$$\ddot{\theta} + \theta - 0.01\tau \sin \theta = 0, \tag{5.21}$$

where the critical load is reached after 100 time units have elapsed. We have also scaled the time according to

$$\tau = \omega_n t, \tag{5.22}$$

and hence the overdots in Eq. (5.21) signify $\dot{\theta} \equiv d\theta/d\tau$. Initial conditions were chosen as $\theta(0) = 0.1$, $\dot{\theta}(0) = 0.0$ to start not too far away from the equilibrium, and the result of a numerical simulation is shown in Fig. 5.4.

Another conceptual aid in understanding this dynamic behavior is to imagine a small ball rolling on the potential-energy surface given by Eq. (5.1). For small

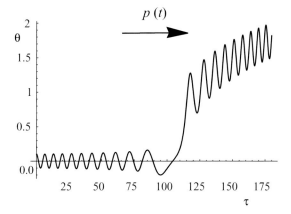

Figure 5.4. A time series evolving through increasing axial load.

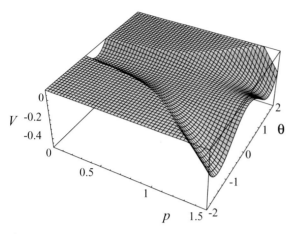

Figure 5.5. Potential energy surface as a function of axial load.

oscillations (for which the linearized equation of motion is appropriate), we see how the evolution of the ball motion in this slowly evolving environment follows the local minima of the underlying potential energy function (see Fig. 5.5). The direction (i.e., positive or negative) followed after criticality is quite arbitrary and rotations to the left, that is, for negative θ, are just as likely for other initial conditions. In this figure, the potential-energy contours are cropped for large positive values to aid the view.

For a conservative system such as this, we can also gain insight from plotting contours of constant total energy. Two such plots are shown in Fig. 5.6 for $p = 0.8$ in part (a) and $p = 1.2$ in part (b), with again the darker shades indicating lower levels of total potential energy. When $p = 1.2$ we have symmetric stable equilibria at $\theta_e = \pm 1.027$. The origin is of course a saddle point at this level of load. Initial conditions close to the stable locations would result in approximately linear vibrations according to the solutions of Eq. (5.20). For initial conditions far away we might

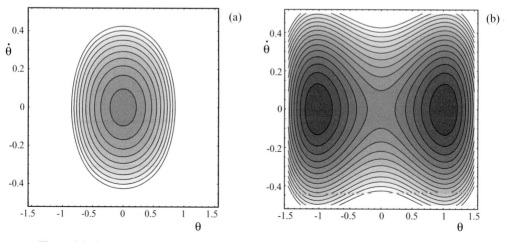

Figure 5.6. Contours of total energy: (a) $p = 0.8$ and (b) $p = 1.2$.

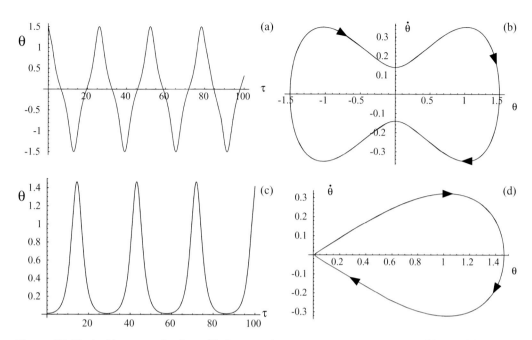

Figure 5.7. Typical large-amplitude oscillations. Both cases correspond to $p = 1.2$: (a) and (b) $\theta(0) = 1.5, \dot{\theta}(0) = 0.0$, (c) and (d) $\theta(0) = 0.01, \dot{\theta}(0) = 0.0$.

expect large-amplitude oscillations that traverse the origin (slowing down when passing over the potential ridge). An example of such a periodic (but *not* simple harmonic) oscillation is shown in Fig. 5.7(a). Indeed, for moderately large amplitudes we would find markedly asymmetric oscillations without a traverse of the upright position. A limited amount of analytic progress can be made in such a situation if we retain higher-order terms in the Taylor series expansion about equilibrium. An example of this type of motion (which is very close to the homoclinic solution starting and finishing at the saddle) is shown in Fig. 5.7(c).

The conservation of energy can be obtained directly from Eqs. (5.1) and (5.2) or we can alternatively use the simple relation

$$\ddot{\theta} = \dot{\theta}\frac{d\dot{\theta}}{d\theta} \tag{5.23}$$

in Eq. (5.5) to separate variables and obtain the velocity as a function of position:

$$\dot{\theta} = \pm\sqrt{C - \theta^2 - 2p\cos\theta}, \tag{5.24}$$

where the constant C depends on the initial conditions. For example, in Fig. 5.7(b) we can use the initial conditions to derive a constant $C = 2.42$, and, using Eq. (5.24), we can confirm, for example, that as the bar passes through its upright position it will be doing so at a nondimensional velocity of 0.141 in either direction. The phase portrait (a plot of velocity versus position) can thus be viewed as a trajectory following one of these contours of constant total energy. The trajectories corresponding to the two preceding examples are also included in Fig. 5.7.

Before leaving this section it is briefly shown how the (nonlinear) natural frequency of large-amplitude vibration can be extracted from the phase trajectories. We can separate variables in Eq. (5.5) and integrate to obtain

$$\int_0^t dt = \int_0^{\theta_{max}} \frac{d\theta}{\sqrt{C - \theta^2 - 2p\cos\theta}}, \tag{5.25}$$

where the period is equal to four times the time it takes to go from 0 to θ_{max}:

$$\tau = 4\int_0^{\theta_{max}} \left(C - \theta^2 - 2p\cos\theta\right)^{-1/2} d\theta. \tag{5.26}$$

Thus taking the initial conditions (and the value of C) corresponding to the trajectory in Fig. 5.7 we can evaluate the preceding integral (numerically) to confirm the (near homoclinic) period of 29.34 units. We shall return to this type of (elliptic integral) approach when we deal with continuous (elastica) systems.

5.2.4 A Note on Inertia

Before moving on it is useful to mention here that in a realistic situation we might expect to increase the axial load in the inverted pendulum model by increasing the mass, that is, $P = Mg$. However, suppose the pendulum arm has some mass m associated with it (that does not change and is not sufficient to cause self-weight buckling). In this case, the energy expressions [from Eqs. (5.1) and (5.2)] change to

$$V = \frac{1}{2}K\theta^2 - MgL(1 - \cos\theta) - \frac{1}{2}mgL(1 - \cos\theta) \tag{5.27}$$

and

$$T = \frac{1}{2}ML^2\dot{\theta}^2 + \frac{1}{6}mL^2\dot{\theta}^2. \tag{5.28}$$

Thus the equation of motion becomes

$$(M + m/3)L^2\ddot{\theta} + K\theta - (M + m/2)gL\sin\theta = 0. \tag{5.29}$$

It is a simple matter to linearize this equation to obtain an expression for the natural frequency (of small-amplitude oscillations)

$$\omega^2 = \frac{K - (M + m/2)gL}{(M + m/3)L^2}, \tag{5.30}$$

which vanishes when the end mass achieves its critical value,

$$M_{cr} = \frac{K}{gL} - \frac{m}{2}. \tag{5.31}$$

The natural frequency and critical mass now provide a suitable means of nondimensionalizing,

$$\Omega^2 = \omega^2 L/g, \quad k = K/(mgL), \quad p = M/m, \tag{5.32}$$

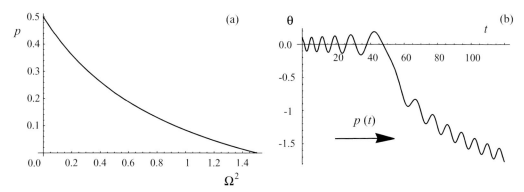

Figure 5.8. (a) The mass–frequency relation for the inverted pendulum model and (b) a numerical time series resulting from a linear increase in mass with time.

which can then be used in the relation between the natural frequency and end load:

$$\omega^2 = \frac{3(2k - 1 - 2p)}{2(1 + 3p)}. \tag{5.33}$$

Given the nondimensionalization, the critical load is now $p_{cr} = k - 0.5$, and a plot of mass (load) versus natural frequency squared is shown in Fig. 5.8(a) for $k = 1$. We can evolve the magnitude of the end mass as a linear function according to $M = 0.01t$, which gives the time series shown in Fig. 5.8(b). The critical mass is reached after 50 time units, after which the (undamped) system starts to oscillate as it follows one of its (nontrivial) postbuckled paths.

5.3 A Discrete-Strut Model

We now move on to consider another rigid-link model, but this time the model is a little more general in the types of behavior it can exhibit. It is also a step closer to the continuous structures we will focus on later. The approach adopted in this section is based on energy considerations rather than the underlying differential equations, and some experimental data are also included. The model under consideration is shown in Fig. 5.9(a). A mechanical model was built to mimic this system by Croll and Walker at University College London [10]. A photographic image is shown in part (b) of Fig. 5.9. It consists of two rigid (massless) links hinged at their supports and in the center where a concentrated mass M is located. Two linear springs provide a restoring force, one against lateral deflection with modulus K, and the other against rotation with modulus C. An axial load of magnitude P acts at the left-hand ends that is unrestrained against horizontal movement. The coordinate Q describes the deflected position of the mass, and we can think of the schematic as providing a plan view; that is, we assume gravity is already taken into account.

The total potential and kinetic energies for this system are given by [5]

$$V = 2C\theta^2 + \frac{1}{2}KL^2 \sin^2\theta - 2PL(1 - \cos\theta) \tag{5.34}$$

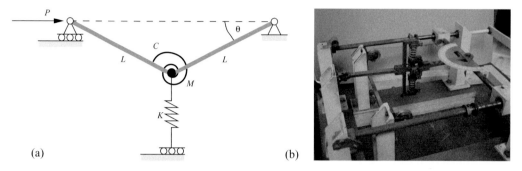

Figure 5.9. (a) Schematic of the two-bar link model and (b) physical realization.

and

$$T = \frac{1}{2}ML^2\dot{\theta}^2. \tag{5.35}$$

Equilibrium solutions are found from

$$\frac{dV}{d\theta} = 4C\theta + KL^2\sin\theta\cos\theta - 2PL\sin\theta = 0. \tag{5.36}$$

We immediately see the trivial solution, $\theta_e = 0$, together with the nontrivial solutions given by

$$\Lambda = \alpha\cos\theta + \frac{(1-\alpha)\theta}{\sin\theta}, \tag{5.37}$$

where

$$\Lambda = P/P_{\text{cr}}, \tag{5.38}$$

$$P_{\text{cr}} = \frac{KL^2 + 4C}{2L}, \tag{5.39}$$

$$\alpha = \frac{KL^2}{KL^2 + 4C}. \tag{5.40}$$

The parameter α is a ratio of spring stiffnesses. For example, if $\alpha = 0$ (i.e., $K = 0$), then we have exactly the same type of equilibrium curves as for the model in the previous section. The stability of the equilibrium paths can again be established from the sign of the second derivative of the total potential-energy function. However, let us assume initially that $\alpha = 1$ (i.e., $C = 0$) so that we have only a lateral (translational) spring acting. The equilibrium expression simplifies to

$$\Lambda = \cos\theta. \tag{5.41}$$

In the presence of initial imperfections, the total potential energy becomes

$$V = \frac{1}{2}KL^2(\sin\theta - \sin\theta_0)^2 - 2PL(\cos\theta_0 - \cos\theta) \tag{5.42}$$

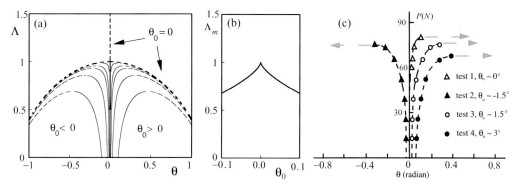

Figure 5.10. (a) Equilibrium paths for the two-bar link model with a lateral spring and initial geometric imperfections. θ_0 ranges from 0.001 closest to the bifurcation to 0.1, (b) imperfection sensitivity and (c) measured data (adapted from [2]).

with the first derivative

$$\frac{dV}{d\theta} = KL^2(\sin\theta - \sin\theta_0)\cos\theta - 2PL\sin\theta. \tag{5.43}$$

Using the nondimensionalization, we obtain

$$\Lambda = \frac{(\sin\theta - \sin\theta_0)\cos\theta}{\sin\theta}. \tag{5.44}$$

In the literature, it is sometimes observed that the trigonometric terms are replaced with their series expansion, and retaining the first few terms results in equilibrium paths [2]:

$$\Lambda = \frac{\theta - \theta_0 - \frac{2}{3}\theta^3 + \frac{1}{2}\theta^2\theta_0}{\theta - \frac{1}{6}\theta^3} \approx 1 - \frac{\theta_0}{\theta} + \frac{\theta_0\theta}{3} - \frac{\theta^2}{2} + \cdots +. \tag{5.45}$$

These nontruncated [Eq. (5.44)], paths have the form shown in Fig. 5.10(a), in which we recognize the characteristic subcritical pitchfork bifurcation. Complementary paths are also present in this example. However, they prove to be unstable and have little to do with a natural loading path starting near the origin and hence are not included in the plot.

For the perfect geometry, it is simple to see that the stability is governed by the coefficient of the second derivative of the potential energy. For the primary path, it is easy to show that equilibrium is unstable if $p > 1$. Similarly for the secondary path it can be shown that the coefficient is always negative and hence the postbuckled behavior is unstable (sometimes called unstable-symmetric branching behavior in the literature [4]). In practice this means that, if the load were gradually increased from zero, the strut model would buckle when $p = 1$ and the system would collapse completely (within the confines of the mathematical modeling). It should be mentioned here that experimental data taken from the system shown in Fig. 5.9(b) very closely match the theory shown in Fig. 5.10(a) [2] and is plotted in Fig. 5.10(c).

For the imperfect geometry the limit of the stability of the paths coincides with the maximum load Λ_m (the horizontal tangency) for a given path. It can be shown that this occurs when $\theta = \theta_0^{1/3}$, and placing this back into Eq. (5.45) results in the cusp geometry shown in Fig. 5.10(b). This displays an important characteristic of some axially loaded structures: the load-carrying capacity of the structure is reduced when initial imperfections are present, i.e., it is *imperfection sensitive*. This type of subcritical behavior may then be viewed as a potentially dangerous consequence when compared with the supercritical behavior described in the previous section.

Following a similar line of reasoning to Section 3.3 we can show that the frequency of small oscillations can be obtained by using Rayleigh's method:

$$\Omega^2 = 1 - \Lambda \cos\theta - \alpha(1 - \cos 2\theta), \tag{5.46}$$

where ω is the effective natural frequency,

$$\Omega = \frac{\omega}{\omega_n}, \tag{5.47}$$

$$\omega_n^2 = \frac{4C + KL^2}{ML^2}, \tag{5.48}$$

and again setting $\alpha = 1$, incorporating initial imperfection θ_0, and simplifying, we get

$$\Omega^2 = (1 - 2\theta^2 + \theta\theta_0) - \Lambda\left(1 - \frac{\theta^2}{2}\right). \tag{5.49}$$

We can then plot Eq. (5.49), incorporating Eq. (5.45) for various initial imperfections, and this is shown in Fig. 5.11 for $\theta_0 = 0.001, 0.01$, and 0.1. Thus we see that the often linear relation between the natural frequency squared and the axial load

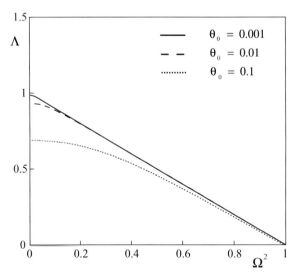

Figure 5.11. The relation between the natural frequency and axial load for the imperfect unstable symmetric model. $\theta_0 = 0.001, 0.01, 0.1$.

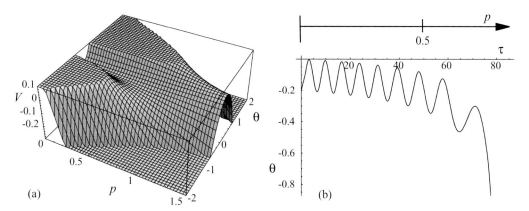

Figure 5.12. (a) The potential-energy surface with an initial imperfection of $\theta_0 = -0.1$ and (b) a typical time series evolving through the p axis, $\theta(0) = -0.2$.

is not true for initially imperfect geometries, and in fact here the natural frequency drops to zero before the critical load is reached.

The potential-energy surface is plotted as a function of load and deflection in Fig. 5.12(a) together with an evolving time series under linearly increasing end load in part (b). We see how the oscillation continues until the limit point is reached and the trajectory slides off to infinity (actually to large oscillations about π but we are not interested in such solutions). According to Fig. 5.10, we expect a maximum load of about $p_m \approx 0.7$ for this imperfection and hence given a ramped load of $p = 0.01t$ we would expect the solution to lose stability after about $\tau = 70$ time units. We observe the decay in natural frequency as a gradual lengthening of the period of oscillation as the instability is approached.

Now, let's assume that $\alpha = 0$ (or $K = 0$), which in fact produces a situation qualitatively similar to the response of the inverted pendulum. In this case, the equilibria (for the initially perfect geometry) are given by

$$\theta = 0, \tag{5.50}$$

for $\theta = 0$ and

$$\Lambda = \theta / \sin \theta \tag{5.51}$$

for $\theta \neq 0$, and a frequency–load relation

$$\Omega^2 = 1 - \Lambda \cos \theta, \tag{5.52}$$

which, for the prebuckled (trivial) equilibrium path gives

$$\Omega^2 = 1 - \Lambda. \tag{5.53}$$

For the postbuckled (nontrivial) equilibrium path, we substitute Eq. (5.51) into Eq. (5.52) to get

$$\Omega^2 = 1 - \theta \cot \theta \approx \frac{\theta^2}{3}. \qquad (5.54)$$

Expanding Eq. (5.51),

$$\Lambda = \frac{\theta}{\sin \theta} \approx 1 + \frac{\theta^2}{6}, \qquad (5.55)$$

and combining Eqs. (5.54) and (5.55) leads to

$$\Omega^2 = 2(\Lambda - 1). \qquad (5.56)$$

Thus we see the interesting result that, as a function of axial load, the postbuckled linear frequency changes at half the rate of the prebuckled frequency, a result also contained in Eq. (5.20). Of course, this relation is true only for the moderately buckled structure because the trigonometric power series was truncated.

It is also mentioned here that these types of link model can be used to illustrate the effect of thermal loading. For example, Croll and Walker [2] showed that an increase in temperature will result in stable-symmetric buckling (supercritical pitchfork bifurcation). It can also be shown that the natural frequency will decay in the usual way as the critical temperature is approached. Thermal buckling is an important consideration for plated structures and will be considered in more detail in a later chapter.

5.4 An Asymmetric Model

The models described in the previous sections of this chapter were symmetric in the absence of initial imperfections. There are a number of structural systems that behave quite differently according to the direction of the deformation [1, 3, 11]. In a buckling context, this is characterized by the asymmetric (or transcritical) point of bifurcation: A structure exhibiting this behavior is shown in Fig. 5.13. The deflection of this SDOF system is described by the coordinate X, the horizontal distance of the top of the (massless) bar. It is convenient to nondimensionalize this by the length of the bar: $x = X/L$. A spring of modulus K provides the restoring force, with x_0 denoting the initial imperfection (and thus $x_0 = X_0/L$). However, the dead load (the point mass) is further offset by a fixed amount l (measured at right angles to the bar and set equal to αx_0).

The equilibrium paths are obtained from the first derivative of the total potential energy and are given by

$$p = \frac{2[(1 - x^2)^{1/2} - (1 - x)^{1/2}(1 + x_0)^{1/2}]}{[x + \alpha x_0(1 - x^2)^{1/2}]}, \qquad (5.57)$$

where $p = P/(KL)$. Some typical paths are shown in Fig. 5.14(a) for a variety of initial imperfections. It is clear that the behavior (including instability behavior) is strongly influenced by the *sign* of the initial imperfection. The stability of these paths

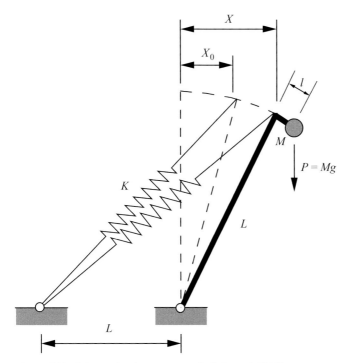

Figure 5.13. Schematic of an asymmetric link model [11].

can be determined from the second derivative of the total potential energy. However, we can use the alternative criterion of requiring real frequencies for stability.
These are obtained from Lagrange's equation and are given by

$$\Omega^2 = (1 + x_0)^{1/2}(1 + x)^{-3/2} - p(1 - x^2)^{-3/2}, \qquad (5.58)$$

where p is determined from Eq. (5.57) and Ω is nondimensionalized with respect
to $\sqrt{(KL^2)/(2M)}$. Some typical frequency–load plots are shown for the same set of

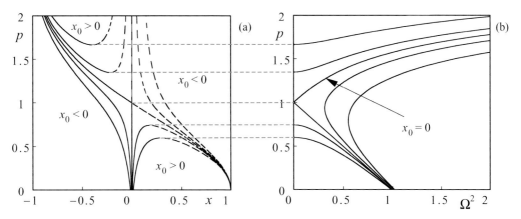

Figure 5.14. (a) Equilibrium paths and (b) frequency–load relation. $\alpha = 5$, $x_0 = 0, \pm 0.005, \pm 0.015$.

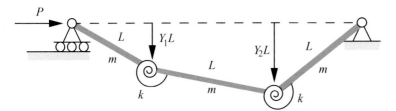

Figure 5.15. A schematic of a three-bar link model.

initial imperfections in Fig. 5.14(b). We see how for negative values of x_0 the natural frequency decays to zero as the system buckles at a limit point. However, when the initial imperfection is in the other direction no instability is encountered and the frequencies start to increase as the structure moves beyond the critical value for the corresponding perfect system. We again note the presence of the complementary equilibrium paths that would not be ordinarily encountered under a monotonic increase in axial load, that is, the natural loading path. The sign of the initial imperfection also has a strong effect on the load-carrying capacity of the structure, and, given the often arbitrary nature of the initial imperfection, this system presents obvious concern in a design context. This type of behavior is sometimes encountered in frame structures. We again note that the axial load is assumed to be an independent parameter (see the discussion in Subsection 5.2.4).

5.5 A Three-Bar Model

Now let us consider the dynamics and stability of a simple mechanism made up of three rigid links of length L and mass per unit length m. They are hinged at their connections, and linear rotational springs of stiffness coefficient k are placed at the two internal joints, as shown in Fig. 5.15 [4]. It is assumed that the structure is in equilibrium in its undeflected (straight) configuration and that an axial load of magnitude P is acting. This model has two degrees of freedom, that is, an equilibrium configuration is determined if two coordinate values are specified. In contrast to the simple examples outlined earlier, there is some flexibility in the way we choose the coordinates to describe the deflected configuration of the system. Suppose we choose the lateral deflections as the coordinates (as shown in the figure). In this case, we can write down the strain energy stored in the springs as

$$U = \frac{1}{2}k[\sin^{-1} Y_1 - \sin^{-1}(Y_2 - Y_1)]^2 + \frac{1}{2}k[\sin^{-1} Y_2 - \sin^{-1}(Y_2 - Y_1)]^2, \qquad (5.59)$$

which can be expanded to give

$$U = \frac{1}{2}k[5Y_1^2 - 8Y_1 Y_2 + 5Y_2^2 + \cdots +]. \qquad (5.60)$$

We can also write the potential energy associated with the movement of the axial load:

$$V = -PL\{3 - (1 - Y_1^2)^{1/2} - (1 - Y_2^2)^{1/2} - [1 - (Y_2 - Y_1)^2]^{1/2}\}, \qquad (5.61)$$

which can also be expanded to give

$$V = -PL[Y_1^2 - Y_1 Y_2 + Y_2^2 + \cdots +]. \qquad (5.62)$$

The kinetic energy for a typical link with displacement a and b at each end is

$$T = \frac{1}{2}m \int_0^L [\dot{a}(1 - x/L) + \dot{b}(x/L)]^2 dx, \qquad (5.63)$$

which, after substitution and adding the effects of all three links, leads to

$$T = \frac{1}{2}mL^2 \frac{2}{3}\left[\dot{Y}_1^2 + \frac{1}{2}\dot{Y}_1\dot{Y}_2 + \dot{Y}_2^2\right]. \qquad (5.64)$$

Thus we have our Lagrangian

$$L = T - U + V_P, \qquad (5.65)$$

and a direct application of Lagrange's equation will lead to the equations of motion by use of the dummy suffix notation

$$T_{ij}\,^e\ddot{Y}_j + V_{ij}\,^e Y_j = 0, \qquad (5.66)$$

and, assuming harmonic oscillations $Y_j = A_j \sin \omega t$, we obtain

$$V_{ij}\,^e A_j - \omega^2 T_{ij}\,^e A_j = 0, \qquad (5.67)$$

thus leading to the characteristic equation

$$\left| V_{ij}\,^e - \omega^2 T_{ij}\,^e \right| = 0. \qquad (5.68)$$

For the specific case at hand, we then have

$$\begin{vmatrix} (5k - 2PL - \omega^2 \frac{2}{3}mL^3) & (-4k + PL - \omega^2 \frac{1}{6}mL^3) \\ (-4k + PL - \omega^2 \frac{1}{6}mL^3) & (5k - 2PL - \omega^2 \frac{2}{3}mL^3) \end{vmatrix} = 0, \qquad (5.69)$$

and, by using

$$p = \frac{PL}{k}, \qquad (5.70)$$

$$\Omega^2 = \frac{\omega^2 mL^3}{k}, \qquad (5.71)$$

we get

$$\Omega^2 = \frac{6}{5}[(8 - 3p) \pm (2p - 7)], \qquad (5.72)$$

and thus the two natural frequencies

$$\Omega_1^2 = \frac{6}{5}(1 - p), \qquad (5.73)$$

$$\Omega_2^2 = 6(3 - p). \qquad (5.74)$$

Each of these eigenvalues corresponds to an eigenvector that describes both the mode of buckling and vibration: The first corresponds to a mode in which both generalized coordinates are equal, the second to a mode in which the coordinates are equal but opposite in sign. From Eqs. (5.73) and (5.74), we can set the natural frequencies equal to zero to obtain the critical loads

$$p_1 = 1 \quad (P_1 = k/L), \tag{5.75}$$

$$p_2 = 3 \quad (P_2 = 3k/L). \tag{5.76}$$

And setting the axial load equal to zero, we get the natural frequencies

$$\Omega_1^2 = 6/5 \quad [\omega_1^2 = 6k/(5mL^3)], \tag{5.77}$$

$$\Omega_2^2 = 18 \quad [\omega_2^2 = 18k/(mL^3)]. \tag{5.78}$$

From Eqs. (5.73) and (5.74), we again see the linear relation between the square of the natural frequency and the axial load. The linearity of this relation occurs because of the equivalence of the buckling modes and the natural modes of vibration, having a finite set of generalized coordinates, and the fact that the frequencies depend on a single parameter. We would not necessarily expect this relation to be exactly linear for the general continuous case, as explained earlier.

This example can also provide a powerful illustration of the utility of choosing principal coordinates. For this two-DOF (2DOF) system, it is a simple matter to expand the determinant of Eq. (5.69), and, of course, there are a myriad of techniques for achieving this numerically for higher-order systems [9]. However, our earlier theory illustrated how coordinate transformations can be useful in the setting up of the equations of motion in going from physical or generalized to principal coordinates. Thompson and Hunt [4] show that the simple transformation

$$u_1 = \frac{Y_1 + Y_2}{2}, \tag{5.79}$$

$$u_2 = \frac{Y_1 - Y_2}{2} \tag{5.80}$$

decouples the equations, enabling the critical loads and natural frequencies to be written immediately [see Eq. (3.43)]. However, this type of direct transformation is usually not obvious *a priori* in a typical analysis. A large variety of numerical algorithms are available to compute this transformation efficiently.

5.6 A Snap-Through Model

In this section, we look at a simple example of snap-through buckling associated with a saddle-node bifurcation [12, 13]. The link model shown in Fig. 5.16 consists of two rigid massless links of length L, with a linear spring K allowing horizontal movement at one end. A point mass M is located at the center where a point load of magnitude P acts in the vertical direction. Again we assume that P acts independently of the mass, and we can also think of Fig. 5.16 as a plan view, such that gravity

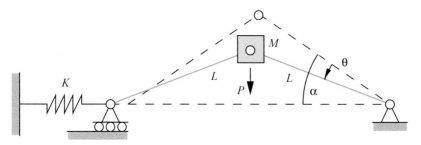

Figure 5.16. Schematic of a simple arch link model.

is not an issue. The generalized coordinate for this SDOF system is again θ, which is measured from the initial rise of the structure that is fixed at $\alpha = \pi/8$ in keeping with earlier work on this model. We can write the total potential energy for this model as

$$V = 2KL^2 \left[\cos(\alpha - \theta) - \cos\alpha\right]^2 - PL\left[\sin\alpha - \sin(\alpha - \theta)\right] \tag{5.81}$$

and a kinetic energy of

$$T = \frac{1}{2}ML^2\dot\theta^2. \tag{5.82}$$

A direct application of Lagrange's equation, and defining

$$p = P/(4KL), \quad \omega_n^2 = 4K/M, \tag{5.83}$$

leads to the nondimensional equation of motion:

$$\ddot\theta + \omega_n^2\left[\cos(\alpha - \theta) - \cos\alpha\right]\sin(\alpha - \theta) - p\cos(\alpha - \theta) = 0. \tag{5.84}$$

The equilibrium path for an initial rise of $\alpha = \pi/8$ is plotted in Fig. 5.17, where it is shown superimposed on contours of the total potential energy. It follows the

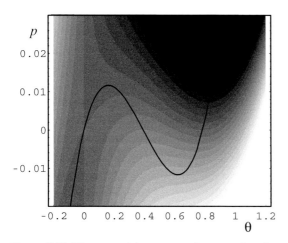

Figure 5.17. The potential-energy surface as a function of axial load with the equilibrium path superimposed.

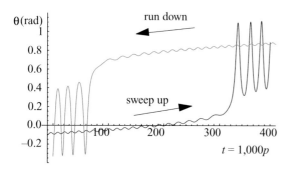

Figure 5.18. Evolution through snap-through illustrating hysteresis. Initial conditions: For both $\dot{\theta} = 0.0$, $\theta(0) = -0.1$ at $t = 0$ and $\theta(0) = 0.8654$ at $t = 400$.

stationary points of the potential energy. The second derivative can be used to de-termine which of these stationary points is a minimum, and we know that a critical point is reached when the equilibrium curve passes through a horizontal tangency. For this system, the limit point is reached when $p = 0.0116$, which corresponds to a deflection of $\theta = 0.166$ rad, that is, when the load reaches this value the struc-ture has deflected to nearly 10 deg and then suddenly snaps through to an inverted position (which is about $\theta = 0.844$ rad, that is, about 25 deg below the horizontal). Note that in this figure the energy is cropped for large values (white) and low values (black) of the potential energy, but in general the darker the shade, the deeper the potential-energy well at that load.

Conducting a numerical integration, we can sweep through the instability by using the following loading function:

$$p = -0.02 + 0.0001t. \tag{5.85}$$

An example is shown in Fig. 5.18, in which an initial condition is used that is 0.01 rad away from stable equilibrium, and $\omega_n = 1$ for convenience. Using the ramp func-tion of Eq. (5.85) thus converts to an anticipated critical time of $t_{cr} = 316$. The loss of stability is abrupt (albeit slightly delayed) under the gradual increase in load. The postbuckled behavior is characterized by relatively large-amplitude oscillation about the inverted position because of the large transient motion initiated by the jump. Note the asymmetric nature of the waveform as the trajectory oscillates along its decidedly asymmetric potential-energy surface. We can conduct a reverse sweep by changing the direction of the load, and again starting from an initial condition adjacent to the equilibrium we again observe the snap back to the original branch, in which case the load evolution has been scaled such that a jump occurs at about $t = 84$. This evolving trajectory is shown in gray in Fig. 5.18, and thus a region of hysteresis is revealed.

The loading described in this example can be considered as "dead" or "force loaded." An alternative, which can often be the most practical approach in a laboratory context, is to use a "displacement-loaded" device. In the former case, it is straightforward to prescribe the load (usually because of gravity) and then measure deflection. In the latter, we prescribe deflection and measure the resulting load (using a load cell for example). In the displacement-controlled approach, the hysteresis is manifested in terms of displacement, and thus the nature of the stability of equilibrium changes [2]. This also brings with it an issue of extensibility, because of a SDOF model, for example, will not be able to oscillate if subject to displacement-controlled loading. The ability to follow equilibrium paths that have bifurcations or turning points is a subject of considerable importance and will be dealt with in more detail in a later section. We note finally that, unlike for symmetric systems possessing a trivial equilibrium solution, this type of limit point buckling is *not* sensitive to initial imperfections. It is affected by the presence of a small initial imperfection (in fact linearly) but not in a disproportionate sense. We shall also look at the analog of this system in a continuous (arch) structure later.

Before leaving this section, we again take a brief look at the relation between the load and the natural frequency. The local stiffness of the force–deflection curve can be obtained from the derivative of the equilibrium condition, that is,

$$\frac{dp}{d\theta} = \frac{\cos\alpha - \cos^3(\alpha - \theta)}{\cos^2(\alpha - \theta)}, \qquad (5.86)$$

which in turn is linearly related to the square of the effective linear natural frequency in the usual way. Hence we can plot the square root of the right-hand side of Eq. (5.86) against load, as shown in Fig. 5.19. An alternative view of *stiffness* would relate load and vertical deflection rather than angle but the relation is very nearly

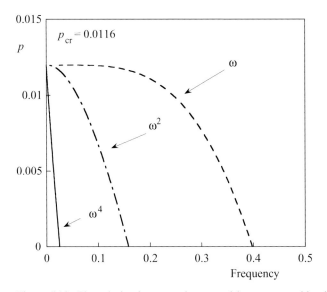

Figure 5.19. The relation between the natural frequency and load for the snap-through model.

the same [2]. It is actually the natural frequency raised to the fourth power that varies linearly with load when close to the instability [14, 15]. This is a typical result in the vicinity of a saddle-node bifurcation and will be touched upon again in the chapter on nondestructive testing.

5.7 Augusti's Model

In this section, we again focus attention on an inverted pendulum model but now replace the torsional spring with a universal joint, that is, a hinge that is not confined to a plane. This is sometimes referred to as Augusti's model in the literature [1, 16, 17]. Thus the deflection of the system needs two coordinates for a complete description. This model shows some interesting behavior when modes interact. Specifically, we outline a bifurcation from nontrivial equilibrium: the secondary bifurcation. The model is shown in Fig. 5.20.

A slender, rigid (but massless) bar of length L is pinned at its base, where rotational springs with constant stiffnesses C_1 and C_2 ($C_2 > C_1$) initially act in perpendicular planes and rotate with the bar. The corresponding angles of rotation with respect to two horizontal, perpendicular axes are $\alpha_1(t)$ and $\alpha_2(t)$, and the angles $\theta_1(t)$ and $\theta_2(t)$ are defined as

$$\theta_1(t) = (\pi/2) - \alpha_1(t), \quad \theta_2(t) = (\pi/2) - \alpha_2(t), \tag{5.87}$$

with $\theta_1(t) = \theta_{10}$ and $\theta_2(t) = \theta_{20}$ when the springs are unstretched. A downward vertical load P is applied at the top of the bar. A concentrated mass M is attached at the top of the bar.

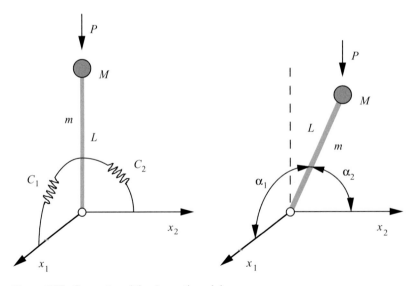

Figure 5.20. Geometry of the Augusti model.

The potential energy V is given by [18]

$$V = \frac{1}{2}C_1(\theta_1 - \theta_{10})^2 + \frac{1}{2}C_2(\theta_2 - \theta_{20})^2$$
$$-PL[(1 - \sin^2\theta_{10} - \sin^2\theta_{20})^{1/2}$$
$$-(1 - \sin^2\theta_1 - \sin^2\theta_2)^{1/2}], \qquad (5.88)$$

and the kinetic energy T is

$$T = \frac{1}{2}ML^2\left[\left(\frac{d\theta_1}{dt}\right)^2\cos^2\theta_1 + \left(\frac{d\theta_2}{dt}\right)^2\cos^2\theta_2\right]$$
$$+ 2ML^2\left\{\frac{[(d\theta_1/dt)\sin 2\theta_1 + (d\theta_2/dt)\sin 2\theta_2]^2}{(1 - \sin^2\theta_1 - \sin^2\theta_2)}\right\}. \qquad (5.89)$$

We can again use Lagrange's equations to obtain the equations of motion. The nonlinear inertia terms in the resulting equations do not affect small vibrations of the system about an equilibrium state. Hence, for simplicity, only the linearized inertia terms are used. The analysis is conducted in terms of the following nondimensional quantities, where Ω is a dimensional vibration frequency:

$$c = C_2/C_1, \quad p = PL/C_1, \quad \tau = t(C_1/ML^2)^{1/2}, \quad \omega = \Omega(ML^2/C_1)^{1/2}, \qquad (5.90)$$

with $c > 1$. It can be shown (from equating the second derivative of the potential energy to zero) that $p = 1$ is the critical buckling load, that is, for the perfect system (the bar is vertical and the springs unstretched). The coupled, nonlinear equations of motion are obtained as

$$\theta_1'' + \theta_1 - \theta_{10} - \frac{p}{2}(1 - \sin^2\theta_1 - \sin^2\theta_2)^{-1/2}\sin\theta_1 = 0, \qquad (5.91)$$

$$\theta_2'' + c\theta_2 - c\theta_{20} - \frac{p}{2}(1 - \sin^2\theta_1 - \sin^2\theta_2)^{-1/2}\sin\theta_2 = 0. \qquad (5.92)$$

Again we focus initially on the underlying equilibria of the perfect system geometry ($\theta_{10} = \theta_{20} = 0$). The four solutions are

$$\theta_1 = 0, \quad \theta_2 = 0,$$
$$\theta_2 = 0, \quad p = \theta_1/\sin\theta_1,$$
$$\theta_1 = 0, \quad p = c\theta_2/\sin\theta_2, \qquad (5.93)$$
$$p = (\theta_1/\sin 2\theta_1)\sqrt{(1 - \sin^2\theta_1 - \sin^2\theta_2)}$$
$$= (c\theta_2/\sin 2\theta_2)\sqrt{(1 - \sin^2\theta_1 - \sin^2\theta_2)}.$$

These curves are plotted in Fig. 5.21(a), with the secondary bifurcation occurring at

$$c\sin 2\theta_1^* = 2\theta_1^*, \quad p = c\cos 2\theta_1^*, \qquad (5.94)$$

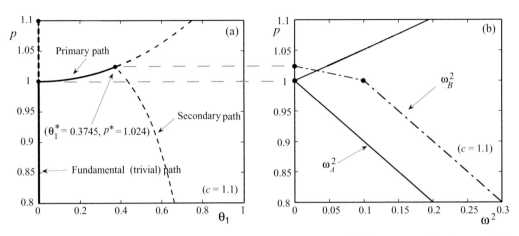

Figure 5.21. (a) Equilibrium paths for the perfect model and (b) corresponding characteristic curves.

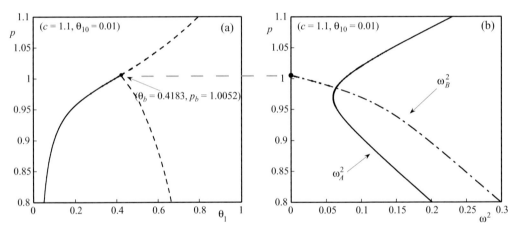

Figure 5.22. (a) Equilibrium paths for the imperfect model and (b) corresponding characteristic curves.

which for the specific case shown (with $c = 1.1$) is at $(\theta_1^*, p^*) = (0.3745, 1.024)$. Shown in part (b) is the dependence of the natural frequencies on the axial load. These are evaluated on the equilibrium paths. For the trivial equilibrium, we have

$$\omega_A^2 = 1 - p, \quad \omega_B^2 = c - p, \tag{5.95}$$

and for the primary (postbuckled, i.e., $p > 1$) branch, we have

$$\omega_A^2 = 1 - p \cos \theta_1 = 1 - \theta_1 / \tan \theta_1, \tag{5.96}$$

$$\omega_B^2 = c - p \cos \theta_1 = c - 2\theta_1 / \sin 2\theta_1. \tag{5.97}$$

For the case in which there is a small amount of initial geometric imperfection, the results shown in Fig. 5.22 are obtained. The initial angle $\theta_{10} = 0.01$ is chosen. The secondary bifurcation now occurs directly from the primary path. It is interesting to note that previous studies of this system have often used truncation to ease some

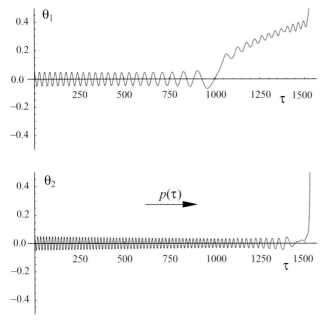

Figure 5.23. A slow sweep through secondary bifurcation.

of the computations, and in that case the load–frequency (squared) relations are exactly linear.

We again close this section by conducting a numerical simulation as the system is swept through the bifurcation(s). Figure 5.23 was based on the following load evolution,

$$p = 0.95 + 0.00005\tau, \tag{5.98}$$

where the initial geometry is perfect and a very small amount of damping was added to the system. We observe a gradual decrease in the θ_1 natural frequency as the system approaches the initial bifurcation at $\tau = 1000$ and then a gradual *increase* in the natural frequency. However, in the lower part of this figure we also see that it is the natural frequency associated with θ_2 that decreases as the secondary bifurcation is approached at $\tau = 1480$.

This system will be revisited a couple of times later in this book: as an example of a path-following algorithm and in cases in which oscillations are not necessarily small.

5.8 Multiple Loads

In all the link model examples so far there has been a single axial load. A mono-tonic increase in this parameter resulted in a linear decay in the natural frequency (squared). However, if more than one independent axial load is present, then the natural frequencies will still decay if either or both of these loads are increased,

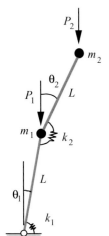

Figure 5.24. A 2DOF model under the action of two axial loads.

but a linear relation is no longer to be expected. To illustrate this effect, consider the link model shown in Fig. 5.24. Following [19, 20], and assuming $k_1 = 2k$, $k_2 = k$, $m_1 = 2m$, $m_2 = m$, we can write the energy terms:

$$U = k\theta_1^2 + \frac{1}{2}k(\theta_2 - \theta_1)^2, \tag{5.99}$$

$$V_P = -\frac{1}{2}LP_1\theta_1^2 - \frac{1}{2}LP_2(\theta_1^2 + \theta_2^2), \tag{5.100}$$

$$T = \frac{1}{2}mL^2(3\dot{\theta}_1^2 + \dot{\theta}_2^2 + 2\dot{\theta}_1\dot{\theta}_2). \tag{5.101}$$

We can nondimensionalize by using the following parameters,

$$\Omega^2 = \frac{\omega^2 mL^2}{k}, \qquad p_i = \frac{P_iL}{k}, \tag{5.102}$$

and by using Lagrange's equations obtain the characteristic equation

$$2\Omega^4 + \Omega^2(p_1 + 4p_2 - 8) + (p_2^2 + p_1p_2 - p_1 - 4p_2 + 2) = 0. \tag{5.103}$$

Setting $\Omega^2 = 0$ gives the two critical load conditions from the quadratic

$$p_2^2 + p_1p_2 - p_1 - 4p_2 + 2 = 0, \tag{5.104}$$

which can be solved for the lowest critical loads acting separately of $p_1 = 2$ and $p_2 = 0.586$. Setting both axial loads equal to zero leads to the natural frequencies $\Omega_1^2 = 0.268$ and $\Omega_1^2 = 3.732$. Thus we can plot the roots of Eq. (5.103) as two surfaces, as shown in Fig. 5.25. The curve on the left is the most relevant given the typical situation of a monotonic increase in the loads, that is, it will be encountered first.

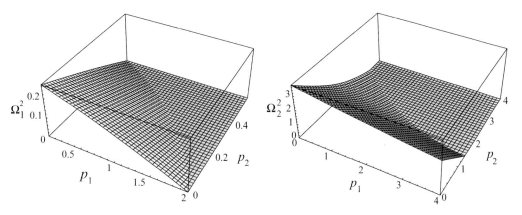

Figure 5.25. The characteristic surface plotting the natural frequency squared in terms of the two axial loads.

5.9 Load-Dependent Supports

It may happen that the stiffness of a structure or its supports is a function of the applied axial load [21, 22]. To illustrate this situation, we briefly return to the inverted pendulum model (see Fig. 5.1), but now we assume a distributed mass such that the moment of inertia of the bar about its base is I. We assume a fixed vertical end load P and ignore gravity. We assume a spring stiffness, rather than a constant torsional stiffness k, that increases linearly with the end load from a baseline value of K_0. The equation of motion is given by

$$\ddot{\theta} - p \sin \theta + k(p)\theta = 0, \qquad (5.105)$$

where the following nondimensional parameters have been used,

$$p = PL/K_0, \quad k(p) = K(pK_0/L)/K_0, \quad t = T\sqrt{(K_0/I)}, \qquad (5.106)$$

and the derivatives in Eq. (5.105) are with respect to the scaled time t.

Equilibrium conditions are obtained from

$$-p \sin \theta + k(p)\theta = 0, \qquad (5.107)$$

and the frequencies of small vibrations about these equilibria are given by

$$\omega^2 = k(p) - p \cos \theta. \qquad (5.108)$$

Now, assume the spring stiffness is related to load in the following way:

$$k(p) = 1 + \gamma p, \qquad (5.109)$$

where γ is taken as positive. Equilibrium paths and load–frequency relations are plotted in Fig. 5.26 for a number of different values of the parameter γ. We see an increase in the critical load and an increase in the natural frequency at a given load as a function of spring stiffness. Further examples of rigid-link models and their stability can be found in Seyranian and Mailybaev [23].

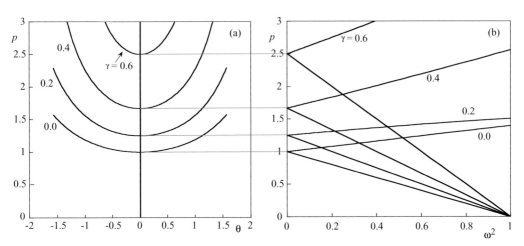

Figure 5.26. (a) Equilibrium paths and (b) load–frequency relation for a link model with a stiffening spring.

5.10 Path Following and Continuation

In a number of places in this chapter, we have had cause to solve nonlinear algebraic equations, and this will be the case throughout this book. The standard technique for solving sets of nonlinear algebraic equations is Newton–Raphson [9, 24] and this is a standard feature in Mathematica [25] and MATLAB [26]. However, the solution path, as a parameter is changed, may be quite complicated (e.g., including turning points), and some difficulty may be encountered. A number of specialized techniques have been developed based on augmenting Newton–Raphson such that a solution path is followed. These are predictor–corrector techniques and work for differential as well as algebraic equations, and are typically called *continuation* methods. Because these techniques typically involve the rates of change of the response as a function of a parameter, they obtain information about stability (based on the evaluation of the Jacobian) without too much difficulty. A popular and efficient algorithm is contained in the software package AUTO.

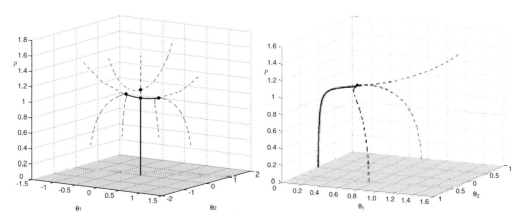

Figure 5.27. Path following results for the Augusti model: (a) geometrically perfect initial configuration and (b) with initial imperfections [29].

Rather than detail specific algorithmic features being given here, readers are referred to Doedel [27] and Doedel et al. [28] for more details. But an example is given based on the solution of the equilibrium equations for the Augusti model from Section 5.7. Using the same parameters as those for Figs. 5.21(a) and 5.22(a), AUTO was used to generate the results shown in Fig. 5.27 [29].

References

[1] J.M.T. Thompson and G.W. Hunt. *A General Theory of Elastic Stability*. Wiley, 1973.
[2] J.G.A. Croll and A.C. Walker. *Elements of Structural Stability*. Wiley, 1972.
[3] G.J. Simitses. *An Introduction to the Elastic Stability of Structures*. Prentice Hall, 1976.
[4] J.M.T. Thompson and G.W. Hunt. *Elastic Instability Phenomena*. Wiley, 1984.
[5] L.N. Virgin. The dynamics of symmetric postbuckling. *International Journal of Mechanical Sciences*, 27:235–48, 1985.
[6] A.N. Kounadis. Nonlinear dynamic buckling of discrete dissipative or nondissipative systems under step loading. *AIAA Journal*, 29:280–9, 1991.
[7] A.N. Kounadis. Nonlinear dynamic buckling and stability of autonomous structural systems. *International Journal of Mechanical Sciences*, 35:643–56, 1993.
[8] I. Elishakoff, S. Marcus, and J.H. Starnes. On vibrational imperfection sensitivity of Augusti's model structure in the vicinity of a nonlinear static state. *International Journal of Non-Linear Mechanics*, 31:229–36, 1996.
[9] W.H. Press, B.P. Flannery, S.A. Teukolsky, and W.T. Vetterling. *Numerical Recipes in Fortran*. Cambridge University Press, 1992.
[10] A.C. Walker, J.G.A. Croll, and E. Wilson. Experimental models to illustrate the nonlinear behavior of elastic structures. *Bulletin of Mechanical Engineering Education*, 10:247–59, 1971.
[11] M.A. Souza. Vibration of thin-walled structures with asymmetric post-buckling characteristics. *Thin-Walled Structures*, 14:45–57, 1992.
[12] P.X. Bellini. The concept of snap-buckling illustrated by a simple model. *International Journal of Nonlinear Mechanics*, 7:634–50, 1972.
[13] D.A. Pecknold, J. Ghaboussi, and T.J. Healey. Snap-through and bifurcation in a simple structure. *Journal of Engineering Mechanics (ASCE)*, 111:909–22, 1985.
[14] L.N. Virgin. Parametric studies of the dynamic evolution through a fold. *Journal of Sound and Vibration*, 110:99–109, 1986.
[15] J.M.T. Thompson and H.B. Stewart. *Nonlinear Dynamics and Chaos*, 2nd ed. Wiley, 2002.
[16] G. Augusti, V. Sepe, and A. Paolone. An introduction to compound and coupled buckling and dynamic bifurcations. In J. Rondal, editor, *Coupled Instabilities in Metal Structures: Theoretical and Design Aspects*. Springer-Verlag, 1998, pp. 1–27.
[17] N. Challamel. Softening branches of a two-degree-of-freedom system induced by spatial buckling. *International Journal of Structural Stability and Dynamics*, 6:493–512, 2006.
[18] L.N. Virgin and R.H. Plaut. Use of frequency data to predict secondary bifurcation. *Journal of Sound and Vibration*, 251:919–26, 2002.
[19] K. Huseyin and J. Roorda. The loading-frequency relationship in multiple eigenvalue problems. *Journal of Applied Mechanics*, 38:1007–11, 1971.
[20] K. Huseyin. *Multiple Parameter Stability Theory and Its Applications*. Oxford University Press, 1986.

[21] R.H. Plaut. Column buckling when support stiffens under compression. *Journal of Applied Mechanics*, 56:484, 1989.

[22] R.H. Plaut. Stability and vibration of a column model with load-dependent support stiffness. *Dynamics and Stability of Systems*, 6:79–88, 1991.

[23] A.P. Seyranian and A.A. Mailybaev. *Multiparameter Stability Theory with Mechanical Applications*. World Scientific, 2003.

[24] T.S. Parker and L.O. Chua. *Practical Numerical Algorithms for Chaotic Systems*. Springer-Verlag, 1989.

[25] S. Wolfram. *The Mathematica Book*. Cambridge University Press, 1996.

[26] MATLAB. User's guide. Technical report, The Math Works, 1989.

[27] E.J. Doedel. *AUTO—Software for continuation and bifurcation problems in ordinary differential equations*. California Institute of Technology, 1986.

[28] E.J. Doedel, A.R. Champneys, T.F. Fairgrieve, Y.A. Kuznetsov, B. Sandstede, and X.J. Wang. Auto97: Continuation and bifurcation software for ordinary differential equations. Technical report, Department of Computer Science, Concordia University, Montreal, Canada, 1997 (available by FTP from ftp.cs.concordia.ca in directory pub/doedel/auto).

[29] H. Chen. Nonlinear analysis of post-buckling dynamics and higher order instabilities of flexible structures. Ph.D. dissertation, Duke University, 2004.

6 Strings, Cables, and Membranes

6.1 Introduction

In transitioning from discrete to continuous systems, we will naturally encounter partial differential equations; that is, the link models of the previous chapter lead to an algebraic eiegnvalue problem whereas the systems considered in this chapter lead naturally to a differential eigenvalue problem, even though we will find utility in approximations leading back to a discrete description. Although the main focus in this book is the behavior of structures subject to compressive axial load, it is instructive to consider systems subject to tensile loads as an introduction. Because these systems tend to be used in a linear context, they typically do not suffer the instability phenomena associated with buckling. However, they do provide a relatively gentle introduction to the behavior and methods of analysis associated with distributed systems subject to axial loading. They also provide a compelling analogy with an everyday example of the relation between axial loading (tension) and natural frequency (pitch): tuning a stringed musical instrument.

6.2 The Stretched String

6.2.1 The Wave Equation

We start by considering the undamped, small-amplitude motion of a stretched string under a tension τ. By considering a small element dx, as shown in Fig. 6.1, we have for horizontal equilibrium

$$\tau(x + dx) \cos\left[\theta(x + dx)\right] - \tau(x) \cos\theta(x) = 0, \tag{6.1}$$

and because for small slopes $\cos\theta \approx 1$, τ is constant. For vertical equilibrium

$$\tau \sin\left[\theta(x + dx)\right] - \tau \sin\theta(x) = \rho dx \frac{\partial^2 w}{\partial t^2}, \tag{6.2}$$

where ρ is the mass per unit length of the string. Expanding the sine terms in Eq. (6.2) gives

$$\tau\left[\theta(x) + dx\frac{\partial\theta(x)}{\partial x} + \cdots +\right] - \tau\left[\theta(x) + \cdots +\right] = \rho dx \frac{\partial^2 w}{\partial t^2}, \tag{6.3}$$

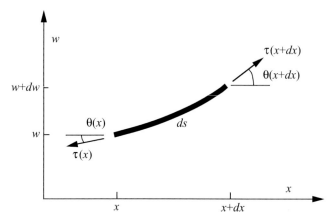

Figure 6.1. Forces acting on a segment of a taut string when undergoing transverse vibration.

and because $\partial\theta/\partial x = \partial^2 w/\partial x^2$ we obtain the governing equation of (lateral motion) for a string:

$$\frac{\partial^2 w}{\partial x^2} = \frac{1}{c_s^2}\frac{\partial^2 w}{\partial t^2}, \tag{6.4}$$

where $c_s = \sqrt{\tau/\rho}$ is a constant, and later to be identified with the speed of lateral motion. It also relates to the velocity of wave propagation along the string, although we are primarily interested in transverse (lateral) effects here. Equation (6.4) is a partial differential equation of the type introduced in Chapter 4, is called the *wave equation*, and occurs in many branches of the physical sciences [1].

A standard approach to solving partial differential equations of this type is based on the separation of variables. This approach will also be used when we consider the dynamic behavior of structures with bending stiffness later. We assume that the displacement of the string can be written as

$$w(x,t) = \phi(x)q(t), \tag{6.5}$$

which, on substitution in Eq. (6.4) leads to

$$\frac{1}{\phi}\frac{d^2\phi}{dx^2} = \frac{1}{q}\frac{1}{c_s^2}\frac{d^2 q}{dt^2}. \tag{6.6}$$

We see that the left-hand side is a function of the position and the right-hand side is a function of time. In this case both sides of the equation must be equal to the same constant, which we set as $-(\omega/c_s)^2$ (with the negative sign and square chosen with forethought). Now we have the two uncoupled ordinary differential equations,

$$\frac{d^2\phi}{dx^2} + \left(\frac{\omega}{c_s}\right)^2\phi = 0, \tag{6.7}$$

$$\frac{d^2 q}{dt^2} + \omega^2 q = 0. \tag{6.8}$$

These linear equations have very familiar solutions

$$\phi(x) = A \sin\left(\frac{\omega}{c_s}\right) x + B \cos\left(\frac{\omega}{c_s}\right) x, \tag{6.9}$$

$$q(t) = C \sin \omega t + D \cos \omega t, \tag{6.10}$$

where the constants (A, B) and (C, D) are obtained from the boundary and initial conditions [2].

Suppose we have a string that is stretched between fixed points. In this case, the boundary conditions can be written as

$$w(0, t) = 0, \tag{6.11}$$

$$w(L, t) = 0. \tag{6.12}$$

Evaluating the spatial part of the solution [Eq. (6.9)] under these circumstances, we get $B = 0$ from Eq. (6.11), and from Eq. (6.12),

$$A \sin\left(\frac{\omega}{c_s}\right) L = 0. \tag{6.13}$$

The solutions to this (in addition to the trivial solution $A = 0$) are

$$\frac{\omega_n L}{c_s} = n\pi, \qquad n = 1, 2, \ldots, \tag{6.14}$$

that is, there are an infinite number of natural frequencies:

$$\omega_n = \frac{n\pi c_s}{L} = \frac{n\pi}{L} \sqrt{\frac{\tau}{\rho}}. \tag{6.15}$$

And thus we have a linear relation between tension τ and the square of the natural frequencies ω_n with corresponding mode shapes,

$$\phi_n(x) = \sin \frac{\omega_n x}{c_s} = \sin \frac{n\pi x}{L}. \tag{6.16}$$

The mode shapes satisfy the conditions of orthogonality described earlier in this book (see Section 4.3).

Thus we have an equation for the lateral displacement of the string given by

$$w(x, t) = \sum_{n=1}^{\infty} q_n(t) \sin \frac{n\pi x}{L} \tag{6.17}$$

$$= \sum_{n=1}^{\infty} (C_n \sin \omega_n t + D_n \cos \omega_n t) \sin \frac{n\pi x}{L}, \tag{6.18}$$

or, using complex notation,

$$w(x, t) = \sum_{n=1}^{\infty} \beta_n e^{i\omega_n t} \sin \frac{n\pi x}{L}, \tag{6.19}$$

where the β_n are complex (and it is the real part that corresponds to physically meaningful solutions). The constants associated with the temporal part of the solution are determined from the initial conditions $w(x, 0)$ and $\partial w(x, 0)/\partial t$ and correspond to the Fourier coefficients.

As an example, consider a taut string that is displaced by an amount H (not too large) at its center and then released (i.e., with an initial velocity of zero). The string has the initial form

$$
\begin{aligned}
w(x, 0) &= \frac{2H}{L}x, & 0 \leq x \leq L/2, \\
w(x, 0) &= \frac{2H}{L}(L - x), & L/2 \leq x \leq L,
\end{aligned}
\tag{6.20}
$$

and evaluating C and D leads to a response

$$
w(x, t) = \frac{8H}{\pi^2} \left(\sin \frac{\pi x}{L} \cos \omega_1 t - \frac{1}{9} \sin \frac{3\pi x}{L} \cos \omega_3 t + \cdots + \right).
\tag{6.21}
$$

That only odd harmonics are excited is due to the symmetric nature of the initial disturbance, which leads to symmetric motion. The resulting triangular wave is dominated by the first (fundamental) mode, as expected.

Thus we see that the boundary conditions influence the mode shapes and natural frequencies and the initial conditions determine the contribution of each mode. We observe our familiar linear relation between the square of the natural frequency and the tension. This closed-form solution would not have been available if the tension in the string had not been constant.

Later in this book we will use a somewhat similar approach for beams and plates. The major differences are that systems with bending stiffness will tend to lead to higher-order differential equations (and hence more boundary conditions) and the forces of interest will primarily be compressive, which allows for bifurcational phenomena to appear.

6.2.2 Traveling-Wave Solution

At this point we might wonder why Eq. (6.4) has the name it does. We now show that an alternative, but equivalent, description can be obtained by writing the general solution in the form

$$
w(x, t) = F_1(x - c_s t) + F_2(x + c_s t).
\tag{6.22}
$$

We can think of the first term on the right-hand side of Eq. (6.22) as representing a wave moving in the positive x direction with constant velocity c_s, with the second term corresponding to a similar wave but moving in the opposite direction. In both cases the functions F represent the (non-changing) shape, or profile, of the wave. Because we have already seen that sinusoidal motion is typical in the vibration of strings, let's consider a wave of the form

$$
w(x, t) = A \sin \frac{2\pi}{\lambda}(x - c_s t),
\tag{6.23}
$$

that is, a wave of amplitude A and wavelength λ traveling in the positive x direction. Introducing the wavenumber $k = 1/\lambda$, we can rewrite Eq. (6.23) as

$$w(x, t) = A \sin (2\pi kx - \omega t), \tag{6.24}$$

in which $\omega = c_s (2\pi)/\lambda$ is the frequency of the wave. However, if we have two equal waves but traveling in opposite directions, we then have

$$w(x, t) = A \sin (2\pi kx - \omega t) + A \sin (2\pi kx + \omega t), \tag{6.25}$$

which can be rewritten as

$$w(x, t) = 2A \sin (2\pi kx) \cos (\omega t). \tag{6.26}$$

Thus we see that the two waves in this case together respresent a standing wave but with an oscillating profile. The two components cancel at those points where $x = n\lambda/2$, and these are called *node* points. If $x = \lambda(2n + 1)/4$, then we have a reinforcement, or antinodes.

To apply this approach to a specific (finite) string, we note that, for example, for a string stretched between two points we will have nodes at the end points, and thus $2kL = r$ where r is an integer, and therefore

$$\omega_r = 2\pi k c_s = r\pi \frac{c_s}{L} = r\pi \sqrt{\frac{\tau}{\rho L^2}} \tag{6.27}$$

with the frequencies obtained previously. We note that, in the presence of damping, the traveling waves will decay, and although it is possible to solve the wave equation with damping, the solution becomes considerably more involved.

6.2.3 Energy Considerations and Rayleigh's Principle

The wave equation can also be obtained by use of Hamilton's principle. The kinetic energy of the string is given by

$$T = \frac{1}{2} \rho \int_0^L \left(\frac{\partial w}{\partial t} \right)^2 dx \tag{6.28}$$

$$= \frac{1}{2} \rho \int_0^L \left(\sum_{n=1}^{\infty} \dot{q} \sin \frac{n\pi x}{L} \right)^2 dx \tag{6.29}$$

$$= \frac{\rho L}{4} \sum_{n=1}^{\infty} \dot{q}_n^2, \tag{6.30}$$

and the potential energy associated with the stretching is given by

$$U = \tau \int_0^L \left(\sqrt{1 + \left(\frac{\partial w}{\partial x} \right)^2} - 1 \right) dx \tag{6.31}$$

$$\approx \frac{1}{2} \tau \int_0^L \left(\frac{\partial w}{\partial x} \right)^2 dx \tag{6.32}$$

$$= \frac{1}{2}\tau \int_0^L \left(\sum_{n=1}^{\infty} \frac{n\pi}{L} q_n \cos \frac{n\pi x}{L} \right)^2 dx \qquad (6.33)$$

$$= \frac{\rho L}{4} \sum_{n=1}^{\infty} \omega_n^2 q_n^2, \qquad (6.34)$$

where we have used the fact that, from Eq. (6.15),

$$\tau = \rho \left(\frac{\omega_n L}{n\pi} \right)^2. \qquad (6.35)$$

We can then invoke Hamilton's principle and, by taking variations in the energy terms, arrive at the wave equation. Assuming there is no external forcing or damping, we also have the conservation of total energy, that is,

$$E = T + U = \frac{\rho L}{4} \sum_{n=1}^{\infty} \omega_n^2 |\beta_n|^2, \qquad (6.36)$$

where the β_n [from Eq. (6.19)] depends on the initial conditions.

We can also use an approximate mode shape (although it is not really necessary in this instance as the exact solution is well known), such as a parabola:

$$w = 4H(t)x(L - x)/L^2. \qquad (6.37)$$

Plugging this into the expression for the kinetic and potential energy expressions [Eqs. (6.28) and (6.31)] and using Rayleigh's method gives

$$\omega^2 = \frac{10\tau}{\rho L^2}, \qquad (6.38)$$

which is less than 1% greater than the exact value [i.e., $\omega^2 = \pi^2 \tau/(\rho L^2)$]. Rayleigh devised a method of improving this result by incorporating an adjustable constant into the displacement function [3].

It is interesting to note that when a string is excited it often exhibits non-planar motion. This type of whirling motion (familiar from a child's skipping rope) is just one type of complicated motion found in the forced string problem. In this chapter, we have assumed that the tension in the string remains constant during motion. However, the tension must fluctuate, and it is the subtle interaction between longitudinal and transverse motion that underlies much of the interesting (nonlinear) behavior. In fact, the greater the tension in the string, the relatively less the tension changes during motion. This problem will be revisited later when we consider large-amplitude vibration.

6.3 A Suspended Cable

A natural extension to the study of the dynamics of a taut string is to consider the behavior of a string that is not taut, but rather sags because of the effect of gravity [4]. Without bending stiffness, it is again the axial load that provides the restoring

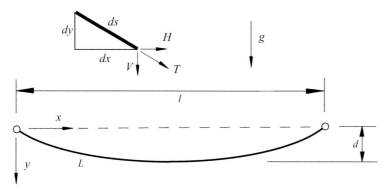

Figure 6.2. The cable geometry and the forces acting on a segment.

force for this important class of practical problems. We focus on symmetric cables having suspension points located at the same vertical elevation, and consider only modes of transverse vibration taking place in-plane [5]. If the cable configuration is *shallow*, then certain simplifying assumptions can be made, but, in general, the sag may have a relatively profound effect on dynamic behavior. Because of the role of gravity, it is convenient to use the coordinate system shown in Fig. 6.2.

Assuming the cable is inextensional, we can write the equilibrium in the vertical direction as

$$\frac{d}{ds}\left(T\frac{dy}{ds}\right) = -mg, \tag{6.39}$$

and in the horizontal direction we have

$$\frac{d}{ds}\left(T\frac{dx}{ds}\right) = 0, \tag{6.40}$$

which can be integrated to give

$$T\frac{dx}{ds} = H, \tag{6.41}$$

in which H is the horizontal component of the cable tension. Thus Eq. (6.39) can be rewritten as

$$H\frac{d^2y}{dx^2} = -mg\frac{ds}{dx}. \tag{6.42}$$

Using horizontal equilibrium and using $ds^2 = dx^2 + dy^2$, we can then write

$$\frac{d}{dx}\left(\frac{dy}{dx}\right) = -\frac{mg}{H}\sqrt{1 + \left(\frac{dy}{dx}\right)^2}. \tag{6.43}$$

The solution to Eq. (6.43) is the well-known catenary, and given that the end points of the cable have the same vertical elevation, we can write the solution in terms of the length-to-span ratio [5, 6],

$$\frac{L}{l} = \frac{\sinh\left[lmg/(2H)\right]}{lmg/(2H)}, \tag{6.44}$$

and the sag-to-length ratio

$$\frac{d}{l} = \frac{H}{lmg}\{\cosh\left[lmg/(2H)\right] - 1\}. \tag{6.45}$$

For a shallow cable (with a small sag), we can expand the hyperbolic functions in Eqs. (6.44) and (6.45) as Taylor series or simplify governing equation (6.42):

$$\frac{d^2y}{dx^2} = -\frac{mg}{H}. \tag{6.46}$$

It can easily be shown that the deflected shape of the cable is then given by

$$y = \frac{l^2 mg}{2H}\frac{x}{l}\left(1 - \frac{x}{l}\right) = 4d\frac{x}{l}\left(1 - \frac{x}{l}\right), \tag{6.47}$$

where the vertical sag in the center of the cable d is given by

$$\frac{d}{l} = \frac{1}{8}\frac{lmg}{H}. \tag{6.48}$$

We can now consider oscillations about equilibrium. In addition to an inertia force (by use of D'Alembert's principle), the added motion $w(x, t)$ will induce a varying horizontal force $h(t)$, and thus the equation of vertical motion is

$$(H + h)\frac{\partial^2}{\partial x^2}(y + w) = -mg + m\frac{\partial^2 w}{dt^2}, \tag{6.49}$$

which, for small-amplitude motion, can be simplified to

$$\frac{\partial^2 w}{\partial x^2} - \frac{m}{H}\frac{\partial^2 w}{dt^2} = \frac{mg}{H}\frac{h}{H}. \tag{6.50}$$

It can also be shown [6] that for inextensional cables (based on linearized theory) $h(t)$ depends on the displacement function $w(x, t)$ such that $\int_0^l w\,dx = 0$.

If we then seek harmonic oscillations of the form

$$w(x, t) = \tilde{w}(x)\cos\omega t, \qquad h(t) = \tilde{h}\cos\omega t, \tag{6.51}$$

then substituting these into Eq. (6.50) and using $\beta^2 = (m\omega^2)/H$ results in

$$\frac{\partial^2 \tilde{w}}{\partial x^2} + \beta^2 \tilde{w} = \frac{mg}{H}\frac{\tilde{h}}{H}. \tag{6.52}$$

Solving Eq. (6.52) by using the boundary conditions $w(0) = w(l) = 0$ results in the following solutions:

- The symmetric modes:

$$\frac{\tilde{w}}{d} = \frac{8C_n}{(\beta_n l)^2}\left[1 - \frac{\cos\left[\beta_n(x - (1/2)l)\right]}{\cos[(1/2)\beta_n l]}\right], \qquad \frac{\tilde{h}}{H} = C_n, \tag{6.53}$$

with $\int_0^l \tilde{w}\,dx = 0$ leading to the lowest root $(1/2)\beta_1 l = 4.493$. Therefore the lowest (symmetric mode) frequency is $\omega_1 = 8.99(H/ml^2)^{1/2}$.

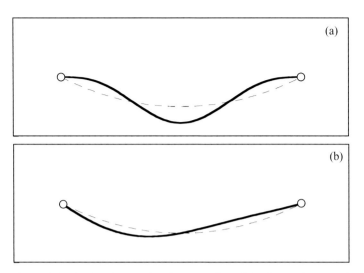

Figure 6.3. The first inextensional symmetric and the first antisymmetric modes.

- The antisymmetric modes:

$$\frac{\tilde{w}}{d} = C_n \sin\left(n\frac{2\pi}{l}x\right), \qquad \omega_n = n\frac{2\pi}{l}\sqrt{\frac{H}{m}} \tag{6.54}$$

with $\tilde{h} = 0$ leading to the lowest (antisymmetric mode) frequency $\omega_1 = 2\pi(H/ml^2)^{1/2}$.

The lowest symmetric and antisymmetric modes are shown in Figs. 6.3(a) and 6.3(b), respectively. In the former case, it is interesting to see that the wavelength of the mode is shorter than the span and the cable straightens out near the supports. In the latter case, we recognize the form from the taut string, as one side moves up and the other side moves down.

It is also interesting to note that these antisymmetric modes would not be influenced by any elasticity of the cable (whereas the symmetric modes may be quite influenced). For example, by introducing the parameter λ, accounting for the elastic flexibility of the cable,

$$\lambda^2 = \frac{EA}{H}\left(\frac{8d}{l}\right)^2, \tag{6.55}$$

we can incorporate this effect into the preceding theory [6]. As the cable flexibility decreases ($\lambda \to \infty$), we approach the inextensible case already considered. At the other end of the spectrum, we have $\lambda \to 0$ and a frequency that tends to decrease, and reaches the value of a taut string $\omega = \pi\sqrt{H/ml^2}$. This effect is summarized in Fig. 6.4. We observe that when $\lambda^2 < 4\pi^2$, the lowest mode is symmetric (with no internal nodes), but at the transition point $\lambda^2 = 4\pi^2$ the in-plane frequencies are equal. This type of modal exchange will also be encountered later in the behavior of a pin-ended column with an elastic restraint at midspan, as well as other structural systems in which axial load is a consideration. When $\lambda^2 > 4\pi^2$, the lowest mode

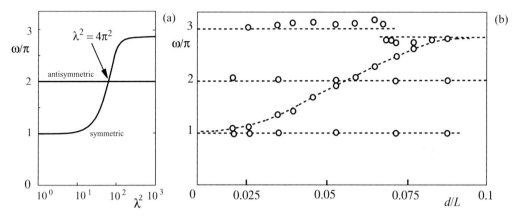

Figure 6.4. Dynamic response of an extensible cable with sag: (a) two lowest natural frequencies and (b) experimental verification (adapted from [7]).

is antisymmetric. Part (b) of Fig. 6.4 also shows this crossover of modes but now as a function of the sag d/l, and some experimental data points are superimposed [6, 7]. The lowest frequency here corresponds to a sway mode. Thus we see that the dynamic response of suspended cables depend on a variety of factors including sag (and hence mass), vertical distance between the support points, flexibility of the cable itself, and so on. The symmetric modes of vibration of the cable are heavily influenced by the cable stiffness (characterized by λ^2) and thus depend on sag and cable flexibility.

6.3.1 The Hanging Chain

The oscillations of a flexible cable suspended vertically from a single fixed point represent a problem that occupies an important place in the historical development of mechanics [8, 9]. It was one of the first systems in which the normal modes of vibration were identified and provided an initial motivation for the development of Bessel functions [10].

In a gravitational field, the *weight* of a vertically suspended slender beam becomes increasingly important, that is, as the length of the beam increases the bending stiffness becomes negligible in comparsion with gravitational effects. The hanging chain has no bending stiffness. At the opposite end of the flexibility spectrum is the rigid-arm pendulum. Rather than the classical analysis being described here (we shall of course focus on beams in bending later) a relatively simple approximate analysis based on Rayleigh's method is shown.

Assuming the origin is placed at the top end of a vertically hanging chain, of length L, a reasonable first-mode shape in terms of lateral deflection w is given by

$$w = Q\left[\frac{x}{L} + \beta\left(\frac{x}{L}\right)^2\right], \tag{6.56}$$

in which x is measured downward from the top and β is a constant to be determined. Note that, in contrast to a hanging beam to be considered later, a linear term is

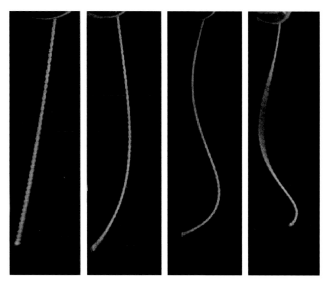

Figure 6.5. The lowest four vibration mode shapes for an experimental hanging beam.

included to allow for a nonzero slope at the clamped end. The kinetic energy is

$$T = \frac{1}{2}m \int_0^L \dot{w}^2 dx = \frac{1}{60}mL\dot{Q}^2(10 + 15\beta + 6\beta^2). \tag{6.57}$$

The potential energy is

$$V = \frac{1}{2}mg \int_0^L \int_0^x w'^2 dx dx = \frac{1}{12}mgQ^2(3 + 4\beta + 2\beta^2), \tag{6.58}$$

and application of Rayleigh's method (see Section 3.3) gives

$$\omega^2 = \frac{5(3 + 4\beta + 2\beta^2)}{10 + 15\beta + 6\beta^2}\left(\frac{g}{L}\right). \tag{6.59}$$

The value of β (= 0.289206) that minimizes the frequency results in $\omega = 1.2025\sqrt{g/L}$. Because of the extremum nature of Rayleigh's method, we know that the lower the frequency estimate is, the closer it will be to the exact answer. However, the exact value (obtained in this case with Bessel functions) is $1.2025\sqrt{g/L}$ [11], and thus the mode shape described in Eq. (6.56) is obviously very accurate. It is interesting to note that for a rigid bar of the same length the frequency is $1.22474\sqrt{g/L}$ [12].

Figure 6.5 shows some experimental snapshots of a hanging axisymmteric chain suspended from a spinning shaft. These mode shapes correspond closely to those obtained with Bessel functions for bending motion with nodal points easily observed.

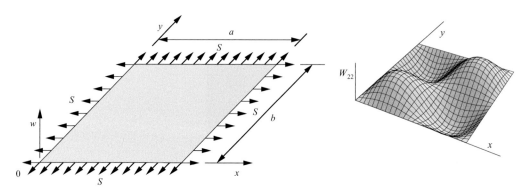

Figure 6.6. (a) A rectangular stretched membrane and (b) the (2,2) mode of vibration.

6.4 A Rectangular Membrane

A stretched membrane may exhibit transverse vibrations in much the same way as the string, but we now need two dimensions to describe the geometry [2, 13]. We take a brief look at the simple case of a rectangular membrane of uniform thickness, with mass density ρ, which is stretched such that the tension S can be assumed to be constant over the membrane. A schematic of the membrane is shown in Fig. 6.6(a).

The lateral deflection is w, and the membrane has length a in the x direction and width b in the y direction. After harmonic motion is assumed, the governing equation is given by

$$\nabla^2 W(x, y) + \beta^2 W(x, y) = 0, \qquad (6.60)$$

where $\beta^2 = \rho \omega^2 / S$ and ∇^2 is the Laplacian [2], that is, $\nabla^2 = \partial^2/\partial x^2 + \partial^2/\partial y^2$, which we shall also make use of in the chapter on plates. We can separate variables, apply the boundary conditions, and obtain the natural frequencies

$$\omega_{mn} = \pi \sqrt{\frac{S}{\rho}\left[\left(\frac{m}{a}\right)^2 + \left(\frac{n}{b}\right)^2\right]}, \qquad m, n = 1, 2, \ldots. \qquad (6.61)$$

The corresponding (normalized) mode shapes are given by

$$W_{mn} = \frac{2}{\sqrt{\rho a b}} \sin\frac{m\pi x}{a} \sin\frac{n\pi y}{b}, \qquad m, n = 1, 2, \ldots. \qquad (6.62)$$

An example of the mode corresponding to the fourth lowest vibration (with frequency $\omega_{22} = 2\pi\sqrt{(S/\rho)(1/a^2 + 1/b^2)}$ is shown in Figure 6.6(b). Hence we again see that the square of the natural frequencies increases linearly with the tension. The lowest frequency for a square membrane is thus given by $\omega = \pi\sqrt{2S/(\rho a^2)}$.

Alternatively, we can write down the kinetic energy associated with the vibration of the membrane as

$$T = \frac{1}{2}\rho \int_0^a \int_0^b \dot{w}^2 dx dy, \qquad (6.63)$$

and the change in potential energy that is due to the tension as the membrane deflects as

$$V = \frac{1}{2}S \int_0^a \int_0^b \left[\left(\frac{\partial w}{\partial x} \right)^2 + \left(\frac{\partial w}{\partial y} \right)^2 \right] dx dy. \tag{6.64}$$

The deflection can be represented by the function

$$w = \sum_{m=1}^{\infty} \sum_{n=1}^{\infty} \phi_{mn}(t) \sin \frac{m\pi x}{a} \sin \frac{n\pi y}{b}, \tag{6.65}$$

which is then used to evaluate the energy terms

$$T = \frac{1}{8}\rho ab \sum_{m=1}^{\infty} \sum_{n=1}^{\infty} \dot{\phi}_{mn}^2 \tag{6.66}$$

and

$$V = \frac{1}{8}Sab \sum_{m=1}^{\infty} \sum_{n=1}^{\infty} \phi_{mn}^2 \left[\left(\frac{m\pi}{a} \right)^2 + \left(\frac{n\pi}{b} \right)^2 \right]. \tag{6.67}$$

We again use Lagrange's equation (2.35) to obtain the equations of motion

$$\ddot{\phi}_{mn} + \frac{S}{\rho}\pi^2 \left[\left(\frac{m}{a} \right)^2 + \left(\frac{n}{b} \right)^2 \right] \phi_{mn} = 0, \qquad m, n = 1, 2, \ldots, \tag{6.68}$$

with a set of natural frequencies obtained previously [Eq. (6.61)]. Later, we shall compare this behavior with that of systems in which there is also bending stiffness, that is, plates.

It turns out that the analysis of membranes depends very much on the shape of the boundary. For example, a circular membrane is more conveniently analyzed by use of polar coordinates and results in Bessel functions [10], and an irregularly shaped membrane would typically require a FEA. The statements made earlier regarding the large-amplitude motion of strings also apply to membranes but we now move on to consider the much wider class of problem in which the structure subject to axial loading possesses bending stiffness and the axial loads are often compressive.

References

[1] R. Courant and D. Hilbert. *Methods of Mathematical Physics*. Wiley Classics Library, 1989.
[2] L. Meirovitch. *Principles and Techniques of Vibrations*. Prentice Hall, 1997.
[3] Lord Rayleigh (John William Strutt). *The Theory of Sound*. Dover, 1945.
[4] H.M. Irvine and T.K. Caughey. The linear theory of free vibrations of a suspended cable. *Proceedings of the Royal Society of London Series A*, 341:299–315, 1974.
[5] S. Krenk. *Mechanics and Analysis of Beams, Columns and Cables*. Springer, 2001.
[6] H.M. Irvine. *Cable Structures*. MIT Press, 1981.
[7] S.E. Ramberg and O.M. Griffin. Free vibration of taut and slack marine cables. *Journal of the Structural Division, Proc. ASCE*, 103:2079–92, 1977.

[8]   H. Lamb. *Higher Mechanics*. Cambridge University Press, 1929.

[9]   D. Yong. Strings, chains, and ropes. *SIAM Review*, 48:771–81, 2006.

[10] G.N. Watson. *A Treatise on the Theory of Bessel Functions*. Cambridge University Press, 1966.

[11] H. Lamb. *The Dynamical Theory of Sound*. Arnold, 1910.

[12] R.D. Blevins. *Formulas for Natural Frequencies and Mode Shapes*. Van Nostrand Rheinhold, 1979.

[13] A.D. Dimarogonas. *Vibration for Engineers*. Prentice Hall, 1996.

7 Continuous Struts

7.1 Introduction

In this chapter, we consider the dynamics of a thin elastic strut including axial effects, arising primarily from one of two situations:

- the axial load is applied externally (including postbuckling), or
- deformation is sufficient to cause coupling between axial and bending behavior (the membrane effect).

In the first section, attention is focused on a traditional approach to setting up the equations of motion (by means of D'Alembert's principle) for the simple case based on engineering beam theory (Euler–Bernoulli) with the addition of axial loads. The resulting partial differential equation of motion is then separated into temporal and spatial ordinary differential equations and the response analyzed for various magnitudes of the axial load [1, 2]. Then an energy approach is used together with Rayleigh's method [3]. In this case, additional terms are retained in the potential energy to allow *postbuckled* effects to be analyzed [4] and the effect of initial geometric imperfections are included. An alternative approach is developed based on a simple application of Hamilton's principle, and in this instance stretching effects are also included and a solution developed by use of Galerkin's method. This approach will be similar to that used in the previous chapter on strings but now bending strain energy enters into the analysis (as well as *compressive* axial loading). In the final part of the chapter we consider the dynamics of struts that are loaded by gravity through self-weight. The next chapter will then continue the study of axially loaded members but with the scope opened to include a wider class of problem.

7.2 Basic Formulation

In this section, we develop the governing equation of motion for a thin, elastic, prismatic beam subject to a constant axial force. In Fig. 7.1(a) a schematic of the beam is shown. It has mass per unit length m, constant flexural rigidity EI, and is subject to an axial load P. The length is L, the coordinate along the beam is x, and the lateral (transverse) deflection is $w(x, t)$. In part (b) is shown an element of the beam between locations x and $x + \Delta x$, which is subject to the D'Alembert forces

$$R(x, t) = m\partial^2 w / \partial t^2. \tag{7.1}$$

(a)

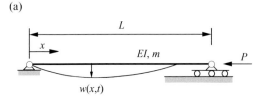

(b)

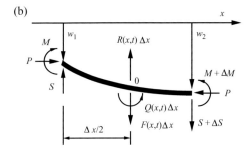

Figure 7.1. A schematic of a thin elastic beam.

At this point, we make the assumption that the angle through which the section rotates $(\Delta w/\Delta x)$ is small and we neglect axial and rotary inertia [5].

Taking moments about 0 we get

$$S\frac{\Delta x}{2} - (S + \Delta S)\frac{\Delta x}{2} + M - (M + \Delta M) + P\Delta w = 0, \tag{7.2}$$

and as $\Delta x \to 0$ we have

$$S - \frac{\partial M}{\partial x} + P\frac{\partial w}{\partial x} = 0. \tag{7.3}$$

Summing forces in the vertical direction we also have

$$(S + \Delta S) - S + F(x, t)\Delta x - R(x, t)\Delta x = 0, \tag{7.4}$$

and again passing to the limit $\Delta x \to 0$, we have

$$\frac{\partial S}{\partial x} + F(x, t) - m\frac{\partial^2 w}{\partial t^2} = 0. \tag{7.5}$$

Differentiating Eq. (7.3) we obtain

$$\frac{\partial S}{\partial x} - \frac{\partial^2 M}{\partial x^2} + P\frac{\partial^2 w}{\partial x^2} = 0, \tag{7.6}$$

and eliminating $\partial S/\partial x$ between Eqs. (7.5) and (7.6) gives

$$m\frac{\partial^2 w}{\partial t^2} - \frac{\partial^2 M}{\partial x^2} + P\frac{\partial^2 w}{\partial x^2} = F(x, t), \tag{7.7}$$

and finally, using the familiar expression from engineering beam theory [6, 7],

$$EI\frac{\partial^2 w}{\partial x^2} = -M, \tag{7.8}$$

we arrive at the governing equation:

$$EI\frac{\partial^4 w}{\partial x^4} + P\frac{\partial^2 w}{\partial x^2} + m\frac{\partial^2 w}{\partial t^2} = F(x, t). \tag{7.9}$$

This linear partial differential equation is a key equation in this book and can be rewritten in shorthand as

$$EIw'''' + Pw'' + m\ddot{w} = F(x,t) \tag{7.10}$$

or

$$EIw_{xxxx} + Pw_{xx} + mw_{tt} = F(x,t), \tag{7.11}$$

where derivatives are signified by primes and subscripts in Eqs. (7.10) and (7.11), respectively, and, given appropriate boundary conditions at $x = 0, L$ and initial conditions, can be solved by use of standard methods as outlined in Chapter 4.

Before we consider the specific aspects of the solution, the utility of nondimensionalizing the governing equation of motion is mentioned. Nondimensionalization is useful because it reduces the number of parameters and allows a more consistent comparison of behavior. In this chapter, we are focused on the unforced problem ($F = 0$), and, by defining the following parameters,

$$\bar{x} = x/L, \quad \bar{w} = w/L, \quad \bar{t} = t\sqrt{EI/(mL^4)}, \quad p = PL^2/(EI), \tag{7.12}$$

we can rewrite Eq. (7.9) as

$$\frac{\partial^4 \bar{w}}{\partial \bar{x}^4} + p\frac{\partial^2 \bar{w}}{\partial \bar{x}^2} + \frac{\partial^2 \bar{w}}{\partial \bar{t}^2} = 0. \tag{7.13}$$

Throughout this book we shall look at solutions to this type of equation and often incorporate nonlinearities into the analysis, in which case a variety of approximate techniques (of the type outlined in Chapter 4) will be utilized. Sometimes the free parameters will be chosen such that they are further normalized (e.g., p might be related to a critical buckling load). However, we start by looking at the simple free-vibration case for which there is good access to analytical solutions.

7.2.1 The Response

We initially consider the free-vibration problem and assume that $F = 0$ and that the motion consists of a function $W(x)$ that varies with $Y(t)$ such that

$$w(x,t) = W(x)Y(t). \tag{7.14}$$

Placing this back into Eq. (7.9) leads to

$$EI\frac{d^4 W}{dx^4}Y + P\frac{d^2 W}{dx^2}Y = -mW\frac{d^2 Y}{dt^2}. \tag{7.15}$$

That is, after dividing by WY (which is not zero if w is not zero), we have an equation separated into spatial (x) and temporal (t) parts, and thus the ratio on each side must be a constant (which we label $-\omega^2$)

$$\frac{m}{Y}\frac{d^2 Y}{dt^2} = -\frac{EI}{W}\frac{d^4 W}{dx^4} - \frac{P}{W}\frac{d^2 W}{dx^2} = -\omega^2. \tag{7.16}$$

Thus we have two *ordinary* differential equations—one in space that we can solve by using the appropriate boundary conditions and that describes the mode shapes (which are half sine waves in the simply supported case); the other in time that we can solve with the appropriate initial conditions and contains the frequency information (simple harmonic motion). It is not uncommon for the vibration of mechanical systems to be dominated by the mode with the lowest frequency, but we consider the full system at first before developing approximate techniques.

7.2.2 The Temporal Solution

We might expect the second-order ordinary differential equation in time to have oscillatory solutions (given positive values of flexural rigidity, etc.). However, we anticipate that the dependence of the form of the temporal solution will depend on the magnitude of the axial load [8, 9]. To be a little more specific, before going on to consider the more general boundary conditions, let us suppose we have ends that are pinned (no deflection and no resistance to rotation), that is, the deflection (w) and bending moment ($-EI\partial^2 w/\partial^2 x$) are zero at $x = 0$ and $x = L$. In the general case, we would assume an exponential form for the solution, but with these relatively convenient boundary conditions, we can take

$$w(x, t) = \sum_{n=1}^{\infty} Y(t) \sin \frac{n\pi x}{L}. \tag{7.17}$$

We can obtain the temporal part of the solution by assuming

$$Y_n(t) = A_n e^{i\omega_n t}, \tag{7.18}$$

and substituting into Eq. (7.9) leads to

$$\sum_{n=1}^{\infty} \left[\left(EI \frac{n^2 \pi^2}{L^2} - P \right) \frac{n^2 \pi^2}{L^2} - m\omega_n^2 \right] A_n \sin \frac{n\pi x}{L} e^{i\omega_n t} = 0. \tag{7.19}$$

Clearly, the term in the square brackets must vanish for a nontrivial solution so that

$$\omega_n^2 = \frac{n^4 EI \pi^4}{mL^4} \left[1 - \frac{PL^2}{n^2 EI \pi^2} \right]. \tag{7.20}$$

If we define the following parameters

$$p_n = n^2 EI \pi^2 / L^2, \quad \bar{\omega}_n^2 = n^4 EI \pi^4 / mL^4, \tag{7.21}$$

Eq. (7.20) becomes

$$\omega_n = \pm \bar{\omega}_n \sqrt{1 - \bar{p}}, \tag{7.22}$$

where $\bar{p} = P/p_n$, and we see that the nature of the solution depends crucially on the discriminant. Making use of the Euler identities, we consider the following four cases, in which A_n and B_n are constants obtained from the initial conditions:

- If $\bar{p} = 0$, then

$$Y_n(t) = A_n \cos \bar{\omega}_n t + B_n \sin \bar{\omega}_n t, \qquad (7.23)$$

  and we observe simple harmonic motion, a familiar result from linear vibration theory.
- If $0 < \bar{p} < 1$, then

$$Y_n(t) = A_n \cos \omega_n t + B_n \sin \omega_n t, \qquad (7.24)$$

  where ω_n is given by Eq. (7.22) and is real, and simple harmonic motion results. Any perturbation will induce oscillatory motion about equilibrium. Assuming no damping, the response neither grows nor decays. This includes the response for a tensile axial load, that is, $\bar{p} < 0$.
- If $\bar{p} = 1$, Eq. (7.22) has a double-zero root, and then the solution can be written as

$$Y_n(t) = A_n + B_n t, \qquad (7.25)$$

  and the motion grows linearly with time (this is a *special* case).
- If $\bar{p} > 1$, the roots of (7.22) are purely imaginary and then

$$Y_n(t) = A_n \cosh \bar{\omega}_n t + B_n \sinh \bar{\omega}_n t, \qquad (7.26)$$

  and the motion grows exponentially with time.

Typical examples of these cases are shown in Fig. 7.2(a) in which the natural frequency in the absence of axial load was taken as unity [10], and hence a natural period of 2π. Also shown in this figure are the stability of equilibrium, part (b), and effective natural frequency (squared), part (c), as a function of axial load.

Let us focus attention on the lowest natural frequency and its corresponding mode ($n = 1$). With no axial load ($\bar{p} = 0$) we obtain $\omega_1 = \bar{\omega}_1$. However, as the axial load increases the natural frequency decreases according to Eq. (7.22), that is, we observe a linear relationship between the magnitude of the axial load and the square of the natural frequency [see Fig. 7.2(c)]. Any nonzero initial conditions result in bounded motion, and we may consider this to be a stable situation (at least in the sense of Lyapunov). When $\bar{p} \to 1$, ω_1 vanishes and the solution ceases to be oscillatory [the linearly increasing (constant-velocity) solution shown in Fig. 7.2]. Any inevitable perturbation will cause the system to become unstable. This type of instability is monotonic because, locally, the deflections grow in one direction (determined by the initial conditions). This type of behavior is sometimes referred to as *divergence*. The higher modes ($n > 1$) will exhibit oscillations (in theory) but the important practical information has been gained, that is, typical behavior is dominated by the lowest mode.

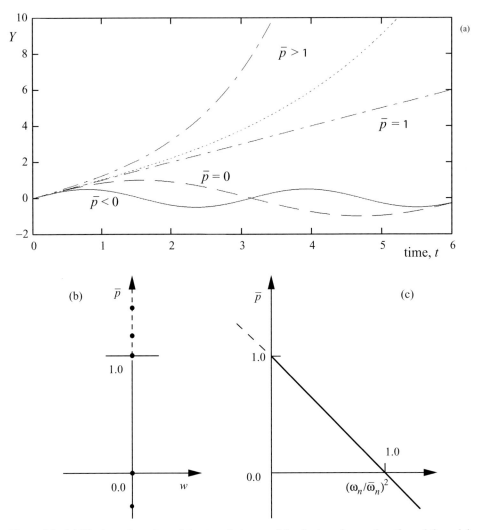

Figure 7.2. (a) The lateral motion of the strut in terms of the first mode as a function of the axial load, (b) stability of equilibrium, and (c) frequency–axial-load relationship.

7.2.3 The Spatial Solution

In the previous subsection, we focused on a simply supported strut that has the advantage of relatively simple mode shapes. In this section, we consider a strut with one end simply supported and the other end clamped. In that case, the lowest natural frequency and its mode shape can still be extracted from the governing differential equation without too much difficulty. For *tensile* axial loads, the system does not suffer instability, and we will see that the lowest natural frequency *increases* with tensile force [8].

Returning to Eq. (7.9) and focusing on the free-vibration problem with an external but constant axial load [$F(x, t) = 0, P \neq 0$], with P positive for tension and negative for compression, we can write a general solution to the spatial part of the solution by assuming $W(x, t) = W(x) \cos \omega t$ and we then have

$$EI \frac{d^4 W(x)}{dx^4} + P \frac{d^2 W(x)}{dx^2} - m\omega^2 W(x) = 0. \tag{7.27}$$

Introducing the nondimensional beam coordinate $\bar{x} = x/L$, then we can write a general solution to the preceding equation in the form

$$W(\bar{x} L) = c_1 \sinh M\bar{x} + c_2 \cosh M\bar{x} + c_3 \sin N\bar{x} + c_4 \cos N\bar{x}, \tag{7.28}$$

in which M and N are given by

$$M = \sqrt{-\Lambda + \sqrt{\Lambda^2 + \Omega^2}}, \qquad N = \sqrt{\Lambda + \sqrt{\Lambda^2 + \Omega^2}}, \tag{7.29}$$

and we have nondimensionalized the axial load and frequency

$$\Lambda = PL^2/(2EI), \quad \Omega^2 = m\omega^2 L^4/(EI). \tag{7.30}$$

Note the slightly different load scaling from that used in Eq. (7.12).

We now apply the boundary conditions. At the left-hand end we have the fully clamped conditions

$$W(0) = 0, \quad dW(0)/d\bar{x} = 0, \tag{7.31}$$

and pinned, or simply supported, at the right-hand end,

$$W(L) = 0, \quad d^2 W(L)/d\bar{x}^2 = 0. \tag{7.32}$$

In fact, we assume that this support is a roller; that is, it does not allow vertical displacement or resistance to bending moments. If the horizontal displacement is suppressed, then membrane effects may induce additional axial loading (a feature to be explored in more detail later in this chapter) and will prove to be an important consideration in the dynamic response of axially loaded plates as well. Plugging these boundary conditions into the general solution [Eq. (7.28)] and applying the condition for nontrivial solutions (i.e., setting the determinant equal to zero) leads to the characteristic equation

$$M \cosh M \sin N - N \sinh M \cos N = 0. \tag{7.33}$$

We can solve this equation numerically by assuming Λ and solving for Ω or vice versa. With $\Lambda = 0$, we have the natural frequency with zero axial load $\Omega_0^2 = 15.42$, in which Eq. (7.33) reduces to $\tanh \sqrt{\Omega} = \tan \sqrt{\Omega}$ [11], and when $\Omega_0 = 0$ we get the buckling load $\Lambda_{cr} = 10.10$.

Figure 7.3 shows the lowest root of Eq. (7.33) as the solid line in terms of

$$\bar{\Lambda} = -\Lambda/\Lambda_{cr}, \qquad \bar{\Omega} = \Omega/\Omega_0. \tag{7.34}$$

Plotting the square of the natural frequency [part (b)] gives almost, but not quite, a straight line. Only tensile loads are plotted in this figure. Also plotted in these figures are some experimental data points (to be described a little later), and the dashed

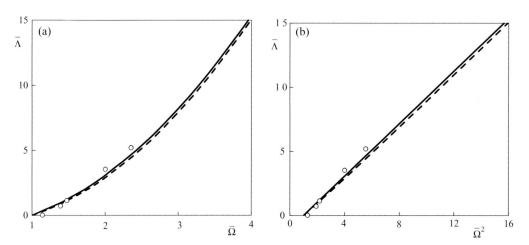

Figure 7.3. The relation between axial load and (a) frequency and (b) frequency squared for a clamped–pinned beam.

line is an upper bound for the frequency developed from energy considerations and based on [1, 12, 13]

$$\bar{\Omega} = \sqrt{1 + \gamma \bar{\Lambda}}, \tag{7.35}$$

where $\gamma = 0.978$ for the clamped–pinned boundary conditions [14].

The corresponding mode shape comes from the smallest root of Eq. (7.33) and the coefficients in Eq. (7.28) are given by

$$c_1 = 1, \quad c_2 = -\tanh M, \quad c_3 = -M/N, \quad c_4 = (M/N)\tan N, \tag{7.36}$$

and is plotted in Fig. 7.4(a) with the mode shape normalized such that its maximum amplitude is unity. Although not apparent in the figure, three curves are actually plotted. Superimposed on the zero-load case (a continuous curve) is the mode shape when the beam is subject to a *compressive* load of one half the elastic critical load. Note that this has an almost negligible effect on the mode shape despite the

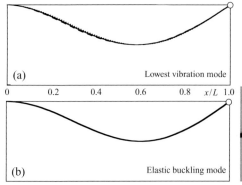

Figure 7.4. The mode shapes correponding to (a) the lowest natural frequency, and (b) the elastic critical load for a clamped–pinned axially loaded beam, and (c) an experimentally buckled beam.

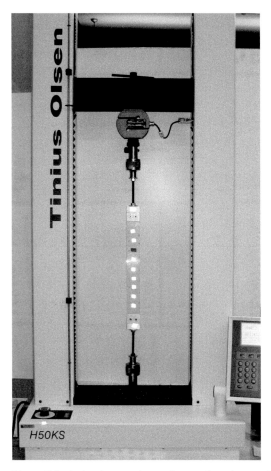

Figure 7.5. A simple experimental setup for a beam with clamped–pinned boundary conditions in a displacement-controlled testing machine.

frequency dropping from $\Omega_0 = 15.42$ for the unloaded case to $\Omega = 10.42$ (and indicated by a dotted curve). Similarly, a tensile force of this same magnitude causes the lowest natural frequency to increase to $\Omega = 19.09$ but also has only a minor influence on the mode shape (dashed curve), at least for the first mode. In general, despite the relatively strong effect of an axial load on the natural frequencies of thin beams (and strings), the effect on mode shapes is relatively minor. For simply supported boundary conditions the mode shapes (sine waves) are not affected at all [15]. We shall see later that this is not necessarily the case for plated structures. Equation (7.33) of course is a transcendental equation and has an infinite number of roots. These correspond to higher modes and will be considered later in this chapter.

Before leaving this section we briefly touch on some simple experimental results. Suppose we have a thin strip of polycarbonate material 44 mm wide by 1.52 mm thick with a length of 318 mm (as shown in Fig. 7.5). This material has a Young's modulus estimated at 2.4 GPa, a density of 1.142×10^3 kg/m^3, and boundary conditions such that it can be considered simply supported at one end and fully clamped at the other. We would expect an elastic critical load in the vicinity of $P_{cr} = 20.2\,EI/L^2 \approx 6\,\mathrm{N}$ [with the buckled mode shape as shown in Fig. 7.4(c)]. We would expect a fundamental natural frequency in the vicinity of $f_1 = 15.4/(2\pi)\sqrt{EI/(\rho AL^4)} \approx 15\,\mathrm{Hz}$. Using these two values to normalize the measured (tensile) axial loads and the measured fundamental frequencies, we obtain the data points shown in Fig. 7.3. Similar tests on struts of other lengths and thicknesses can also be suitably nondimensionalized, and the results confirm the almost linear (stiffening) effect of the (tensile) loads on the square of the natural frequencies. It should be mentioned here that in a practical testing situation there is likely to be a little membrane effect that is due to finite-amplitude oscillations as well as possible friction at the pinned end. This is especially the case when the axial loading is compressive.

7.3 Rayleigh's Quotient

In Section 4.3 it was shown, in general terms, how continuous systems typically led to the differential eigenvalue problem and Rayleigh's quotient was introduced as a useful approximate technique. We now apply this approach to analyze the free vibrations of a prismatic axially loaded beam. In this case the loading is conservative and the boundary conditions are homogeneous. We introduce the functional (the Lagrangian)

$$\mathcal{L} = \frac{1}{2}m\int_0^L \dot{w}^2 dx + \frac{1}{2}P\int_0^L w'^2 dx - \frac{1}{2}EI\int_0^L w''^2 dx, \qquad (7.37)$$

and, rather than using Lagrange's equations to obtain the governing equation of motion directly [Eq. (7.13)], we use the calculus of variations [16] and follow the approach described in Section 4.3 to arrive at expressions for stable (harmonic) behavior and hence the critical load and lowest natural frequency [10]:

$$\omega^2 < \frac{\int_0^L \frac{1}{2}(EIW''^2 - PW'^2)dx}{\int_0^L \frac{1}{2}mW^2 dx}, \qquad (7.38)$$

$$P < \frac{\int_0^L \frac{1}{2}(EIW''^2 - m\omega^2 W)dx}{\int_0^L \frac{1}{2}W'^2 dx}. \qquad (7.39)$$

For the pin-ended beam, we confirm the linear relation between the axial load and the square of the natural frequency [Fig. 7.2(c)].

7.4 Rayleigh–Ritz Analysis

Following on from the discussion in Chapter 3 we realize that there are situations in which it is convenient to use approximate analyses based on energy, especially when nonlinearities need to be considered. We will typically move from a differential eigenvalue context to an algebraic eigenvalue context. In this section, we make use of a Rayleigh–Ritz approach and open the scope to include the behavior of beams with initial curvature and those subject to axial loads that may exceed the elastic critical value, that is, postbuckling [17–19]. The Rayleigh–Ritz approach improves on the Rayleigh quotient by allowing more freedom with assumed displacement functions.

In using conventional beam theory in the previous section, we made the standard assumption that the bending moment and (an approximate form of) curvature were *linearly* related (and with flexural rigidity EI). This allowed a relatively simple analytic solution to be found. However, a number of approximate techniques have been developed that are relatively easy to apply and can be used in somewhat more complicated situations, including large deflections (e.g., in the initial postbuckled regime).

If we do not restrict ourselves to small deflections and slopes, it can be shown that the curvature ψ is related to the lateral deflection by

$$\psi = w''(1 - w'^2)^{-1/2}, \tag{7.40}$$

where the prime denotes differentiation with respect to the arc length x. It can be shown that this is roughly equivalent to the (Cartesian) curvature expression more familiar from standard geometry (see Naschie [20] for a discussion). It can also be shown (based on inextensional beam theory) that the total strain energy (see Chapter 2) stored in bending is

$$U = \frac{1}{2}EI \int_0^L \psi^2 \, dx. \tag{7.41}$$

Placing ψ in the preceding expression and expanding, we obtain

$$U = \frac{1}{2}EI \int_0^L (w''^2 + w''^2 w'^2 + \cdots +) \, dx. \tag{7.42}$$

Similarly, we can relate the end shortening to the lateral deflection

$$\xi = L - \int_0^L (1 - w'^2)^{1/2} dx, \tag{7.43}$$

which in turn leads to the potential energy of the load:

$$V_P = -\frac{1}{2}P \int_0^L \left(w'^2 + \frac{1}{4}w'^4 + \cdots + \right) dx. \tag{7.44}$$

Given a form for w these two expressions [(7.42) and (7.44)] can then be added to the appropriate kinetic energy expression to give the Lagrangian.

For simply supported boundary conditions we assume a solution (buckling mode) of the form

$$w = Q(t) \sin \frac{\pi x}{L}, \tag{7.45}$$

which can then be used to evaluate the strain and end-shortening energies to give

$$V(Q) = U + V_p = \frac{1}{2}EI \left(\frac{\pi}{L}\right)^4 \frac{L}{2}Q^2 + \frac{1}{2}EI \left(\frac{\pi}{L}\right)^6 \frac{L}{8}Q^4 + \cdots +$$

$$- P \left(\frac{1}{2} \left(\frac{\pi}{L}\right)^2 \frac{L}{2}Q^2 + \frac{1}{2} \left(\frac{\pi}{L}\right)^4 \frac{3L}{32}Q^4 + \cdots + \right), \tag{7.46}$$

and a kinetic-energy expression

$$T = \frac{1}{2}m \int_0^L \dot{w}^2 dx = \frac{1}{2}m \left(\frac{L}{2}\right) \dot{Q}^2, \tag{7.47}$$

where terms including the first nonlinear potential-energy contribution have been retained (appropriate for moderately large-deflection theory). For example, if P is greater than the Euler critical value, then the potential energy takes the form of two minima separated by a hilltop, that is, a twin-well form of potential energy.

By use of the nondimensionalization

$$\Lambda = \frac{P}{EI(\pi/L)^2}, \quad \Omega^2 = \frac{\omega^2}{EI/m(\pi/L)^4}, \tag{7.48}$$

application of Rayleigh's method gives

$$\Omega^2 = 1 + \frac{3}{2} \left(\frac{\pi}{L}\right)^2 Q^2 - \Lambda - \frac{9}{8} \Lambda \left(\frac{\pi}{L}\right)^2 Q^2. \tag{7.49}$$

For the trivial equilibrium path ($Q = 0$), we get

$$\Omega^2 = 1 - \Lambda. \tag{7.50}$$

Nontrivial equilibrium paths ($dV/dQ = 0$) are given by

$$\Lambda = 1 + \frac{\pi^2}{8} \left(\frac{Q}{L}\right)^2, \tag{7.51}$$

and evaluating Eq. (7.49) about the postbuckled paths (and ignoring higher-order terms) leads to

$$\Omega^2 = 2(\Lambda - 1). \tag{7.52}$$

The load–frequency relation again results in an interesting scaling between prebuckled and postbuckled behavior:

$$\frac{(d\Lambda/d\Omega)_p}{(d\Lambda/d\Omega)_f} = -2, \tag{7.53}$$

a result also found for the discrete model [Eqs. (5.53) and (5.56)].

A similar (but less accurate) analysis can be based on assumed polynomial buckled and vibration mode shapes. For example, by assuming a simple parabola as the fundamental mode shape, we would arrive at a natural frequency of 120 (as opposed to π^4) and a critical load of 12 (as opposed to π^2). That these values are higher than (or equal to) the exact values is always true for Rayleigh–Ritz analysis.

The Role of Initial Imperfections. In cases in which appreciable axial loading is present and especially in buckling, it is well established that small geometric imperfections may have a profound effect on behavior [21–24] and this was an effect observed for link models earlier. In terms of the linear theory, the potential energy becomes [25]

$$V = \frac{1}{2}EI \int_0^L (w'' - w_0'')^2 dx - \frac{P}{2} \int_0^L \left(w'^2 - w_0'^2\right) dx, \tag{7.54}$$

with an associated equation of motion

$$EI(w'''' - w_0'''') + Pw'' + m\ddot{w} = 0. \tag{7.55}$$

Note that here both the initial imperfection w_0 and the lateral deflection w are measured from the straight configuration (alternatively, the lateral deflection can be measured from the initially deflected configuration, and this leads to the equation of motion in a slightly different form [26]). Again assuming a simply supported strut with an initial deflection (i.e., prior to the application of any load) of the form $w_0 = Q_0 \sin(\pi x/L)$, we find the static response is given by

$$w = \left(\frac{Q_0}{1 - \alpha}\right) \sin(\pi x/L), \tag{7.56}$$

where $\alpha = P/P_E$ and P_E is the Euler load, and thus the maximum lateral deflection (at the strut midpoint) is given by

$$Q_{max} = \frac{Q_0}{1 - \alpha} = \frac{Q_0}{1 - P/P_{cr}}, \tag{7.57}$$

that is, deflections grow unlimited when the elastic critical load for the underlying geometrically perfect strut is approached [27].

In terms of the large-deflection theory, the governing equation of motion is related in form to Eq. (3.56), i.e., a skewed pitchfork bifurcation, and evaluating the various energy expressions leads to the equilibrium paths

$$\Lambda = 1 + \frac{\pi^2}{8}\left(\frac{Q}{L}\right)^2 - \left(\frac{Q_0}{L}\right)\left(\frac{L}{Q}\right), \tag{7.58}$$

where Q is the central displacement and the frequency expression is given by

$$\Omega^2 = 2(\Lambda - 1) + 3\left(\frac{Q_0}{L}\right)\left(\frac{L}{Q}\right). \tag{7.59}$$

We see that these degenerate into the perfect case.

Plotting the results for typical initial imperfections leads to Fig. 7.6. We see that, because there is no distinct instability, the natural frequency simply reaches a minimum (greater than zero) in the vicinity of the critical load for the underlying perfect system (a supercritical pitchfork bifurcation). The different dashes in this figure represent levels of initial imperfection incremented by 0.01 from 0 to 0.05. The complementary equilibrium paths (see Fig. 5.3) are not plotted in part (a), nor are results for a negative initial imperfection. Monitoring the natural frequency for different axial loads can be extrapolated as a nondestructive evaluation of the buckling load (see Chapter 11). This aspect of the interaction of dynamics and stability will be revisited later.

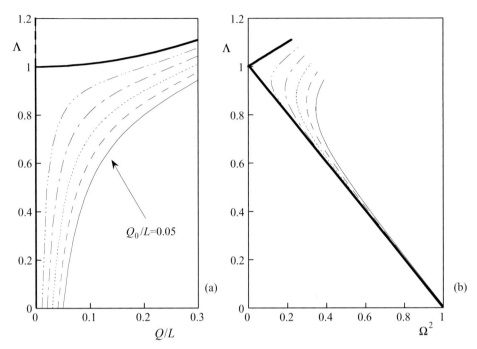

Figure 7.6. (a) Equilibrium paths for a slightly curved axially loaded strut and (b) corresponding frequency–load relation.

7.5 A Galerkin Approach

In this section, we conduct an alternative analysis of a simply supported strut. Here, we apply Hamilton's principle followed by a single-mode Galerkin procedure, as outlined in Chapter 4. Again axial load is included in the analysis and moderately large deflections (corresponding to a degree of stretching) are allowed, that is, we focus attention on a system in which no translation (in-plane or out-of-plane) is allowed at the supports [28].

From Eq. (2.28), we restate Hamilton's principle in the form

$$\int_{t_1}^{t_2} (\delta T - \delta U + \delta W)\, dt = 0. \tag{7.60}$$

The strain energy consists of bending and stretching terms [2]:

$$\delta U = \int_0^L \left[N_x \delta\left(\frac{\partial u}{\partial x}\right) + N_x \frac{\partial w}{\partial x} \delta\left(\frac{\partial w}{\partial x}\right) + EI\left(\frac{\partial^2 w}{\partial x^2}\right) \delta\left(\frac{\partial^2 w}{\partial x^2}\right) \right] dx. \tag{7.61}$$

The kinetic energy is given by

$$\delta T = \int_0^L m \frac{\partial w}{\partial t} \delta\left(\frac{\partial w}{\partial t}\right) dx, \tag{7.62}$$

and the work done by the external load is given by

$$\delta W = \int_0^L \left[P\delta\left(\frac{\partial u}{\partial x}\right) + P\frac{\partial w}{\partial x} \delta\left(\frac{\partial w}{\partial x}\right) \right] dx. \tag{7.63}$$

Integrating by parts and applying the boundary conditions (pinned supports at either end) leads to the equation of motion (in the lateral direction) of

$$m\frac{\partial^2 w}{\partial t^2} + EI\frac{\partial^4 w}{\partial x^4} - \frac{\partial}{\partial x}\left[(N_x - P)\frac{\partial w}{\partial x}\right] = 0. \tag{7.64}$$

The second term consists of axial effects from both the external applied load P and
membrane effects, that is, coupling between bending and stretching, N_x, based on a
truncation of the end shortening and given by

$$N_x = \frac{EA}{2L}\int_0^L \left(\frac{\partial w}{\partial x}\right)^2 dx, \tag{7.65}$$

in which A is the cross-sectional area of the member.

The deflection w and distance along the beam x were previously nondimensionalized according to Eqs. (7.12) but, given the stretching effect (and the fact that
these effects occur for relatively small deflections), it is more convenient to scale the
lateral deflection by the beam thickness

$$W = w/h, \quad \bar{x} = x/L, \tag{7.66}$$

which enables Eq. (7.64) to be rewritten as

$$\frac{\partial^4 W}{\partial \bar{x}^4} - \frac{12h}{EL^2}\left(\frac{L}{h}\right)^4 \frac{\partial}{\partial \bar{x}}\left[(N_{\bar{x}} - P)\frac{\partial W}{\partial \bar{x}}\right] + \frac{12mh}{E}\left(\frac{L}{h}\right)^4 \frac{\partial^2 W}{\partial t^2} = 0. \tag{7.67}$$

We now scale the in-plane loads and time by using

$$p = P\left(\frac{L^2}{EI}\right), \tag{7.68}$$

$$N_{\bar{x}} = N_x\left(\frac{L^2}{EIA}\right), \tag{7.69}$$

$$\tau = t\sqrt{\frac{EI}{mL^4}}, \tag{7.70}$$

leading to the final nondimensional equation of motion,

$$\frac{\partial^2 W}{\partial \tau^2} + \frac{\partial^4 W}{\partial \bar{x}^4} - \frac{\partial}{\partial \bar{x}}\left[(N_{\bar{x}} - p)\frac{\partial W}{\partial \bar{x}}\right] = 0, \tag{7.71}$$

with the stretching–bending coupling from

$$N_{\bar{x}} = 6\int_0^1 \left(\frac{\partial W}{\partial \bar{x}}\right)^2 d\bar{x}. \tag{7.72}$$

The solution of Eq. (7.71) is facilitated by the fact that the nonlinear term involves the product of the derivative of W and a definite integral that is a constant for
a given W. We can conduct a single-mode analysis of this system by assuming that

$$W(\bar{x}, \tau) = A(\tau)\sin \pi\bar{x}, \tag{7.73}$$

and, placing this in Eqs. (7.71) and (7.72), we get

$$\ddot{A} \sin \pi \bar{x} + A\pi^4 \sin \pi \bar{x} - (p + 3A^2)(-A\pi^2 \sin \pi \bar{x}) = 0, \qquad (7.74)$$

and thus

$$\ddot{A} + A(\pi^4 - p\pi^2) + 3\pi^2 A^3 = 0. \qquad (7.75)$$

For small-amplitude motion we can neglect A^3, although this case will be studied in more detail in the final chapter. Hence (from Fig. 16.8), we expect to see the stiffening effect that is due to the immovable ends, even when the deflection is not especially large (recall that the deflections were scaled by the beam thickness in this case). Although exact solutions are available by elliptic integrals and large-deflection analysis, such considerations are left to later chapters. In the absence of axial load, and for small-amplitude motion, we simply have a harmonic oscillation with a frequency of

$$\Omega_n = \pi^2, \qquad (7.76)$$

which in dimensional terms is our familiar fundamental (bending) frequency

$$\omega_n = \pi^2 \sqrt{\frac{EI}{mL^4}}. \qquad (7.77)$$

Again we can examine the condition $\omega_n^2 \to 0$ to obtain the critical value of the axial load,

$$p = \pi^2, \qquad (7.78)$$

which is, of course, the Euler load

$$P_E = EI \left(\frac{\pi}{L}\right)^2. \qquad (7.79)$$

This is the load at which the beam would buckle (assuming no initial imperfections) if the compressive load were slowly increased from zero. It should be mentioned that if there were absolutely no imperfection or disturbance then the equilibrium would not become unstable. Again, the linear relation between the axial load and the square of the natural frequency is confirmed. The fundamental mode of vibration however, although usually dominating the motion, would also be accompanied by higher modes (for arbitrary initial conditions), a subject we turn to next.

7.6 Higher Modes

Although the lowest critical load (and its corresponding mode shape) dominates a typical buckling problem, the higher modes in vibration play a significant role.

Returning to an earlier beam example considered (clamped–pinned) we can extract other roots from the characteristic Eq. (7.33). The subsequent table shows

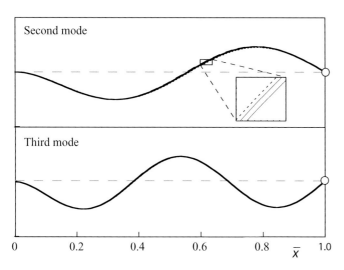

Figure 7.7. The second and third vibration modes for a clamped–pinned strut. The inset shows more clearly the curves in the presence of tensile and compressive axial forces.

how the natural frequencies corresponding to higher modes also change with axial load (tensile loads now negative):

	$P = 0$	$P = 0.5P_{cr}$	$P = -0.5P_{cr}$
Ω_1	15.42	10.42	19.09
Ω_2	49.94	44.95	54.52
Ω_3	104.25	99.11	109.14

Figure 7.7 shows the corresponding mode shapes. Again we see that, despite the influence of axial load on natural frequencies, the effect on mode shape is again minor (this will not necessarily be the case for more complicated structures like plates and shells) [29].

For the approximate analysis we return to the simply supported case and add a second term to the assumed shape:

$$W(\bar{x}, \tau) = A_1(\tau) \sin \pi\bar{x} + A_2(\tau) \sin 2\pi\bar{x}. \tag{7.80}$$

Using Eq. (7.80) to evaluate Eqs. (7.71) and (7.72) now leads to the pair of equations

$$\ddot{A}_1 + 3\pi^2 A_1 (A_1^2 + A_2^2) + A_1(\pi^4 - p\pi^2) = 0, \tag{7.81}$$

$$\ddot{A}_2 + 12\pi^2 A_2 (A_1^2 + A_2^2) + A_2(16\pi^4 - 4p\pi^2) = 0. \tag{7.82}$$

If we further assume that deflections are small, then the equations decouple and we immediately obtain the second lowest critical load ($p = 4\pi^2$) and the second lowest mode of vibration ($\Omega = 4\pi^2$), both of which correspond to a full sine wave. That the linearized equations are uncoupled is a consequence of the assumed modes actually corresponding to the normal modes, as mentioned before. For a general

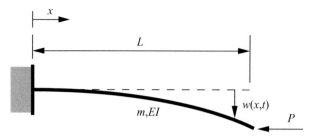

Figure 7.8. Schematic of a cantilevered strut.

structure, this will not typically be the case. There are other (physical) ways in which
the equations of motion can couple, and these will be considered in more detail in
the next chapter.

Consider a cantilevered strut, of the type shown in Fig. 7.8. Here, the strut is
subject to an axial end load, and we analyze the way in which the natural frequencies
depend on the magnitude of the axial load by using an approximate Rayleigh–Ritz
method. In choosing the assumed buckling and vibration modes, we wish to have a
function that resembles the true modes as closely as possible. Even for a relatively
simple structure (like the cantilever under consideration) the exact modes can be
quite complicated.

The cubic polynomial

$$w(x, t) = C(t)x^2 + D(t)x^3 \tag{7.83}$$

satisfies the conditions of zero displacement and slope when $x = 0$. We use this to
evaluate the strain energy in bending,

$$U = \frac{1}{2}EI \int_0^L w''^2 dx = 2EIL(C^2 + 3CDL + 3D^2L^2), \tag{7.84}$$

the potential energy of the end load,

$$V_P = -P\frac{1}{2}\int_0^L w'^2 dx = -\frac{PL^3}{30}(20C^2 + 45CDL + 27D^2L^2), \tag{7.85}$$

and the kinetic energy

$$T = \frac{1}{2}m \int_0^L \dot{w}^2 dx = \frac{m}{210}(21\dot{C}^2L^5 + 35\dot{C}\dot{D}L^6 + 15\dot{D}^2L^7). \tag{7.86}$$

Again, in the vicinity of equilibrium and for linear vibrations we expect the total potential and kinetic energies to be quadratic functions of the generalized coordinates
and velocities, respectively, and thus

$$T = \frac{1}{2}T_{ij}\dot{q}_i\dot{q}_j, \tag{7.87}$$

$$U + V_P = \frac{1}{2}V_{ij}^E q_i q_j, \tag{7.88}$$

where use is made of the dummy suffix notation [3], that is, any subscript occurring
more than once in a product is summed over all values.

Evaluating the partial derivatives leads to the second variation of the strain energy,

$$U_{ij} = EIL \begin{bmatrix} 4 & 6L \\ 6L & 12L^2 \end{bmatrix}, \qquad (7.89)$$

the work done by the axial load

$$V_{P_{ij}} = -PL^3 \begin{bmatrix} (4/3) & (3/2)L \\ (3/2)L & (9/5)L^2 \end{bmatrix}, \qquad (7.90)$$

and the kinetic energy

$$T_{ij} = mL^5 \begin{bmatrix} (1/5) & (1/6)L \\ (1/6)L & (1/7)L^2 \end{bmatrix}. \qquad (7.91)$$

We note that these matrices would have been of infinite dimension and diagonal if the exact mode shapes (eigenfunctions) had been used. We next make use of the following definitions,

$$\Omega^2 = \frac{mL^4}{EI}\omega^2, \qquad p = \frac{PL^2}{EI}, \qquad (7.92)$$

and, using the characteristic determinantal equation,

$$|U_{ij} - V_{P_{ij}} - \omega^2 T_{ij}| = 0, \qquad (7.93)$$

we obtain the characteristic equation

$$12 - 5.2p - 0.97\Omega^2 + 0.05p\Omega^2 + 0.15p^2 + 0.000794\Omega^4 = 0, \qquad (7.94)$$

from which we can readily extract the roots. In the absence of an axial load ($p = 0$), we have a lowest root of $\Omega^2 = 12.6$. This compares with an exact value of 12.36. By setting the natural frequency to zero we obtain a lowest root of $p = 2.49$, which compares with the exact value of $\pi^2/4$ (2.47). The other roots correspond to the higher of the two modes, although these are less accurate. It is interesting to note that using a single generalized coordinate [e.g., with $D = 0$ in Eq. (7.83)] leads to a critical load of $p = 3$ and a natural frequency of $\Omega = 4.47$.

Improved Accuracy. There are a number of ways in which the accuracy of approximate methods can be improved. A couple of examples are given here for a clamped–clamped beam. We already have the general expressions for the various energy terms, which depend on the assumed buckling and vibration modes. Taking more terms leads not only to estimates of the higher modes but also to increasing accuracy in the estimate of the fundamental mode. For a beam clamped against deflection and rotation at both ends, use can be made of the Duncan polynomials [30], defined as

$$S_m(\bar{x}) \equiv \frac{\sqrt{4m+1}}{(2m)!} \frac{d^{2m-2}}{d\bar{x}^{2m-2}} \left[\bar{x}^{2m}(1-\bar{x})^{2m}\right], \qquad (7.95)$$

in which m is an integer and d stands for derivative. This satisfies the boundary conditions at both ends. In the Ritz procedure, use is then made of the shape function to evaluate the energy integrals. These values have been tabulated for various S functions and their derivatives (and location along the beam). Hence for a single mode we can assume

$$w = S_1(\bar{x})q_1(t) \tag{7.96}$$

and then evaluate the energy terms by using the Duncan polynomial to get

$$U = \frac{1}{2}EI \int_0^1 S_1''^2(\bar{x})q^2 d\bar{x} = \frac{1}{2}EI(1.0)\,q^2, \tag{7.97}$$

$$V_P = -\frac{1}{2}P \int_0^1 S_1'^2(\bar{x})q^2 d\bar{x} = \frac{1}{2}P(-0.02380952)q^2, \tag{7.98}$$

$$T = \frac{1}{2}m \int_0^1 S_1^2(\bar{x})\dot{q}^2 d\bar{x} = \frac{1}{2}m(0.001984127)\dot{q}^2. \tag{7.99}$$

Using Lagrange's equation or Rayleigh's method leads directly to a natural frequency:

$$\omega^2 = \frac{EI - P(0.02380952)}{m(0.001984127)}. \tag{7.100}$$

Note that in the absence of an axial load we get a natural frequency coefficient of $\omega^2|_{P=0} = 22.45$, and setting $\omega = 0$ in the preceding equation gives a critical load coefficient of $P = 4.255\pi^2$, both of which compare reasonably well with the exact answers of 22.37 and $4\pi^2$, respectively [1].

Adding a second term, i.e.,

$$w = S_1(\bar{x})q_1(t) + S_2(\bar{x})q_2(t), \tag{7.101}$$

and making more extensive use of the Duncan polynomial tables results in the determinant,

$$\begin{vmatrix} 1 - 0.0238\alpha - 0.001984\beta & 0.005324\alpha + 0.000161\beta \\ 0.005324\alpha + 0.000161\beta & 1 - 0.0000649\alpha - 0.0000749\beta \end{vmatrix} = 0, \tag{7.102}$$

where $\alpha = P/(EI)$ and $\beta = m\omega^2/(EI)$. Solving this equation leads to the lowest root for the critical load corresponding to $4.002\pi^2$, and in the absence of axial load we now have a natural frequency coefficient of 22.37, both of which are very accurate. However, it should be mentioned here that the Duncan polynomials are not orthogonal and hence may lead to problems when higher modes are considered within the FEM for example [31], but can be appropriately modified [32]. The Rayleigh–Ritz method requires assumed mode shapes that satisfy the geometric boundary conditions (which for a beamlike structure are the deflections and slopes), but choosing functions satisfying the nongeometric boundary conditions tends to improve accuracy.

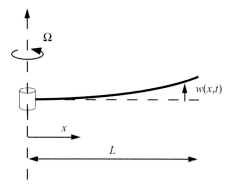

Figure 7.9. A rotating cantilever beam.

7.7 Rotating Beams

An example of slender beams subject to tensile axial loads can be found in rotor blades [33–35]. Centrifugal forces that are due to high rates of rotation typically lead to stiffening effects [36, 37]. Consider the rotating (cantilever) beam shown in Fig. 7.9. We can write the governing equation of motion in the usual form [38],

$$\frac{\partial^2}{\partial x^2}\left(EI\frac{\partial^2 w}{\partial x^2}\right) - \frac{\partial}{\partial x}\left(P\frac{\partial w}{\partial x}\right) + m\frac{\partial^2 w}{\partial t^2} = 0, \qquad (7.103)$$

but now the axial load P (in tension) is given by

$$P = \int_x^L m\Omega^2 x \, dx. \qquad (7.104)$$

We see that this is similar to Eq. (7.27). Separating variables by using $w = W(x)Y(t)$ and setting the constant equal to $\lambda^2\Omega^2$, we can write Eq. (7.103) (for a uniform beam) in terms of two separated variables,

$$EI\frac{d^4 W}{dx^4} - \frac{d}{dx}\left(P\frac{dW}{dx}\right) - m\lambda^2\Omega^2 W = 0, \qquad (7.105)$$

and

$$\frac{d^2 Y}{dt^2} + \lambda^2\Omega^2 Y = 0, \qquad (7.106)$$

and scaling by using $\bar{w} = W/L$, $\bar{x} = x/L$, and $\psi = \Omega t$, we can then write

$$EI\frac{d^4\bar{w}}{d\bar{x}^4} - L^2\frac{d}{d\bar{x}}\left(P\frac{d\bar{w}}{d\bar{x}}\right) - m\lambda^2\Omega^2 L^4\bar{w} = 0, \qquad (7.107)$$

and

$$\frac{d^2 Y}{d\psi^2} + \lambda^2 Y = 0. \qquad (7.108)$$

Boundary conditions appropriate for a cantilever beam are zero deflection and slope at the hub (clamped) end and zero bending moment and shear force at the free end. Equation (7.107) can be attacked in a variety of ways, but it is recognized that because the axial force is not a constant [Eq. (7.104)] in this case recourse to

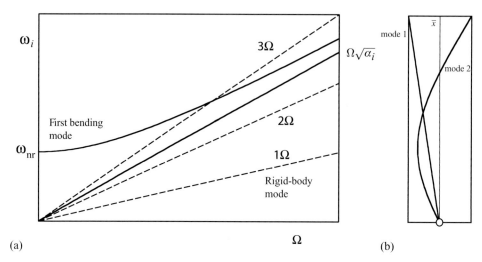

Figure 7.10. (a) A "spoke" diagram showing the relation between natural frequencies and rate of rotation and (b) the first two mode shapes of a pinned–free rotating beam using an assumed solution.

approximate-solution techniques is required. We have already seen the general influence of tensile axial loads on the natural frequencies of lateral vibrations. In applications to rotor blades it is interesting to note that a common configuration is to use a hinge at the root of the beam, and hence the somewhat unusual boundary conditions of pinned–free are encountered [34]. In this case, the lowest mode is a rigid-body rotation (a *flapping* motion). The cantilever is often termed a hingeless blade.

Here, a useful approach is briefly mentioned (related to Rayleigh–Ritz) that is due to Southwell [33, 39]. He showed that a relation between the rotating and nonrotating frequencies could be established as

$$\omega_i^2 = \omega_{nr}^2 + \alpha_i \Omega^2, \tag{7.109}$$

where ω_{nr} is the natural frequency of the nonrotating blade and the Southwell coefficient α_i is given by

$$\alpha_i = \frac{\int_0^1 m\bar{x} \left[\int_0^{\bar{x}} (d\bar{w}_i/d\bar{x})^2 d\bar{x} \right] d\bar{x}}{\int_0^1 m\bar{w}_i^2 \, d\bar{x}}, \tag{7.110}$$

where $\bar{w}_i = \bar{w}_i(x)$ is an assumed mode shape, satisfying the boundary conditions. Although α_i is not strictly constant, the mode shape changes very little with rotation speed. Equation (7.109) is plotted in Fig. 7.10(a). It can seen that for $\Omega \to 0$ we get the lowest bending mode, which for a pinned–free cantilever is $15.418\sqrt{EI/(mL^4)}$ and for a clamped–free cantilever is $3.516\sqrt{EI/(mL^4)}$. As the rate of rotation gets large we observe an asymptotic relation $\omega_i \to \sqrt{\alpha_i}\Omega$.

A specific example of this approach was given in Bramwell [33]. For a hinged-free blade the deflection was expressed as

$$\gamma = c_1\gamma_1 + c_2\gamma_2, \tag{7.111}$$

in which

$$\gamma_1(x) = x, \quad \gamma_2(x) = 10x^3/3 - 10x^4/3 + x^5. \tag{7.112}$$

It was shown that the two roots resulting from this analysis correspond to the frequencies $\omega = \Omega$ (flapping motion) and $\omega = 2.757\Omega$, and the (normalized) mode shapes are plotted in Fig. 7.10(b). Given this basic shape, use can be made of Southwell's method [Eq. (7.110)] to show how the frequencies change with the speed of rotation according to Fig. 7.10(a).

In a practical (rotor blade) sense, an important issue is that as the rate of rotation changes then so do the natural frequencies, and hence the possibility of resonance occurring within the operating range of the rotor is an important issue [40]. Also, in a practical context it is likely that a rotating beam may be tapered. A convenient method of handling this situation is the use of a dynamic stiffness approach [41, 42]. More details on this type of numerical approach will be introduced in the chapter on frames. Here, the following table [43] summarizes the effect of rotation rate on the lowest three frequencies of clamped–free and pinned–free uniform cantilevers in which $\eta = \Omega/\sqrt{EI/mL^4}$.

η		Clamped–free	Pinned–free
0	ω_1	3.5160	0.000
	ω_2	22.0345	15.4182
	ω_3	61.6972	49.9649
1	ω_1	3.6816	1.000
	ω_2	22.1810	15.6242
	ω_3	61.8418	50.1437
3	ω_1	4.7973	3.000
	ω_2	23.3203	17.1807
	ω_3	62.9850	51.5498

In terms of transient analysis by use of time-marching numerical integration, it is important to include appropriate gyroscopic and Coriolis effects in these types of systems with velocity-dependent loads [44]. In some cases, the governing equations of motion can be solved exactly [43], and a number of the approximate techniques introduced in Chapter 4 can be used for various helicopter applications [34]. It is also interesting to note that, although the primary interest here has been focused on bending modes, the frequencies associated with other modes may actually decrease with the rate of rotation [45].

A considerable amount of research has been conducted on modal interaction in rotor blades that is due to elastic and inertial coupling [46–48]. Clearly, the effects of fluid loading (including forward flight) are complicated, but suffice it to say here that there are practical situations in which there is elastic coupling between flapping and lagging motion [49, 50]. Southwell's method can be applied and flap–lag dynamic

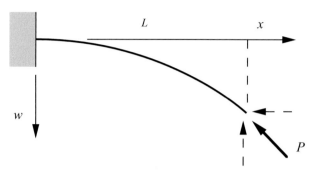

Figure 7.11. A cantilever strut loaded tangentially.

interaction obtained. This is an important design consideration for helicopters, and stability boundaries have been developed to take into account the various parameters of the problem [51]. Given the periodic nature of rotating systems certain special mathematical techniques can be used including Floquet theory (see Chapter 14) [52]. Turbomachinery tends to operate at extremely high rates of revolution and their blades can experience considerable stiffening effects. For example, tip speeds can be close to Mach 1; that is, a small turbine with a radius of a few centimeters might operate at 100,000 rpm, whereas a large commercial jet engine might operate in the vicinity of 1500 rpm [53]. Use is made of Campbell diagrams (also known as waterfall plots and spectrograms) to keep the natural frequencies away from the harmonics of the rotor speed (and thus avoid resonance—see Chapter 15). Some experimental verification of the stiffening effect in rotating turbine blades can be found in Fan et al. [54]. Circular plates are sometimes designed so that their natural frequencies can be tuned as a function of the rate of spinning [55]. An example of modal coupling in a more conventional structural setting will be given in the next chapter.

7.8 A Strut with a Tangential Load

If only the buckling load is sought in a strut problem then conventional static analysis will often suffice. However, there are some circumstances in which a dynamic analysis is needed, and we now take a quick look at the case of a cantilevered strut under the action of an axial load that changes direction such that it follows the slope at the end of the strut. This example, which concerns nonconservative forces, occupies a rather central position in the development of structural stability and is often referred to as Beck's problem [56–58]. A schematic of this situation is shown in Fig. 7.11. Here, the forcing is nonconservative because energy is not conserved in this case because of the boundary conditions at the tip, although there are some issues concerning the manner in which the system might develop into the shape shown in Fig. 7.11 [10]. We can write down the governing equation of motion,

$$w'''' + pw'' + \mu \ddot{w} = 0, \qquad (7.113)$$

where the axial load and mass are nondimensionalized with respect to the flexural rigidity, that is,

$$p = \frac{P}{EI}, \quad \mu = \frac{m}{EI}. \tag{7.114}$$

We can then assume that

$$w(x, y) = A_0 e^{\lambda x} e^{i\omega t}, \tag{7.115}$$

and thus we obtain the characteristic equation

$$\lambda^4 + p\lambda^2 - \mu\omega^2 = 0. \tag{7.116}$$

The key difference between this problem and that of a constant direction load is the application of the boundary conditions at the free end. In addition to the requirement for zero displacement and slope at the fixed end and zero bending moment at the free end, we must now also satisfy zero shear force $EIw'''(L, t) = 0$. This approach is often referred to as the kinetic approach and is suitable for nonconservative problems of this type. In fact, a static approach would suggest that no nontrivial equilibrium positions exist [58].

The general solution of the spatial part of Eq. (7.113) is given by

$$W(x) = B_1 \sin \alpha x + B_2 \cos \alpha x + B_3 \sinh \beta x + B_4 \cosh \beta x, \tag{7.117}$$

where

$$\alpha = \sqrt{\frac{\sqrt{4\mu\omega^2 + p^2} - p}{2}}, \tag{7.118}$$

$$\beta = \sqrt{\frac{\sqrt{4\mu\omega^2 + p^2} + p}{2}}, \tag{7.119}$$

and satisfying the boundary conditions requires (for a nontrivial solution)

$$\begin{vmatrix} 0 & 1 & 0 & 1 \\ \alpha & 0 & \beta & 0 \\ -\alpha^2 \sin \alpha L & -\alpha^2 \cos \alpha L & \beta^2 \sinh \beta L & -\beta^2 \cosh \beta L \\ -\alpha^3 \cos \alpha L & \alpha^3 \sin \alpha L & \beta^3 \cosh \beta L & \beta^3 \sinh \beta L \end{vmatrix} = 0. \tag{7.120}$$

This leads directly to the characteristic equation

$$(p^2 + 2\mu\omega^2) + 2\mu\omega^2 \cosh \alpha L \cos \beta L + p\sqrt{\mu\omega^2} \sinh \alpha L \sin \beta L = 0. \tag{7.121}$$

A plot of (ω versus p) from Eq. (7.121) is shown in Fig. 7.12. We see that when the axial load is zero we have the two lowest frequencies for a cantilever ($\omega_1 = 3.516$ and $\omega_2 = 22.03$, indicated by the black data points in Fig. 7.12). However, as the axial load is increased, the lowest frequency increases and the second lowest falls, coming together when $p = 20.1/L^2$. For values of p above this critical value, solutions become complex, that is, oscillating motions with increasing amplitude. The critical value of $P_{cr} = 2.03\pi^2 EI/L^2$, corresponds to a critical load that is about eight

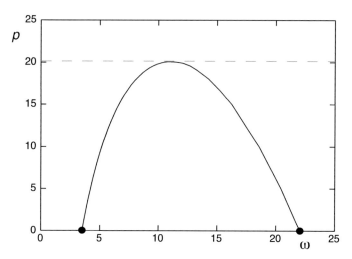

Figure 7.12. Frequency versus load for a cantilever subject to a follower force.

times greater than the critical load for the equivalent case with a purely vertical end load, i.e., $P_{cr} = (\pi^2/4)EI/L^2$. This phenomenon is often referred to as *coupled-mode flutter*. Although this may not represent a very practical case, it does reinforce the need to view instability within a dynamical context in some circumstances [57]. This problem can also be solved with a FEM (see Fig. 9.10). A tangential load that is distributed along the length of the column is investigated in Leipholz [58], and Bazant and Cedolin [10] describe an analogous system with a concentrated end mass.

7.9 Self-Weight

Returning to conservative problems, consider the column shown in Fig. 7.13. It has height H, constant bending stiffness EI, and constant weight G per unit length. We

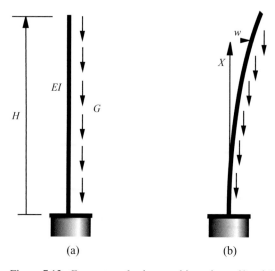

Figure 7.13. Geometry of column subjected to self-weight.

would expect this column to buckle under its own weight at a critical height, followed by a gradual droop corresponding to a stable-symmetric (or supercritical)
bifurcation. Again, we would also expect the lowest natural frequency to reduce
with axial load (column height), but consideration of dynamics is left until the next
section.

The equilibrium equation is

$$EIw''''(x) + G[(H - x)w'(x)]' = 0. \tag{7.122}$$

To put the analysis in nondimensional terms, define

$$a = \left(\frac{EI}{G}\right)^{1/3}, \quad \bar{x} = \frac{x}{a}, \quad \bar{w} = \frac{w}{a}, \quad h = \frac{H}{a}. \tag{7.123}$$

(The lengths are not nondimensionalized by H, because the height is the parameter
of interest.) This leads to

$$\bar{w}''''(\bar{x}) + [(h - \bar{x})\bar{w}'(\bar{x})]' = 0. \tag{7.124}$$

The boundary conditions are $\bar{w}(0) = \bar{w}'(0) = \bar{w}''(h) = \bar{w}'''(h) = 0$. The critical
nondimensional height is $h_{cr} = 1.986$ [26, 59]. This system will be subject to a large-
deflection analysis based on the elastica in a later chapter.

Approximate values of the critical height can be obtained with the use of the
Rayleigh–Ritz method, as described earlier. The potential energy U is given by

$$U = \frac{1}{2}\int_0^h (w'')^2 dx - \frac{1}{2}\int_0^h (h - x)(w')^2 dx. \tag{7.125}$$

Making U stationary for the kinematically admissible function

$$w(x) = Qx^c, \tag{7.126}$$

where $c > 1$, leads to the approximate critical height $h_{cr} = 2.289$ if $c = 2$, and the
value $h_{cr} = 2.143$ for the minimizing choice $c = 1.747$. If

$$w(x) = Q[1 - \cos(cx/h)], \tag{7.127}$$

we obtain $h_{cr} = 2.025$ for $c = \pi/2$ (corresponding to the buckling mode for a cantilever with an axial end load, as used in [60]), and $h_{cr} = 2.003$ for the optimal value
$c = 1.829$. Finally, the two-term approximation

$$w(x) = Q_1 x^2 + Q_2 x^3 \tag{7.128}$$

furnishes the excellent approximation $h_{cr} = 1.991$.

For a circular cross section of radius R, we can also use a Rayleigh–Ritz analysis
(based on a simple polynomial displacement function) to obtain the critical height
for uniform pole:

$$h_{cr} = 1.26\left(\frac{E}{\rho}R^2\right)^{1/3}, \tag{7.129}$$

in which ρ is the specific weight. For a uniform taper $[r = (h - x)R_0/h]$ in which
R_0 is the radius at the base of the cantilever the coefficient in Eq. (7.129) changes

to 2.17. It is interesting to note that for trees an appropriate rule of taper is $\{r = [(h-x)/h]^{3/2}R_0\}$, and the coefficient becomes 2.60 so we see a somewhat optimal design in nature (at least in terms of vertical loading) [61, 62].

Using (7.127), we can also show that including an initial imperfection as well as the next term in the Taylor series expansion of the potential energy leads to the familiar form of supercritical pitchfork bifurcation for the postbuckled static response. Furthermore, by computing the kinetic energy of the system, and then using Lagrange's equations (or Rayleigh's quotient), we also expect the lowest natural frequency to increase after buckling (as shown in some experimental data later in the chapter) [60].

7.9.1 A Hanging Beam

A related problem to self-weight buckling is the behavior of long vertical pipes that are subject to *nonuniform* axial loads that, for example, might be due to the combined effects of gravity and internal hydrostatic pressure [63]. A practical example would be the behavior of drill strings in a well bore [64, 65]. Again various end conditions are possible, but we shall focus attention on the specific case of a vertical beam, fully fixed at its top end and completely free at the bottom. Hydrostatic pressure is assumed to vary linearly with distance from the top, that is, a submerged column, and the effect of gravity is included in the following analysis. At the end of this subsection we will show that for very long and slender beams the behavior tends toward that of the hanging chain encountered in the previous chapter.

The governing equation of motion is based on the usual assumptions of linearly elastic material and small deflections, and hence we still have the same form as that of Eq. (7.64):

$$EI\frac{\partial^4 w}{\partial x^4} - \frac{\partial}{\partial x}\left[(G - G_m)x\frac{\partial w}{\partial x}\right] + m\frac{\partial^2 w}{\partial t^2} = 0, \tag{7.130}$$

in which G is the weight of the beam (per unit length) and G_m is the weight of the fluid column displaced (also per unit length). Again we assume harmonic motion of the form $w = W(x)\sin \omega t$, and thus

$$\frac{d^4 W}{d\bar{x}^4} - \alpha \bar{x}\frac{d^2 W}{d\bar{x}^2} - \alpha\frac{dW}{d\bar{x}} - \lambda^4 W = 0, \tag{7.131}$$

in which

$$\lambda^4 = m\omega^2 L^4/(EI), \quad \alpha = (G - G_m)L^3/(EI), \quad \bar{x} = x/L. \tag{7.132}$$

In the following analysis, the parameter α [in which the component $(G - G_m)$ can be thought of as the *traction*] will be set and a solution given for λ.

For the specific case under consideration, the boundary conditions consist of fixed at the top end $(W(1) = dW/d\bar{x}(1) = 0)$ and free at the lower end $[d^2 W/d\bar{x}^2(0) = d^3 W/d\bar{x}^3(0) = 0]$. Details of the solution technique can be found in Huang and Dareing [64], with Fig. 7.14 showing the three lowest frequencies as a function of α.

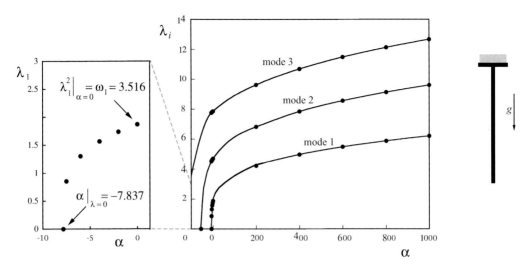

Figure 7.14. Natural frequencies of a hanging beam as a function of α (adapted from Huang and Dareing [64]).

It can be shown that the lowest natural frequency (mode 1) drops to zero when the traction reaches a level of $\alpha = -7.8373$, that is, buckling occurs (see the expanded view on the left-hand side). In the case of a hanging beam with no hydrostatic pressure but under the action of gravity alone, we can effectively reduce the bending stiffness of the system by allowing α to go to very large values (i.e., very large tensions). This tends toward the behavior of the hanging chain observed in the previous chapter; for example, with $\alpha = 1000$, the lowest natural-frequency coefficient approaches 1.22. This compares with the value 1.2026 from Eq. (6.59) and would have been very closely matched if the upper support were pinned rather than fixed. References [64] and [15] give many examples of different boundary conditions as well as the case in which the traction does not tend to zero at one end.

We can also conduct a brief Rayleigh–Ritz analysis by assuming a simple parabola for the lowest mode: $W = Cx^2$. In this case, Rayleigh's quotient leads to

$$\omega^2 = \frac{1/3mgL^3 + 4EI}{1/5mL^4}, \tag{7.133}$$

from which we obtain the frequencies of $4.47\sqrt{EI/mL^4}$ (recall that the exact coefficient is 3.516) and $1.29\sqrt{g/L}$ for the cases of zero weight and zero bending stiffness, respectively, and buckling at $\alpha = -12$.

7.9.2 Experiments

A cantilever under the action of self-weight loading provides a relatively simple context for experimental verification of some of the behavior described in this section. Consider first a vertically mounted (built-in end at the bottom), slender elastic rod, of axisymmetric (circular) cross section, whose length can be increased such that (self-weight) buckling is induced. A cantilever test was conducted while the

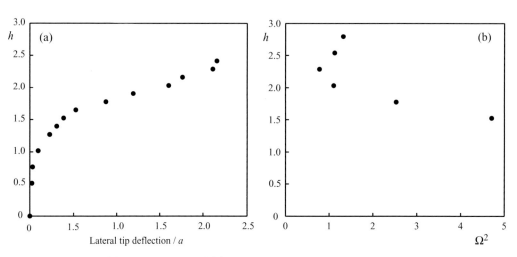

Figure 7.15. (a) Tip deflection and (b) fundamental frequency for a slender rod.

beam was mounted in a horizontal configuration to determine the flexural rigidity, and this estimate suggested a critical elastic buckling length in the vicinity of 20 cm, based on $1.986(EI/W)^{1/3}$. The lateral deflection was measured as a function of height, and the results are plotted in Fig. 7.15(a). The supercritical nature of the bifurcation is apparent. In part (b) of this figure, the lowest natural frequency is also plotted as a function of height. The minimum is achieved in the vicinity of the critical height: It does not drop to zero in practice because of the inevitable presence of a geometric imperfection (in fact, for this type of axisymmetric cross section there is no obvious preferred direction for postbuckled deflection in the perfect case). Also, frequencies become increasingly difficult to measure (for small-amplitude vibration) near the buckling length because of the increasing effect of damping. At least, increasing in terms of a damping ratio. It can also be argued that the damping force has less effect because the velocities are decreasing. The postbuckled equilibria and frequencies (which start to increase in the postbuckling range) will be revisited from a theoretical viewpoint in a later chapter within the context of an elastica analysis.

Another method of illustrating the effect of gravity is to conduct tests on a *double* cantilever, that is, a thin rod clamped at its center point and oriented in the vertical direction. As the rod becomes more slender, the difference between the "upright" and "downward" natural frequencies becomes more apparent. A number of thin polycarbonate strips were fabricated such that a range of the nondimensional parameter α [see Eqs. (7.132)] could be examined. The hub was clamped to an electromagnetic shaker and the system subject to a broadband, random excitation. A laser vibrometer was then used to acquire velocity data from discrete locations along both beams, and subsequent signal processing used to obtain frequency-response data.

As an example, and to provide a baseline, consider a *horizontal* polycarbonate strip with cross-sectional dimensions of 25.4 mm × 4.67 mm, Young's modulus

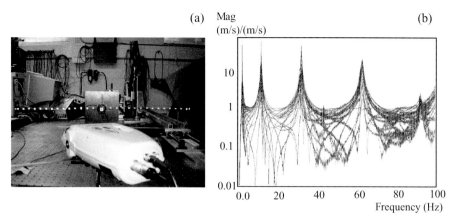

Figure 7.16. A thin prismatic cantilever: (a) experimental configuration and (b) frequency response.

$E = 1.93$ GPa [measured from a laterally (static) loaded cantilever test, and within the range of the manufacturer's specifications], mass per unit length $m = 0.131$ kg/m, and length $L = 0.737$ m. Figure 7.16(a) shows the experimental arrangement, with part (b) showing the superimposed frequency response extracted from 30 evenly spaced locations along the entire length. The four lowest measured natural frequencies (in hertz) are 1.812, 11.34, 31.71, and 62.35, which compare with analytical values [based on Eq. (7.130)] of 1.836, 11.51, 32.22, and 63.13. This represents good correlation, with typical differences between 1% and 2%. The effect of damping is again neglected in the analysis.

Now, if we take the same system and rotate it 90 deg we get the frequency separation (mode splitting) shown in Fig. 7.17. At first glance from Fig. 7.17(a) it appears that there is no difference from the results shown in Fig. 7.16(b). However, on closer inspection, each peak is revealed as two adjacent peaks [just the fundamental frequencies are shown in Fig. 7.17(b)]. For this beam $|\alpha| = 1.23$. For larger values of α the up or down orientation has a greater effect. For example, when we reduce the

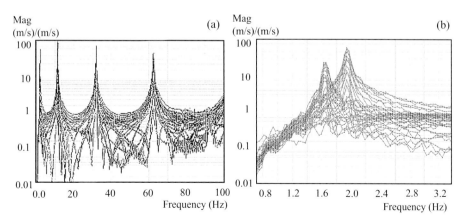

Figure 7.17. Normalized frequency-response spectrum for a cantilever in a gravitational field, $|\alpha| = 1.23$: (a) lowest few frequencies and (b) blowup of the lowest frequency.

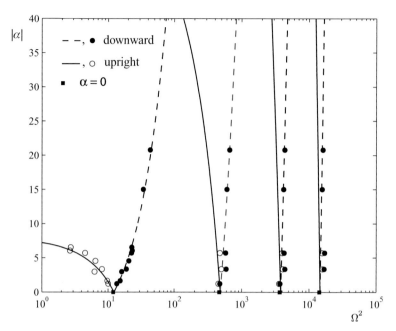

Figure 7.18. The four lowest frequencies versus $|\alpha|$.

thickness of the strip to 2.38 mm and the length to 0.66 m, while holding everything else constant, we then have $|\alpha| = 5.71$, and the two peaks separate to approximately 1.65 and 1.95 Hz.

For the upright cantilever we have the result that buckling occurs when $\alpha = -7.837$ [61], and the trivial equilibrium loses its stability. Thus the results shown in Fig. 7.17(b) (for which $|\alpha| = 1.23$) correspond to a cantilever whose length is $(1.23/7.837)^{1/3} = 55\%$ of its buckling length.

For the case $\alpha = 20.78$, only the downward (trivial) configuration is stable, and for this case the FE package, ANSYS, was used for the analytical results. For this case, the frequency of the mode one oscillation is greater than the zero-weight case by 90%, the second frequency by 17%, the third by 6.5%, and the fourth by 3.5%.

The results are shown graphically in Fig. 7.18 for both analytical and experimental frequencies versus α. The lowest frequency, as expected, drops to zero when $|\alpha| = -7.837$.

7.10 Thermal Loading

Before leaving this chapter we briefly consider the situation in which the axial load is produced through a thermal gradient. We start by looking at a simply supported beam that is not allowed to move axially at its ends (and thus generates axial forces). If the beam, with coefficient of linear thermal expansion β, is subject to a constant thermal load, the governing equation of motion becomes

$$m\ddot{w} + EIw'''' + AE(\beta T)w'' = 0, \qquad (7.134)$$

and thus we obtain a natural frequency that is a function of the temperature change:

$$\omega^2 = \left(\frac{\pi}{L}\right)^4 \left(\frac{EI}{m}\right)\left[1 - (\beta T)\left(\frac{L}{r\pi}\right)^2\right],\qquad(7.135)$$

where $r = \sqrt{I/A}$ is the radius of gyration. In the absence of a temperature gradient we recover the natural frequency of a regular simply supported beam. Again we have a natural frequency that drops to zero as the critical buckling temperature is approached, but increases if the beam is cooled:

$$T_{cr} = \left(\frac{r\pi}{L}\right)^2 \frac{1}{\beta}.\qquad(7.136)$$

Of course, this solution depends on the ends of the beam being prevented from moving, but we basically obtain Euler buckling. The issue of thermal loading is a very complicated one, especially for nonsimple structures, and the material will often have some temperature dependence (certainly at high temperatures) as well as nonuniform thermal loading. We will return to the question of thermal loading later with regard to plates [66, 67].

7.11 Other Effects

The main focus in this book is on the interaction of buckling and dynamics with application to elastic slender structures. In more stocky axially loaded members, some of the thin beam-theory assumptions need to be reexamined. Specifically, shear deformation and rotary inertia effects may be important [68]. Timoshenko beam theory incorporates these effects by relaxing the restriction that plane sections remain normal to the deflected beam axis during deformation [69]. An additional term then appears in the governing equation of motion, and the shear strain tends to diminish the lowest critical load. However, for relatively slender structures (a similar argument applies to plates) the shear strain has a negligible effect. Clearly, for stocky heavily loaded axial structures, plasticity becomes an issue, but again this does not fall within the scope of material in this book.

Energy dissipation may have an important influence on dynamic behavior. For example, in a linear system, it is the presence of damping that limits the response at resonance. Viscoelastic effects may be important [70]. Damping was introduced earlier in general terms relating to linear-viscous damping. In structural mechanics, this type of damping arises because of transverse displacement as well as straining of the beam material. The former contributes a term proportional to velocity in the equation of motion, the latter a term that may take the form $c_s I \partial^5 w/\partial x^4 \partial t$. The vibration of axially loaded beams with certain distributed masses has been considered in Cox [71].

The following chapter will broaden the scope a little. Slender beamlike elements will again be considered but now a number of other factors will be included in the analysis.

References

[1]   A. Bokaian. Natural frequencies of beams under compressive axial loads. *Journal of Sound and Vibration*, 126:49–65, 1988.

[2]   L. Meirovitch. *Principles and Techniques of Vibrations*. Prentice Hall, 1997.

[3]   J.M.T. Thompson and G.W. Hunt. *Elastic Instability Phenomena*. Wiley, 1984.

[4]   J. Szabo, Z. Gaspar, and T. Tarnai. *Post-Buckling of Elastic Structures*. Elsevier, 1986.

[5]   W. Nowacki. *Dynamics of Elastic Systems*. Chapman & Hall, 1963.

[6]   S.P. Timoshenko and D.H. Young. *Elements of Strength of Materials*. Van Nostrand Reinhold, 1977.

[7]   R.E.D. Bishop and W.G. Price. The vibration characteristics of a beam with an axial force. *Journal of Sound and Vibration*, 59:237–44, 1978.

[8]   A.E. Galef. Bending frequencies of compressed beams. *Journal of the Acoustical Society of America*, 44:643, 1968.

[9]   V. Birman, I. Elishakoff, and J. Singer. On the effect of axial compression on the bounds of simple harmonic motion. *Israel Journal of Technology*, 20:254–8, 1982.

[10] Z.P. Bazant and L. Cedolin. *Stability of Structures*. Oxford University Press, 1991.

[11] D. Young and R.P. Felgar. Tables of characteristic functions representing normal modes of vibration of a beam. Technical report, The University of Texas, Publication No. 4913, 1949.

[12] N.G. Stephen. Beam vibration under compressive axial load—upper and lower bound approximation. *Journal of Sound and Vibration*, 131:345–50, 1989.

[13] A. Bokaian. Natural frequencies of beams under tensile axial loads. *Journal of Sound and Vibration*, 142:481–98, 1990.

[14] G.J. Simitses. *Dynamic Stability of Suddenly Loaded Structures*. Springer-Verlag, 1989.

[15] R.D. Blevins. *Formulas for Natural Frequencies and Mode Shapes*. Van Nostrand Rheinhold, 1979.

[16] R. Courant and D. Hilbert. *Methods of Mathematical Physics*. Wiley Classics Library, 1989.

[17] G.W. Housner and W.K. Tso. Dynamic behavior of supercritically loaded struts. *ASCE Journal of Engineering Mechanics*, 88:41–65, 1962.

[18] J. Roorda. *The Instability of Imperfect Elastic Structures*. Ph.D. dissertation, University of London, 1965.

[19] R. Schmidt. Initial postbuckling of columns by the Rayleigh–Ritz method. *Industrial Mathematics*, 36:67–76, 1986.

[20] M.S. El Naschie. *Stress, Stability and Chaos in Structural Engineering: An Energy Approach*. McGraw-Hill, 1990.

[21] S.M. Dickinson. The lateral vibration of slightly bent slender beams subject to prescribed axial end displacement. *Journal of Sound and Vibration*, 68:507–14, 1980.

[22] I. Elishakoff, V. Birman, and J. Singer. Influence of initial imperfections on nonlinear free vibrations of elastic bars. *Acta Mechanica*, 55:191–202, 1985.

[23] T.M. Atanackovic. Stability of a compressible elastic rod with imperfections. *Acta Mechanica*, 76:203–22, 1989.

[24] L. Tomski and S. Kukla. Free vibrations of a certain geometrically nonlinear system with initial imperfection. *AIAA Journal*, 28:1240–5, 1989.

[25] C.L. Dym. *Stability Theory and Its Applications to Structural Mechanics*. Noordhoff International, 1974.

[26] S.P. Timoshenko and J.M. Gere. *Theory of Elastic Stability*, 2nd ed. McGraw-Hill, 1961.

[27] H.G. Allen and P.S. Bulson. *Background to Buckling*. McGraw-Hill, 1980.

[28] S. Woinowsky-Krieger. The effect of an axial force on the vibration of hinged bars. *Journal of Applied Mechanics*, 17:35–36, March 1950.

[29] F.J. Shaker. Effect of axial load on mode shapes and frequencies of beams. Technical report, NASA Lewis Research Center Report TN-8109, 1975.

[30] W.J. Duncan. Galerkin's method in mechanics and differential equations. *Reports and Memoranda No. 1798*, 1937.

[31] S. Karunamoorthy, D.A. Peters, and D. Barwey. Orthogonal polynomials for energy methods in rotary wing structural dynamics. *Journal of the American Helicopter Society*, 38:93–8, 1993.

[32] Y.R. Wang. *The effect of wake dynamics on rotor eigenvalues in forward flight*. Ph.D. dissertation, Georgia Institute of Technology, 1992.

[33] A.R.S. Bramwell. *Helicopter Dynamics*. Arnold, 1976.

[34] W. Johnson. *Helicopter Theory*. Princeton University Press, 1980.

[35] V.T. Nagaraj. Relationship between fundamental natural frequency and maximum static deflection for rotating Timoshenko beams. *Journal of Sound and Vibration*, 201:404–6, 1997.

[36] C.L. Lee. Limit cycle measurements from a cantilever beam attached to a rotating body. *AIAA Journal*, 36:1540–1, 1998.

[37] C.D. Eick and M.P. Mignolet. Vibration and buckling of flexible rotating beams. *AIAA Journal*, 33:528–38, 1995.

[38] M. Behzad and A.R. Bastami. Effect of centrifugal force on natural frequency of lateral vibration of rotating shafts. *Journal of Sound and Vibration*, 274:985–95, 2004.

[39] R.V. Southwell and B.S. Gough. On the free transverse vibrations of airscrew blades. Technical Report 766, Rep. Mem. Aeronautical Research Council, 1921.

[40] T.G. Carne, D.R. Martinez, and S.R. Ibrahim. Modal identification of rotating blade system. Technical report, AIAA-83-0815, 1983.

[41] S.M. Hashemi, M.J. Richard, and G. Dhatt. A new dynamic finite element (dfe) formulation for lateral free vibrations of Euler–Bernoulli spinning beams using trigonomoetric shape functions. *Journal of Sound and Vibration*, 220:601–24, 1999.

[42] J.R. Banerjee. Free vibration of centrifugally stiffened uniform and tapered beams using the dynamic stiffness method. *Journal of Sound and Vibration*, 233:857–75, 2000.

[43] S. Naguleswaran. Lateral vibration of a centrifugally tensioned uniform Euler–Bernoulli beam. *Journal of Sound and Vibration*, 176:613–24, 1994.

[44] G.H. Argyris and H-P. Mlejnek. *Dynamics of Structures*. North-Holland, 1991.

[45] J.R. Banerjee, H. Su, and D.R. Jackson. Free vibration of rotating tapered beams using the dynamic stiffness method. *Journal of Sound and Vibration*, 298:1034–54, 2006.

[46] R.A. Ormiston and D.H. Hodges. Linear flap–lag dynamics of hingeless Helicopter rotor blades in hover. *Journal of the American Helicopter Society*, 17:2–14, 1972.

[47] D.H. Hodges and E.H. Dowell. Nonlinear equations of motion for the elastic bending and torsion of twisted nonuniform rotor blades. Technical Report, NASA TN D-7818, 1974.

[48] A. Bazoune. Survey on modal frequencies of centrifugally stiffened beams. *Shock and Vibration Digest*, 37:449–69, 2005.

[49] D.H. Hodges and R.A. Ormiston. Stability of elastic bending and torsion of uniform cantilevered rotor blades in hover. *AIAA*, 1973. Paper No. 73-405.

[50] E.H. Dowell, J. Traybar, and D.H. Hodges. An experimental–theoretical correlation study of non-linear bending and torsion deformations of a cantilever beam. *Journal of Sound and Vibration*, 50:533–44, 1977.

[51] A.T. Frederick and I. Chopra. Assessment of transient analysis techniques for rotor stability testing. *Journal of the American Helicopter Society*, 35:39–50, 1990.

[52] C.E. Hammond. An application of Floquet theory to prediction of mechanical instability. *Journal of the American Helicopter Society*, 4:14–23, 1974.

[53] N.A. Cumpsty. *Jet Propulsion*. Cambridge University Press, 2003.

[54] Y.C. Fan, M.S. Ju, and Y.G. Tsuei. Experimental study on vibration of a rotating blade. *Journal of Engineering for Gas Turbines and Power*, 116:672–7, 1994.

[55] R.G. Parker and C.D. Mote. Tuning of the natural frequency spectrum of a circular plate by in-plane stress. *Journal of Sound and Vibration*, 145:95–110, 1991.

[56] M. Beck. Die Knicklast des einseitig eingespannten, tangential gedruckten Stabes. *Zeitschrift für Angewandte Mathematik und Physik*, 3:225–8, and 476–7, 1952.

[57] H. Ziegler. *Principles of Structural Stability*. Blaidsell, 1968.

[58] H.H.E. Leipholz. *Stability Theory*. Wiley, 1987.

[59] C.Y. Wang. A critical review of the heavy elastica. *International Journal of Mechanical Sciences*, 28:549–59, 1986.

[60] L.N. Virgin. Free vibrations of imperfect cantilever bars under self-weight loading. *Proceedings of the Institution of Mechanical Engineering*, 201:345–7, 1987.

[61] A.G. Greenhill. Determination of the greatest height consistent with stability that a vertical pole or mast can be made, and of the greatest height to which a tree of given proportions can grow. *Proceedings of the Cambridge Philosophical Society*, 4:65–73, 1881.

[62] A.G. Pugsley. Limits to size set by trees. *The Structural Engineer*, 66:322–3, 1988.

[63] D.F. Pilkington and J.B. Carr. Vibrations of beams subjected to end and axially distributed loading. *Journal of Mechanical Engineering Science*, 12:70–2, 1970.

[64] T. Huang and D.W. Dareing. Buckling and frequencies of long vertical pipes. *Journal of the Engineering Mechanics Division, ASCE*, 95:167–81, 1969.

[65] C.P. Sparks. Transverse modal vibrations of vertical tensioned risers: A simplified analytical approach. *Oil and Gas Science and Technology*, 57:71–86, 2002.

[66] D.J. Johns. *Thermal Stress Analysis*. Pergamon, 1965.

[67] B.A. Boley and J.H. Weiner. *Theory of Thermal Stresses*. Wiley, 1960.

[68] E. Sevin. On the elastic bending of columns due to dynamic axial forces including effects of axial inertia. *Journal of Applied Mechanics*, 27:125–31, 1960.

[69] H. Abramovich. Natural frequencies of Timoshenko beams under compressive axial loads. *Journal of Sound and Vibration*, 157:183–9, 1992.

[70] S.C. Chuang and J.T.-S. Wang. Vibration of axially loaded damped beams on viscoelastic foundation. *Journal of Sound and Vibration*, 148:423–35, 1991.

[71] H.L. Cox. Vibration of axially-loaded beams carrying distributed masses. *Journal of the Acoustical Society of America*, 30:568–71, 1958.

8 Other Column-Type Structures

The previous chapter focused attention on the behavior of axially loaded prismatic thin beams in which the external loading consisted primarily of loads applied at the end of the member or in which significant axial effects were induced (e.g., the membrane, or stretching, effect). In this chapter, the scope of the analysis is opened to include a wider variety of situations in which axial loading and dynamic effects are considered, but in which lateral loading, for example, plays an important role, or in which columns with variable cross-sectional properties are considered. This will then lead naturally into the chapter on frames, in which boundary conditions, for example, do not necessarily fall into simple categories. We will also typically have to rely more on approximate techniques. Furthermore, there are cases in which the geometry of the structure means that a system's vibration and stability characteristics may depend on a number of parameters not considered in the previous chapter that focused on relatively straightforward prismatic members. We start this chapter by looking at a couple of cases in which this happens.

8.1 A Beam on an Elastic Foundation

It is not uncommon for a beam to have some kind of continuous support along its length. We can think of this as an elastic foundation and assume the foundation stiffness is linear [1, 2]. A practical example of this might be the sleepers under a railroad track, where a significant axial-loading effect is caused by thermal expansion [3, 4]. Referring to the schematic shown in Fig. 8.1 and again assuming the ends of the beam are pinned, we can extend the analysis of the previous chapter.

The regular beam of the previous chapter resulted in a partial differential equation that was conveniently solved by separating variables:

$$w(x, t) = \sum_{i=1}^{n} Y_i(t) W_i(x), \qquad (8.1)$$

and using the Fourier expansion, based on simply supported boundary conditions,

$$W_i = \sin \frac{i\pi x}{L}, \qquad (8.2)$$

resulted in an infinite set of uncoupled ordinary differential equations in terms of these modes.

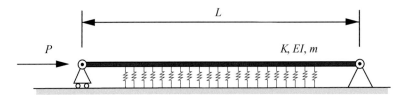

Figure 8.1. A schematic of a thin elastic beam restrained by a linearly elastic foundation.

The incorporation of a linear-elastic foundation results in additional strain energy stored in the foundation:

$$U_K = \frac{1}{2} K \int_0^L w^2 dx \qquad (8.3)$$

$$= \frac{1}{2} K \int_0^L q_i{}^2 \sin^2 \frac{i\pi x}{L} dx \qquad (8.4)$$

$$= \frac{1}{2} K q_i{}^2 \frac{L}{2}. \qquad (8.5)$$

Therefore, we obtain the natural frequencies:

$$\omega_i^2 = \frac{1}{m} \left[EI \left(\frac{i\pi}{L} \right)^4 + K - P \left(\frac{i\pi}{L} \right)^2 \right]. \qquad (8.6)$$

Clearly, we recover the results from the last chapter when we set $K = 0$, but, depending on the stiffness of the elastic foundation, we see the possibility of a frequency other than the first (corresponding to a half-sine wave) dropping to zero first under the action of increasing P. Introducing the following nondimensional parameters,

$$\Omega^2 = \frac{\omega^2}{\frac{EI}{m} \left(\frac{\pi}{L} \right)^4}, \quad p = \frac{P}{EI \left(\frac{\pi}{L} \right)^2}, \quad k = \frac{K}{EI \left(\frac{\pi}{L} \right)^4}, \qquad (8.7)$$

we obtain nondimensional equations for each mode:

$$1 + k - p = \Omega^2 \ : \ i = 1,$$

$$16 + k - 4p = \Omega^2 \ : \ i = 2,$$

$$81 + k - 9p = \Omega^2 \ : \ i = 3, \qquad (8.8)$$

and so on. Without the elastic foundation, we observe the familiar relation between the axial load and the square of the natural frequency. However, the elastic foundation has an interesting effect on the critical loads; for example, we see that when $k = 4$ the lowest two buckling loads are the same ($p = 5$). For $4 < k < 36$, the critical value of p is $4 + (k/4)$ and the corresponding mode has two half-sine waves. In general, if $(n - 1)^2 n^2 < k < n^2 (n + 1)^2$, the critical value of p is $n^2 + k/(n)^2$ and the governing buckling mode has n half-sine modes [5].

Suppose we fix $k = 12$. The relationships in Eqs. (8.8) are plotted in Fig. 8.2. For this specific foundation stiffness, we have (in the absence of axial loads) frequencies $\Omega_1^2 = 13$, $\Omega_2^2 = 28$, and $\Omega_3^2 = 93$. The lowest three buckling loads (i.e., when the lowest natural frequency is zero) are $p_2 = 7, p_3 = 10.33$, and $p_1 = 13$. We see how

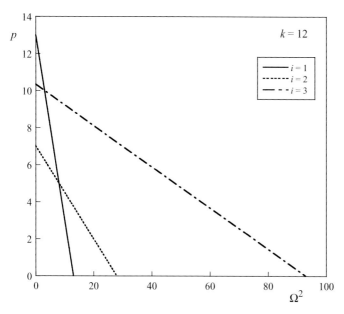

Figure 8.2. The interaction of axial load and natural frequencies for a pinned beam resting on a foundation with stiffness $k = 12$.

these modes have changed order, such that the $i = 2$ mode actually becomes critical first. Of course, as we would expect, both the nondimensional critical load and the lowest natural frequency are considerably higher than the $k = 0$ case (i.e., unity).

8.2 Elastically Restrained Supports

It may happen that the actual boundary conditions do not correspond exactly to the classification of pinned, fixed, and so on. In this case, an elastic spring can be incorporated into the analysis such that when the torsional stiffness is set equal to zero we obtain the pinned or simply supported case, and when it is infinite we have the fully fixed boundary condition [6]. This allows for a range of intermediate values that can be used to reflect varying degrees of partial restraint [7, 8].

Because the elastic end constraints affect only the boundary conditions we still have the familiar equation governing the small-amplitude, harmonic, transverse vibrations of a beam given by

$$\frac{d^4W}{dx^4} + p\frac{d^2W}{dx^2} - \Omega^2 W = 0, \tag{8.9}$$

in which

$$W = w/L, \quad \bar{x} = x/L, \quad p = PL^2/EI, \quad \Omega^2 = \rho A\omega^2 L^4/EI. \tag{8.10}$$

Again, the solution is given by

$$W = C_1 \sinh \alpha\bar{x} + C_2 \cosh \alpha\bar{x} + C_3 \sin \beta\bar{x} + C_4 \cos \beta\bar{x}, \tag{8.11}$$

with

$$\alpha^2 = \sqrt{(p^2/4) + \Omega^2} + p/2,$$
$$\beta^2 = \sqrt{(p^2/4) + \Omega^2} - p/2. \tag{8.12}$$

Note the similarity between Eqs. (8.12) and (7.29), with the slight difference that is due to the definition of nondimensional load used in Eqs. (7.30) compared with Eqs. (8.10). However, with torsional end constraint the boundary conditions become

$$W(0) = W(1) = 0$$

$$d^2W/d\bar{x}^2 - \sigma_1 dW/d\bar{x} = 0 \quad \text{at} \quad \bar{x} = 0, \tag{8.13}$$

$$d^2W/d\bar{x}^2 + \sigma_2 dW/d\bar{x} = 0 \quad \text{at} \quad \bar{x} = 1,$$

where

$$\sigma_1 = k_1 L/EI, \qquad \sigma_2 = k_2 L/EI, \tag{8.14}$$

and the k's are the spring stiffness at the left- and right-hand ends, respectively.

Application of the preceding conditions leads to the characteristic equation

$$\left[(\alpha^2 + \beta^2)^2 + \sigma_1\sigma_2(\alpha^2 - \beta^2)\right]\sinh\alpha\sin\beta$$

$$- 2\sigma_1\sigma_2\alpha\beta(\cosh\alpha\cos\beta - 1) \tag{8.15}$$

$$+ (\sigma_1 + \sigma_2)(\alpha^2 + \beta^2)(\alpha\cosh\alpha\sin\beta - \beta\sinh\alpha\cos\beta) = 0. \tag{8.16}$$

We consider a specific case of a beam, subject to a tensile axial force, with the left-hand end pinned ($\sigma_1 = 0$) and the right-hand end subject to a torsional spring (σ_2). Thus we can examine the dynamics of the beam ranging from zero-end rotational stiffness (i.e., a pinned–pinned beam) to infinite-end rotational stiffness (i.e., a pinned–clamped beam) to compare with some of the results of the previous chapter. Figure 8.3 shows how the lowest two natural frequencies vary with tensile axial load for different levels of end torsional restraint. For zero axial load, we observe the frequency coefficients $\Omega_1 = \pi^2$ and $\Omega_2 = 4\pi^2$ for the pinned–pinned case (i.e., $\sigma_2 = 0$). The natural frequencies increase with tensile force as expected. For pinned–fixed boundary conditions ($\sigma_2 = \infty$), we obtain the frequencies $\Omega_1 = 15.4$ and $\Omega_2 = 50.0$. These cases are indicated by the circles. Two intermediate cases are also shown: for $\sigma_2 = 5$ and $\sigma_2 = 20$. In those cases in which a structural component makes up one element of a larger structure, or framework, then the actual boundary conditions will typically depend on the stiffness provided by adjacent members, and this will be the focus of the next chapter.

8.3 Beams with Variable Cross Section

In this section, we make use of the Rayleigh–Ritz approach to obtain the axial load versus frequency relation for a simply supported beam with a square cross section ($h \times h$) and a linear taper, as shown in Fig. 8.4. The column length is L, Young's modulus E, and density ρ. The width (and depth) at any distance x along the length

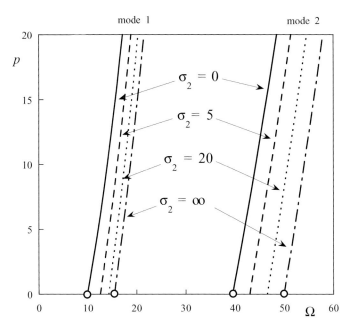

Figure 8.3. Variation of the lowest two natural frequencies with tensile axial load for varying end constraint.

of the column is given by

$$h(x) = h(0)\left[1 + \frac{\alpha - 1}{L}x\right], \tag{8.17}$$

in which $\alpha = h(0)/h(L)$. From this, we can compute the area and second moment of area:

$$A(x) = A(0)\left[1 + x(\alpha - 1)/L\right]^2,$$
$$I(x) = I(0)\left[1 + x(\alpha - 1)/L\right]^4, \tag{8.18}$$

where $A(0) = h(0)^2$ and $I(0) = h(0)^4/12$ are the area and second moment of area at the left-hand (smaller) end. Given simply supported boundary conditions we can assume a half-sine wave as the fundamental mode (we know this is exact for a prismatic beam), that is, $w(x) = C\sin \pi x/L$. The energy expressions for strain energy in

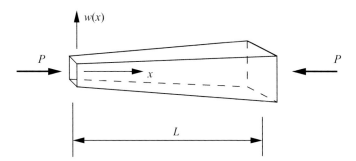

Figure 8.4. A schematic of a thin elastic beam whose size is a linear function of its length.

bending, potential energy of the loading, and kinetic energy are given by

$$U = \frac{1}{2}\int_0^L EI(x){w''}^2 dx, \tag{8.19}$$

$$V_P = \frac{1}{2}\int_0^L P{w'}^2 dx, \tag{8.20}$$

$$T = \frac{1}{2}\int_0^L \rho A(x)\dot{w}^2 dx. \tag{8.21}$$

These expressions can be evaluated for the assumed mode shape, which results in

$$U = \frac{1}{4}EI(0)LC^2\left(\frac{\pi}{L}\right)^4 [1 + 2(\alpha - 1) + (2 - 3/\pi^2)(\alpha - 1)^2$$
$$+ (1 - 3/\pi^2)(\alpha - 1)^3 + (1/5 - 1/\pi^2 + 3/(2\pi^4))(\alpha - 1)^4], \tag{8.22}$$

$$V_P = \frac{1}{4}LPC^2\left(\frac{\pi}{L}\right)^2, \tag{8.23}$$

$$T = \frac{1}{4}L\rho A(0)\dot{C}^2[1 + (\alpha - 1) + (1/3 - 1/(2\pi^2))(\alpha - 1)^2]. \tag{8.24}$$

We can then make use of Lagrange's equation or Rayleigh's method to obtain the natural frequency in the usual way. We note that this result subsumes the case of a prismatic (constant-cross-section) beam in which $\alpha = 1$. However, in general, we get

$$\bar{\omega}^2 = \frac{[1 + 2(\alpha - 1) + 1.696(\alpha - 1)^2 + 0.696(\alpha - 1)^3 + 0.317(\alpha - 1)^4 - \bar{p}]}{[1 + (\alpha - 1) - 0.283(\alpha - 1)^2]}, \tag{8.25}$$

where the natural frequency and axial load are nondimensionalized according to

$$\bar{\omega}^2 = \frac{\omega^2}{EI(0)/m(\frac{\pi}{L})^4}, \qquad \bar{p} = \frac{P}{EI(0)(\frac{\pi}{L})^2}. \tag{8.26}$$

For the prismatic beam ($\alpha = 1$), we get the exact coefficients from the previous chapter.

The effect of a varying cross section is shown in Fig. 8.5. The linearity of the $\bar{p}$ versus $\bar{\omega}^2$ relation in part (a) is a consequence of the single-mode assumption. By increasing the value of α, we are stiffening the beam, and hence both the critical load and natural frequency increase [9, 10]. For example for a beam whose cross-sectional dimension at $x = L$ is twice that at $x = 0$, that is, $\alpha = 2$ results in a nondimensional critical load of approximately 5.47, and a natural frequency (squared) in the absence of axial load of about 3.3. The variable cross section does, of course, render the sine wave an approximate mode shape [11]. More terms in the Rayleigh–Ritz procedure can be used. Auciello [12] did this by using higher-order orthogonal polynomial functions. The result of using (very accurate) six terms in just such an expansion is also shown in Fig. 8.5(b), and specifically for $\alpha = 2$ in part (c).

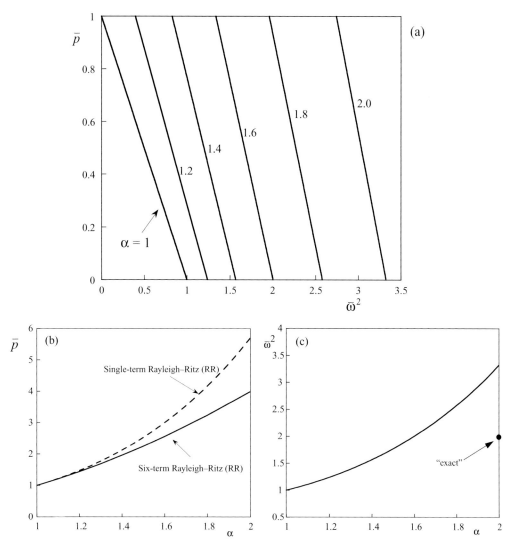

Figure 8.5. (a) The frequency–load relation for various tapered columns, (b) variation of the critical load with taper (with a sample "exact" result for $\alpha = 2$ superimposed), and (c) variation of natural frequency with taper when the axial load is zero.

As the degree of taper increases, the single-mode approximation breaks down [13].

The analysis is simplified somewhat if it is assumed that it is the second moment of area that varies linearly with length rather than the cross-sectional dimension, or if a beam is wedged shape then the area will vary linearly with length and the second moment of area will vary as the cube of the length, and so on. Tapered columns that are loaded by their self-weight can also be handled, although the mass distribution must also be taken into account. Of course, the height to which a tree might grow is a nice example of this, although it is the lateral loading caused by wind that ultimately

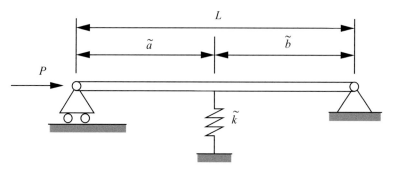

Figure 8.6. A schematic of a thin elastic beam restrained by a linear spring located close to midspan.

limits height [14–17]. It has been shown [18] that many trees have a natural taper of the form

$$R = R(0) \left(\frac{h-x}{h} \right)^{3/2}, \tag{8.27}$$

where R is the radius, $R(0)$ is the radius at the base, and h is the height. Stepped columns that have discrete changes in their cross-sectional properties can also be handled in this manner.

8.4 Modal Coupling

Given a simply supported beam, it is clear that the half-sine wave mode would be suppressed if lateral deflection at the center of the beam were prevented, such that a complete, rather than a half, sine wave would then be the lowest buckling mode and produce the fundamental mode of vibration [19]. Suppose a spring is attached to the beam very close to, but not at, the center of the beam, as shown in Fig. 8.6. In this case, in which the spring is offset from the midspan point, the modes are coupled [20]. Again the beam has the usual parameters, which in nondimensional terms are given by

$$x = \tilde{x}/L, \quad a = \tilde{a}/L, \quad b = \tilde{b}/L = 1 - a, \quad w = \tilde{w}/L, \tag{8.28}$$

$$\gamma = \sqrt{PL^2/EI}, \quad k = \tilde{k}L^3/EI, \quad \omega = \tilde{\omega}\sqrt{mL^4/EI}.$$

Note that we have started off using the tilde expression for *dimensional* quantities in order to simplify the notation. Denoting w_1 and w_2 as the lateral deflections on the left- and right-hand spans, respectively, we use D'Alembert's principle and assume harmonic motion to obtain

$$w_1'''' + \gamma^2 w_1'' - \omega^2 w_1 = 0 \tag{8.29}$$

for $0 < x < a$, and

$$w_2'''' + \gamma^2 w_2'' - \omega^2 w_2 = 0 \tag{8.30}$$

for $a < x < 1$.

The boundary conditions are the same as before, namely, zero deflection and bending moment at either end, but we also have the additional geometric requirements concerning continuity at the point where the spring is located. They are

$$w_1(0) = 0, \quad w_1''(0) = 0,$$
$$w_2(1) = 0, \quad w_2''(1) = 0, \tag{8.31}$$

$$w_1(a) = w_2(a), \qquad w_1'(a) = w_2'(a),$$
$$w_1''(a) = w_2''(a), \quad w_2'''(a) = w_2'''(a) + kw_1(a). \tag{8.32}$$

It is apparent that the general solution to the governing equation(s) will depend on the parameters in a manner similar to that of the previous chapter. It is convenient to categorize the response according to the influence of spring stiffness on the buckling and natural frequencies of the system.

We first consider the buckling behavior, which again we conveniently obtain by setting $\omega^2 = 0$. In this case, the general solution to Eqs. (8.29) and (8.30) can be written as

$$w_j(x) = A_j \cos \gamma x + B_j \sin \gamma x + C_j x + D_j, \quad j = 1, 2, \tag{8.33}$$

where

$$k = \frac{\gamma^3 \sin \gamma}{\gamma ab \sin \gamma - \sin \gamma a \sin \gamma b}. \tag{8.34}$$

Solving this system results in the relation between nondimensional axial load and spring stiffness shown in Fig. 8.7(a). For zero spring stiffness we have the lowest two critical loads given by $\gamma/2\pi = 0.5, 1.0$ corresponding to the symmetric (half-sine wave) and the antisymmetric (full-sine wave) modes obtained in the previous chapter.

As the stiffness of the lateral spring is increased (but still located exactly at the center of the beam), we observe that the lowest buckling load starts to increase, and when the spring stiffness reaches $k = 16\pi^2$ the first two critical loads coincide. On subsequent increase in spring stiffness, the formerly second buckling mode is now the lowest and the beam will buckle into a complete sine wave with the spring enforcing a nodal point at midspan. This crossover is sometimes called the *ideal stiffness* because the spring (bracing) has done its job in increasing the buckling capacity and any additional bracing would be surplus.

If the spring is now located slightly offset from the beam midspan then we observe that some interesting coupling taking place. For example, when $a = 0.525$, the modes *veer* in the vicinity of where the modes crossed in the previous case. This is also shown in Fig. 8.7. Veering also takes place in interactions between certain higher modes [21].

Next, consider the effect that the spring stiffness (and axial load) has on the dynamics of the system. Analytically, we make a further division here according to the sign of ω^2:

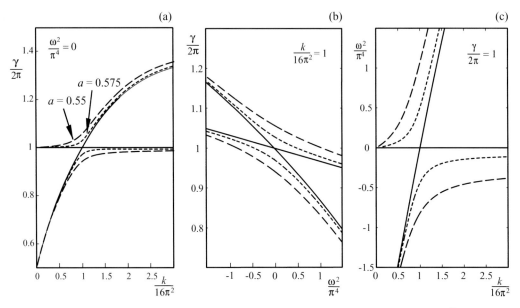

Figure 8.7. Curve veering: The solid lines represent the centrally placed spring, $\bar{a} = \bar{b} = 0.5$, with the offset cases as shown.

- If $\omega^2 > 0$ then the general solution to Eqs. (8.29) and (8.30) is given by

$$w_j(x) = A_j \cos \sigma x + B_j \sin \sigma x + C_j \cosh \lambda x + D_j \sinh \lambda x, \, j = 1, 2, \qquad (8.35)$$

where

$$\sigma = \left[(\gamma^2 + \sqrt{\gamma^4 + 4\omega^2})/2 \right]^{1/2}, \qquad (8.36)$$

$$\lambda = \left[(-\gamma^2 + \sqrt{\gamma^4 + 4\omega^2})/2 \right]^{1/2}. \qquad (8.37)$$

Again these expressions are related to the general solutions given in Eqs. (7.29) and (8.12). Applying the boundary and transition conditions leads to the characteristic equation

$$k = \frac{\sigma\lambda(\sigma^2 + \lambda^2) \sin \sigma \sinh \lambda}{\sigma \sin \sigma \sinh \lambda a \sinh \lambda b - \lambda \sinh \lambda \sin \sigma a \sin \sigma b}. \qquad (8.38)$$

- If $\omega^2 < 0$, then the general solution is given by

$$w_j(x) = A_j \cos \sigma x + B_j \sin \sigma x + C_j \cos \eta x + D_j \sin \eta x, \quad j = 1, 2, \qquad (8.39)$$

where

$$\eta = \left[(\gamma^2 - \sqrt{\gamma^4 + 4\omega^2})/2 \right]^{1/2}. \qquad (8.40)$$

The characteristic equation is then given by

$$k = \frac{\sigma\eta(\sigma^2 - \eta^2) \sin \sigma \sin \eta}{\sigma \sin \sigma \sin \eta a \sin \eta b - \eta \sin \eta \sin \sigma a \sin \sigma b}. \qquad (8.41)$$

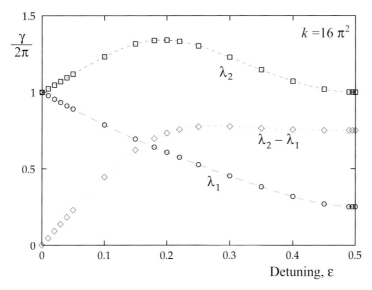

Figure 8.8. Separation of computed eigenvalues as a function of the detuning parameter.

It should be pointed out here that this case corresponds to exponentially grow-ing (unstable) motions. For loads greater than the critical value, we would expect the system to follow its (stable) postbuckled path, for which there would be real fre-quencies. Suppose we now hold k fixed at $16\pi^2$. For a centrally located spring, we get simple crossing when $\omega^2/\pi^4 = 0$ and $\gamma/2\pi = 1$, together with veering as the spring location is again slightly offset from the center, as shown in Fig. 8.7(b).

Finally, we set the axial load at $\gamma = 2\pi$ and observe the relation between the spring stiffness and natural frequencies. This is shown in Fig. 8.7(c). We can also plot the separation of the eigenvalues as a function of the detuning parameter $\epsilon = a - 0.5$ as shown in Fig. 8.8 when the spring stiffness $k = 16\pi^2$.

8.5 Flexural–Torsional Buckling and Vibration

So far we have concentrated on the modeling of slender beams such that a 1D de-scription was adequate. Another class of axially loaded structures again involves the interaction of vibration and instability but where the cross-sectional properties are such that additional factors need to be considered. Thin-walled bars of an open section fall into this class, and they are commonly used in a (lateral) load-carrying configuration (e.g., I-section beams and columns [22]). We encounter torsional and coupled flexural–torsional behavior, in addition to flexural buckling and vibration. In this section, we take a brief look at the dynamics associated with this type of behavior. Much of the underlying development concerning the relation among dis-placements, stresses, and strains is omitted, but the reader is referred to Timoshenko and Gere [23] for a thorough derivation.

Structural members of open cross sections tend to have relatively low torsional stiffness and may buckle because of twisting. This effect will often cause a reduction

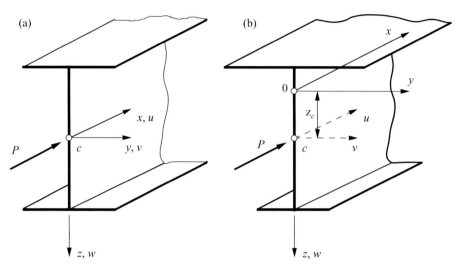

Figure 8.9. Thin-walled I-beam subject to an axial load: (a) a doubly symmetric cross section, (b) a beam with one axis of symmetry.

in the load-carrying capacity. We focus attention on a specific type of axial–torsional buckling and vibration associated with a simply supported, (doubly symmetric) I-beam by using an energy approach. The geometry is shown in Fig. 8.9(a), and for simplicity, we initially assume that the axial load P acts through the centroid of the section (which coincides with the shear center) [24]. The beam has length L and the usual physical properties EI and m.

We can write the strain energy for a prismatic beam as

$$U = \int_0^L \frac{1}{2} \left(EI_z w''^2 + EI_y v''^2 + E\Gamma \theta''^2 + GJ\theta'^2 \right) dx. \tag{8.42}$$

In this equation, the first two terms are familiar from our earlier bending theory, the third term is the strain energy that is due to longitudinal warping resistance (Γ is the warping moment of inertia of the cross section), and the final term is due to St. Venant's torsion (where GJ is the torsional stiffness, i.e., a product of the shear modulus G and the polar moment of inertia J) [25, 26]. Both of these deformations are related to an angle of twist θ measured about the centroid. The potential energy associated with the end load is given by

$$V_P = -\int_0^L \frac{1}{2} P \left(v'^2 + w'^2 + \frac{I_0}{A}\theta'^2 \right) dx, \tag{8.43}$$

and, finally, the kinetic energy is given by

$$T = \int_0^L \frac{1}{2} m \left(\dot{v}^2 + \dot{w}^2 + \frac{I_0}{A}\dot{\theta}^2 \right) dx. \tag{8.44}$$

Given the simply supported boundary conditions (and assuming the twist is zero at both ends for simplicity), we can take the displacements to be

$$w = C_1(t) \sin \frac{n\pi x}{L}, \tag{8.45}$$

$$v = C_2(t) \sin \frac{n\pi x}{L}, \tag{8.46}$$

$$\theta = C_3(t) \sin \frac{n\pi x}{L}. \tag{8.47}$$

Focusing attention on the lowest ($n = 1$) mode and evaluating Eqs. (8.42) and (8.43) by using these shapes, we obtain the total potential-energy function:

$$U + V_P = \frac{\pi^2}{4L} \left\{ C_1^2 \left(\frac{\pi^2 E I_z}{L^2} - P \right) + C_2^2 \left(\frac{\pi^2 E I_y}{L^2} - P \right) \right.$$
$$\left. + C_3^2 \left[\frac{A}{I_0} \left(GJ + \frac{E\Gamma \pi^2}{L^2} \right) - \frac{I_0}{A} P \right] \right\}. \tag{8.48}$$

The kinetic energy is evaluated as

$$T = \frac{1}{4} mL \left(\dot{C}_1^2 + \dot{C}_2^2 + \frac{I_0}{A} \dot{C}_3^2 \right). \tag{8.49}$$

We can then use Lagrange's equations to obtain the equations of motion:

$$mL\ddot{C}_1 + \frac{\pi^2}{L} \left(\frac{\pi^2 E I_z}{L^2} - P \right) C_1 = 0, \tag{8.50}$$

$$mL\ddot{C}_2 + \frac{\pi^2}{L} \left(\frac{\pi^2 E I_y}{L^2} - P \right) C_2 = 0, \tag{8.51}$$

$$mL\ddot{C}_3 + \frac{\pi^2}{L} \left[\frac{A}{I_0} \left(GJ + \frac{\pi^2 E\Gamma}{L^2} \right) - P \right] C_3 = 0. \tag{8.52}$$

The uncoupled nature of the equations is a consequence of the symmetry of the cross section and loading. The natural frequencies in the three modes are thus given by

$$\omega_w^2 = \frac{1}{m} \left[-P \left(\frac{\pi}{L} \right)^2 + E I_y \left(\frac{\pi}{L} \right)^4 \right], \tag{8.53}$$

$$\omega_v^2 = \frac{1}{m} \left[-P \left(\frac{\pi}{L} \right)^2 + E I_z \left(\frac{\pi}{L} \right)^4 \right], \tag{8.54}$$

$$\omega_\theta^2 = \frac{A}{m I_0} \left[-P \frac{I_0}{A} \left(\frac{\pi}{L} \right)^2 + E\Gamma \left(\frac{\pi}{L} \right)^4 + GJ \left(\frac{\pi}{L} \right)^2 \right]. \tag{8.55}$$

The presence of the axial load has the familiar relation with the natural frequencies. Using our dynamic definition of criticality, we immediately see the (pure flexure) Euler loads in the w and v directions:

$$P_w = E I_z \left(\frac{\pi}{L} \right)^2, \tag{8.56}$$

$$P_v = E I_y \left(\frac{\pi}{L} \right)^2. \tag{8.57}$$

The third critical condition corresponds to pure torsional instability:

$$P_\theta = \frac{A}{I_0} \left[E\Gamma \left(\frac{\pi}{L}\right)^2 + GJ \right].$$
(8.58)

Which of the three critical loads is lowest clearly depends on certain features of the cross section. For regular structural sections it is usual that the lowest mode corresponds to flexure about the weaker axis, and this is why the Euler load plays an important role in design [23].

In the absence of axial loads, we get the natural frequencies:

$$\bar{\omega}_w^2 = \frac{EI_z}{m} \left(\frac{\pi}{L}\right)^4,$$
(8.59)

$$\bar{\omega}_v^2 = \frac{EI_y}{m} \left(\frac{\pi}{L}\right)^4,$$
(8.60)

$$\bar{\omega}_\theta^2 = \frac{A}{mI_0} \left[E\Gamma \left(\frac{\pi}{L}\right)^4 + GJ \left(\frac{\pi}{L}\right)^2 \right].$$
(8.61)

For these uncoupled modes, we can further nondimensionalize to confirm the linear relation between the axial load and the square of the natural frequencies:

$$\left(\frac{\omega_w}{\bar{\omega}_w}\right)^2 = 1 - \frac{P}{P_w},$$
(8.62)

$$\left(\frac{\omega_v}{\bar{\omega}_v}\right)^2 = 1 - \frac{P}{P_v},$$
(8.63)

$$\left(\frac{\omega_\theta}{\bar{\omega}_\theta}\right)^2 = 1 - \frac{P}{P_\theta}.$$
(8.64)

So far the approach has followed lines similar to that of our previous analyses. However, if the beam is subject to end moments or the cross section does not exhibit any symmetry, then the situation becomes considerably more involved, with coupled modes appearing. Without going into too much detail, we briefly examine the behavior of a beam with only one axis of symmetry, such that the shear center and the centroid are separated by a vertical amount z_c, as shown in Fig. 8.9(b). In this case, an additional term appears in the potential energy associated with the axial load, which causes coupling between the v and θ modes of the form

$$V_{\text{offset}} = 2C_2 C_3 P z_c.$$
(8.65)

Using Lagrange's equations then leads to the condition for coupled natural frequencies in nondimensional terms:

$$\begin{vmatrix} (\omega_v^2 - \omega^2) & z_c[\omega^2 - \frac{P}{m}\left(\frac{\pi}{L}\right)^2] \\ z_c[\omega^2 - \frac{P}{m}\left(\frac{\pi}{L}\right)^2] & \frac{z_c^2}{\lambda}(\omega_\theta^2 - \omega^2) \end{vmatrix} = 0,$$
(8.66)

where

$$\lambda = \frac{1}{1 + I_0/(z_c^2 A)}.$$
(8.67)

Expanding this determinant leads to the characteristic equation

$$(\omega_\theta^2 - \omega^2)(\omega_v^2 - \omega^2) - \lambda\left[\omega^2 - \frac{P}{m}\left(\frac{\pi}{L}\right)^2\right]^2 = 0, \tag{8.68}$$

and by using Eqs. (8.63) and (8.64) we can express this as

$$\left[\bar\omega_\theta^2\left(1 - \frac{P}{P_\theta}\right) - \omega^2\right]\left[\bar\omega_v^2\left(1 - \frac{P}{P_v}\right) - \omega^2\right] - \lambda\left[\omega^2 - \frac{P}{m}\left(\frac{\pi}{L}\right)^2\right]^2 = 0. \tag{8.69}$$

Setting the axial load to zero results in the standard expression for the natural frequency of coupled flexural–torsional vibration:

$$\omega_{v\theta}^2 = \frac{(\bar\omega_\theta^2 + \bar\omega_v^2) \mp \sqrt{(\bar\omega_\theta^2 - \bar\omega_v^2)^2 + 4\lambda\bar\omega_\theta^2\bar\omega_v^2}}{2(1-\lambda)}. \tag{8.70}$$

We obtain the critical load in this coupled flexural–torsional mode by setting the frequency equal to zero in Eq. (8.69), which leads to

$$(P_{cr} - P_\theta)(P_{cr} - P_v) - P_{cr}^2\lambda = 0, \tag{8.71}$$

which, on evaluation, gives

$$P_{cr} = \frac{1}{2(1-\lambda)}\left[(P_\theta + P_v) \pm \sqrt{(P_\theta + P_v)^2 - 4(1-\lambda)P_\theta P_v}\right]. \tag{8.72}$$

Thus we conclude that a beam with a single axis of symmetry may buckle in a mode that couples displacement about the weaker principal axis and twisting, and this critical load may be lower than the purely flexural mode for the doubly symmetric section. With $z_c = 0$, we have $\lambda = 0$ and recover buckling in a purely flexural or purely torsional mode. The coupling causes the linear relation between axial load and frequency squared to be lost. Coupling between modes is also an important consideration in the dynamics of rotating beams with specific reference to helicopter rotor blades, as discussed in the previous chapter [27].

For a section with no axis of symmetry, some coupling will always be present (between both flexural modes and the torsional mode). However, even with just a single assumed mode for each coordinate, to determine the relation between natural frequencies and axial load, a 3×3 determinant must be solved, and this is best achieved numerically.

Lateral buckling may be handled in an analogous way. In this case a beam is loaded vertically on its top flange, say, and may buckle sideways because of the combined action of bending and twisting [23, 28–31].

8.6 Type of Loading

So far, other than in Beck's problem, the external load has been assumed to remain constant in direction and magnitude. This is implicitly based on a dead-loading scheme, that is, the axial force is a result of a mass in a gravitational field, although the mass of the end load is assumed not to influence the oscillations of the structural

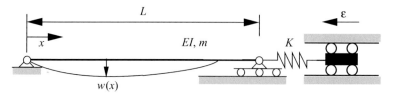

Figure 8.10. A continuous thin strut subject to semirigid loading by a prescribed end shortening.

member. Although this is a common enough means of imparting an axial load to a slender structure, it is also possible to load by use of a prescribed end displacement, and many testing machines fall into this category. Clearly, this also has a relation with thermal loading described in the previous chapter. This may have a major effect on instability behavior, especially in the postbuckled regime. Here, a brief introduction is made to semirigid loading and some of the analysis is repeated, but this time the external axial load is replaced with an imposed end shortening ϵ acting through a spring of stiffness K, as shown in Fig. 8.10. Assuming the beam is inextensional, we now have a potential-energy function (prior to buckling) of the form

$$V = \frac{1}{2}EI\left(\frac{\pi}{L}\right)^4 \frac{L}{2}Q^2 + \frac{1}{2}K\left(\frac{1}{2}\left(\frac{\pi}{L}\right)^2 \frac{L}{2}Q^2 - \epsilon\right)^2, \qquad (8.73)$$

based on the mode shape from Eq. (7.45). Thus we obtain a natural frequency

$$\omega^2 = \frac{EI}{m}\left(\frac{\pi}{L}\right)^4 - \frac{K\epsilon}{m}\left(\frac{\pi}{L}\right)^2, \qquad (8.74)$$

which reverts to the standard case when $\epsilon = 0$ (assuming the spring is initially unstressed). As $\omega^2 \to 0$, we also obtain the critical end shortening $\epsilon_{cr} = EI/K(\pi/L)^2$. This analysis can be carried through to the postbuckled range, and many of the earlier relations between the frequency and the load experienced by the strut apply.

Displacement controlled loading can be especially influential in the testing of structures liable to exhibit snap-through behavior at a limit point [32]. It can be shown that, although this type of loading will generally have a stabilizing effect, we cannot be as conclusive for MDOF softening systems, for which snapdown is a challenging problem [33]. We can also consider the case in which the ends of the beam are moved toward each other at a certain rate [34].

8.7 A Continuous Arch

We have already seen in the chapter on link models how an initially deflected structure might snap through if subjected to a lateral load. Arches resist loads primarily through compression, and hence instability is again an issue. The *rise* of the arch is an additional parameter to consider. To keep the analysis simple, we consider an arch of the type shown in Fig. 8.11(a) [35, 36].

Here we make a nondimensional study of the deflection and natural frequencies as functions of the loading. The initial (unloaded) shape of the arch $y_0(x)$ is

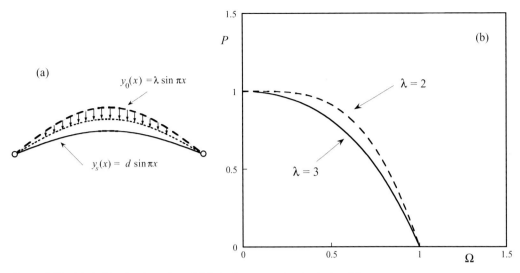

Figure 8.11. (a) A thin elastic arch and (b) the frequency-load relation [2].

sinusoidal with amplitude λ and is subject to a sinusoidal distributed lateral load $q(x)$ of amplitude p. The arch deflects to a position $y_s(x)$ that is also sinusoidal with amplitude d, where d is the largest root of the equation

$$d^3 + (1 - \lambda^2)d + p - \lambda = 0. \tag{8.75}$$

The rise of the arch λ has an important influence on the behavior. First consider the case in which $\lambda = 2$. In this case, we obtain a fundamental natural frequency

$$\omega_f = \sqrt{3(d^2 - 1)}, \tag{8.76}$$

which is associated with a vibration mode shape consisting of a half-sine wave and a critical load $p_{cr} = 4$. The relation between ω_f and p is shown in Fig. 8.11(b) as the dashed curve. The frequency ratio is defined as $\Omega = \omega_f/\omega_0$, where ω_0 is the frequency of the unloaded arch (i.e., when $p = 0, d = \lambda$), and $P = p/P_{cr}$. The actual instability mechanism here is a saddle-node bifurcation, which is not uncommon for shallow arches [37, 38].

Next we consider an arch that is less shallow with $\lambda = 3$. The frequency is now given by

$$\omega_f = 2\sqrt{(d^2 - 5)}, \tag{8.77}$$

corresponding to a full-sine-wave vibration mode shape and a critical load $p_{cr} = 9.71$. The relation between Ω and P is also shown in Fig. 8.11 as the solid curve. The loss of stability now corresponds to a bifurcation point, and this type of instability is more typical for deeper arches [39].

In earlier chapters, we saw how the relation between frequency (squared) and axial load was linear. This is not the case here (the form of this relation will be investigated further in Chapter 11), and in fact in the limit point, or saddle-node, case it is the fourth power of the frequency that drops linearly to zero as the critical point is reached. This topic will be revisited later in terms of nondestructive testing. Dynamic buckling (caused by a suddenly applied load and treated in Chapter 13) is a class of problem that has also received attention over the years [35, 39–41].

References

[1]   M. Hetenyi. *Beams on Elastic Foundation*. University of Michigan Press, 1946.

[2]   R.H. Plaut and E.R. Johnson. The effect of initial thrust and elastic foundation on the vibration frequencies of a shallow arch. *Journal of Sound and Vibration*, 78:565–71, 1981.

[3]   B.A. Boley and J.H. Weiner. *Theory of Thermal Stresses*. Wiley, 1960.

[4]   D.J. Johns. *Thermal Stress Analysis*. Pergamon, 1965.

[5]   N. Challamel, C. Lanos, and C. Casandjian. Localization in the buckling or in the vibration of a two-span weakened column. *Engineering Structures*, 28:776–82, 2006.

[6]   C.L. Amba-Rao. Effect of end conditions on the lateral frequencies of uniform straight columns. *Journal of the Acoustical Society of America*, 42:900–1, 1967.

[7]   M. Chi, B.G. Dennis, and J. Vossoughi. Transverse and torsional vibrations of an axially-loaded beam with elastically constrained ends. *Journal of Sound and Vibration*, 96:235–41, 1984.

[8]   P.J. Wicks. Compound buckling of elastically supported struts. *Journal of Engineering Mechanics*, 113:1861–69, 1987.

[9]   D. Zhou and Y.K. Cheung. The free vibration of a type of tapered beam. *Computer Methods in Applied Mechanics and Engineering*, 188:203–16, 2000.

[10] D. Zhou and Y.K. Cheung. Vibrations of tapered Timoshenko beams in terms of static Timoshenko beam functions. *Journal of Applied Mechanics*, 68:596–602, 2001.

[11] M. Eisenberger. Buckling loads for variable cross section bars in a nonuniform thermal field. *Mechanics Research Communications*, 19:259–66, 1992.

[12] N.M. Auciello. On the transverse vibrations of non-uniform beams with axial loads and elastically restrained ends. *International Journal of Mechanical Sciences*, 43:193–208, 2001.

[13] J. Jaroszewicz and L. Zoryi. Investigation of the effect of axial loads on the transverse vibrations of a vertical cantilever with variable parameters. *International Applied Mechanics*, 36:1242–51, 2000.

[14] G.W. Housner and W.O. Keightley. Vibrations of linearly tapered cantilever beams. *Journal of the Engineering Mechanics Division, ASCE*, 88:95–123, 1962.

[15] R.P. Goel. Transverse vibration of tapered beams. *Journal of Sound and Vibration*, 47:1–7, 1976.

[16] F.W. Williams and J.R. Banerjee. Flexural vibrations of axially-loaded beams with linear or parabolic taper. *Journal of Sound and Vibration*, 99:121–38, 1985.

[17] L.N. Virgin. Free vibrations of imperfect cantilever bars under self-weight loading. *Proceedings of the Institution of Mechanical Engineering*, 201:345–7, 1987.

[18] A.G. Pugsley. Limits to size set by trees. *The Structural Engineer*, 66:322–3, 1988.

[19] R.H. Plaut, K.D. Murphy, and L.N. Virgin. Curve and surface veering for a braced column. *Journal of Sound and Vibration*, 187:879–85, 1995.

[20] W.J. Supple. Coupled branching configurations in the elastic buckling of symmetric structural systems. *International Journal of Mechanical Sciences*, 9:97–112, 1967.

[21] A.H. Chilver. Coupled modes of elastic buckling. *Applied Mechanics Reviews*, 15:15–28, 1967.

[22] E.P. Popov. *Introduction to Mechanics of Solids*. Prentice Hall, 1968.

[23] S.P. Timoshenko and J.M. Gere. *Theory of Elastic Stability, 2nd ed.* McGraw-Hill, 1961.

[24] T.M. Roberts. Natural frequencies of thin-walled bars of open cross-section. *Journal of Engineering Mechanics*, 113:1584–93, 1987.

[25] R. Parnes. *Solid Mechanics in Engineering*. Wiley, 2001.

[26] H.W. Haslach and R.W. Armstrong. *Deformable Bodies and Their Material Behavior.* Wiley, 2004.

[27] D.H. Hodges and E.H. Dowell. Non-linear equations of motion for elastic bending and torsion of twisted non-uniform rotor blades. Technical Report, NASA TN D-7818, 1974.

[28] R.S. Barsoum and R.H. Gallagher. Finite element analysis of torsional and flexural–torsional stability problems. *International Journal for Numerical Methods in Engineering*, 2:335–52, 1970.

[29] W.F. Chen and E.M. Liu. *Structural Stability: Theory and Implementation*. Elsevier, 1987.

[30] F. Mohri, L. Azrav, and M. Potier-Ferry. Vibration analysis of buckled thin-walled beams with open sections. *Journal of Sound and Vibration*, 275:434–46, 2004.

[31] S.P. Machado and V.H. Cortinez. Free vibration of thin-walled composite beams with static initial stresses and deformations. *Engineering Structures*, 29:372–82, 2007.

[32] J.M.T. Thompson and G.W. Hunt. *Elastic Instability Phenomena*. Wiley, 1984.

[33] H.B. Harrison. Post-buckling behaviour of elastic circular arches, Part 2. *Proceedings of the Institution of Civil Engineers*, 65:283–98, 1978.

[34] N.J. Hoff. Buckling and stability: The forty-first Wilbur Wright memorial lecture. *Journal of the Royal Aeronautical Society*, 58:3–52, 1954.

[35] W.E. Gregory and R.H. Plaut. Dynamic stability boundaries for shallow arches. *Journal of Engineering Mechanics*, 108:1036–50, 1982.

[36] R.H. Plaut and L.N. Virgin. Use of frequency data to predict buckling. *Journal of Engineering Mechanics*, 116:2330–5, 1990.

[37] H.L. Schreyer and E.F. Masur. Buckling of shallow arches. *Journal of Engineering Mechanics, ASCE*, 92:1–19, 1966.

[38] L.W. Rehfield. Nonlinear flexural oscillations of shallow arches. *AIAA Journal*, 12:91–3, 1974.

[39] R.E. Fulton and F.W. Barton. Dynamic buckling of shallow arches. *Journal of Engineering Mechanics, ASCE*, 97:865–77, 1971.

[40] N.J. Hoff and V.G. Bruce. Dynamic analysis of the buckling of laterally loaded flat arches. *Journal of Mathematical Physics*, 32:276–88, 1954.

[41] A.M. Liapunov. *Stability of Motion (Collected Papers)*. Academic, 1966.

[42] D.L.C. Lo and E.F. Masur. Dynamic buckling of shallow arches. *Journal of Engineering Mechanics, ASCE*, 102:901–17, 1976.

9 Frames

So far, we have focused attention on the dynamic behavior of individual slender structural members. There are, of course, many practical situations in which a number of members are connected to form a truss or frame. Often, such systems are analyzed as moment frames designed to resist loads in bending. However, there are occasions in which significant axial loading occurs, and thus we augment the standard methods of structural dynamics to account for this situation. We start with revisiting the partial differential equation description of the dynamics of a slender beam with axial loading. However, we now allow for varying degrees of elastic restraint at the boundaries, because, in a typical framework, the stiffness of a joint depends (in a nonsimple way) on the effects of the members contributing to that joint [1, 2]. In Section 8.2, we developed some general expressions for beams with elastically restrained ends. Initially, we follow the standard approach in which a characteristic equation is developed from the governing partial differential equation. However, this is rarely a practical approach for a frame, and hence the chapter then proceeds to introduce the dynamic stiffness method [3]. This is a systematic, FE technique that provides a powerful, matrix-based approach to solving problems in structural dynamics. We focus on frames consisting of prismatic beam members, and, because of the way in which most frames are designed to resist lateral as well as axial loads, any postbuckling, as such, will be encountered as the growth of large deflections during loading [4].

9.1 A Beam with General Boundary Conditions

In Section 7.2, the governing equation for harmonic lateral vibrations of slender axially loaded beams was introduced and then extended the analysis to include the effect of elastic end restraint in Section 8.2. Assuming harmonic motion, we return to these general expressions here:

$$EI\frac{d^4W(x)}{dx^4} + P\frac{d^2W(x)}{dx^2} - m\omega^2 W(x) = 0. \tag{9.1}$$

Again, the mode shapes are given by

$$W(\bar{x}) = c_1 \sinh M\bar{x} + c_2 \cosh M\bar{x} + c_3 \sin N\bar{x} + c_4 \cos N\bar{x}, \tag{9.2}$$

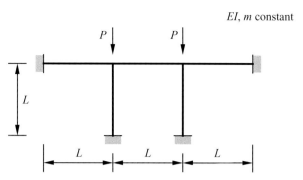

Figure 9.1. A symmetrically loaded plane frame. Prismatic members with the same flexural rigidities [7].

with $\bar{x} = x/L$, and where M and N are given by

$$M = \sqrt{-\Lambda + \sqrt{\Lambda^2 + \Omega^2}},$$

$$N = \sqrt{\Lambda + \sqrt{\Lambda^2 + \Omega^2}}. \tag{9.3}$$

The axial-load and natural-frequency parameters can be conveniently nondimensionalized by use of

$$\Lambda = PL^2/2EI, \quad \Omega^2 = m\omega^2 L^4/EI, \tag{9.4}$$

that is, the same as in Eqs. (7.30) and slightly different from Eqs. (8.10). Now, instead of applying one of the simple boundary conditions, for example, clamped ($W = dW/dx = 0$), or relating the end displacement to an elastic torsional spring constant $[d^2W/dx^2 - (k_1 L)/(EI)dW/dx = 0]$, we have displacement and rotations at the ends that depend on the applied shear forces and moments according to standard beam theory (see Section 7.2) [5]. In the absence of axial loads and inertia effects, these conditions lead to the familiar slope-deflection equations [6]. However, now the coefficients in these relations depend on involved combinations of trigonometric and hyperbolic functions of M and N. Certain conditions of continuity are then used to ensure, for example, that the moments at a common joint balance. Rather than write out the general expressions, we illustrate the process with an example.

Consider the frame shown in Fig. 9.1 [7]. This is a plane frame with built-in end supports, and thus has only two nodal DOFs: the rotations at the two joints where the loads are applied. We assume that EI and m are constant throughout. Solving the eigenvalue associated with the governing equations of motion gave the results shown in Fig. 9.2. The lowest natural frequency in both symmetric and asymmetric modes is plotted as a function of the axial loading. In typical fashion, we establish the end points of these interaction curves. For the lowest symmetric mode of vibration without axial loads, we obtain a natural frequency of $\Omega \approx 13.5$. As the effective natural frequency tends to zero, we obtain a critical load coefficient of $2\Lambda_{cr} \approx 30.6$. This latter value is not unreasonable because, when viewing one of the columns, we might

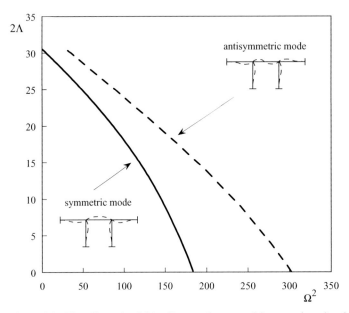

Figure 9.2. The effect of axial loading on the natural frequencies of a plane frame.

expect the upper boundary condition to be roughly intermediate between fully fixed (in which case the coefficient of $2\Lambda_{cr} \to 4\pi^2$) and pinned ($\approx 2\pi^2$). Higher modes can also be obtained from the other roots of the characteristic equation. Later, we will confirm this result by using finite-element analysis (FEA) and study the effect of making one or two minor changes.

9.2 The Stiffness Method

Although the preceding section furnished an exact solution (at least within the confines of linear beam theory), it was somewhat cumbersome. In most practical structures, we will need a more systematic approach, better suited to computational methods, and this is where the FEM has proven to be so powerful [8, 9]. We focus here on those aspects of the stiffness method that address issues relating to the dynamics and stability of thin prismatic beams. The standard stiffness method (which neglects any coupling between axial and bending effects) is covered widely in the literature [10]. In similarity with the next chapter on plates, only a relatively brief outline of the basic theory is given here. For more in-depth descriptions of FE procedures, see Zienkiewicz and Taylor [9], Cook [11], Bathe [12].

The FEM is a systematic numerical technique in which a continuum is divided into small elements, the properties of which can be couched in terms of behavior (e.g., deflections and slopes) at the nodes. Columns and beams make up the "elements" of the system, with joints providing a natural selection for "nodes," although in practice a single structural element tends to be subdivided into many subelements. Here we focus on a regular thin-beam element in which the bending stiffness is influenced by the presence of an axial load and how this affects the dynamics of a plane frame of which it is a part [13].

We still suppose that the response of a beam can be expressed as

$$w(x, t) = W(x)Y(t). \tag{9.5}$$

Then we still have the potential energy of flexural deformations given by

$$V = \frac{1}{2}EI \int_0^L [W''(x)]^2 dx. \tag{9.6}$$

Furthermore, we also assume that the transverse deflections can be described by a cubic polynomial,

$$W(x) = a_1 w_1 + a_2 \theta_1 + a_3 w_2 + a_4 \theta_2, \tag{9.7}$$

where

$$a_1 = 1 - 3(x/L)^2 + 2(x/L)^3,$$
$$a_2 = x[1 - 2(x/L) + 2(x/L)^2], \tag{9.8}$$
$$a_3 = (x/L)^2[3 - 2(x/L)],$$
$$a_4 = x^2/L[-1 + (x/L)],$$

and subscripts 1 and 2 (on the nodal deflections) in Eq. (9.7) refer to the left $x = 0$ and right $x = L$ ends of the beam, respectively. That is, we assume that the nodal deflections (w, θ) and forces are related by a displacement, or shape, function (which satisfies the boundary conditions).

This expression can then be used to obtain the strains and the principle of virtual work used to evaluate the force corresponding to coordinate i that is due to a unit displacement of coordinate j, given by

$$k_{ij} = EI \int_0^L W_i''(x) W_j''(x) dx, \tag{9.9}$$

and hence the element stiffness matrix,

$$[K] = \frac{EI}{L^3} \begin{bmatrix} 12 & 6L & -12 & 6L \\ 6L & 4L^2 & -6L & 2L^2 \\ -12 & -6L & 12 & -6L \\ 6L & 2L^2 & -6L & 4L^2 \end{bmatrix}. \tag{9.10}$$

This matrix can also be derived from the traditional slope-deflection equations. It is symmetric and positive-definite.

Similarly, we recall that the strain energy that is due to (constant) axial loading on a beam of length L is given by

$$V_p = -\frac{P}{2} \int_0^L [W'(x)]^2 dx, \tag{9.11}$$

and, following the same arguments and using the same shape functions, we obtain the consistent geometric influence coefficients

$$k_{Gij} = P \int_0^L W_i'(x)W_j'(x)dx, \qquad (9.12)$$

and hence the geometric stiffness matrix [14]:

$$[K_G] = \frac{P}{30L} \begin{bmatrix} 36 & 3L & -36 & 3L \\ 3L & 4L^2 & -3L & -L^2 \\ -36 & -3L & 36 & -3L \\ 3L & -L^2 & -3L & 4L^2 \end{bmatrix}. \qquad (9.13)$$

A consistent mass-matrix formulation, by use of the same deflected shape function, Eq. (9.7), leads to the mass matrix

$$[M] = \frac{\rho AL}{420} \begin{bmatrix} 156 & 22L & 54 & -13L \\ 22L & 4L^2 & 13L & -3L^2 \\ 54 & 13L & 156 & -22L \\ -13L & -3L^2 & -22L & 4L^2 \end{bmatrix}. \qquad (9.14)$$

Thus, assuming harmonic motion in the usual way, we can arrive at the eigenvalue problem for a prismatic beam:

$$([K] + \lambda[K_G] - \omega_i^2[M])\, W_i = 0 \qquad (9.15)$$

in which λ is the load factor, that is, the factor by which P must be increased for buckling to occur. (Compression loads are negative in this formulation.)

We note at this point that it is also possible to use simplified displacement functions to derive the geometric stiffness and mass matrices (e.g., crudely lumping the distributed mass at the nodes in some way) [15, 16]. Damping can also be introduced, and again the proportional damping assumption, that is, $[C] = a[M] + b[K]$, provides compelling ease of analysis (damping is typically small in structural applications).

Some Simple Beam Examples. Consider a cantilever beam, clamped at one end and free at the other and subject to an axial load. For the two allowable DOFs at the free end, we can immediately write the eigenvalue problem:

$$\left| \frac{EI}{L^3} \begin{bmatrix} 12 & -6L \\ -6L & 4L^2 \end{bmatrix} + \frac{P}{30L} \begin{bmatrix} 36 & -3L \\ -3L & 4L^2 \end{bmatrix} - \frac{\omega^2 \rho AL}{420} \begin{bmatrix} 156 & -22L \\ -22L & 4L^2 \end{bmatrix} \right| = 0. \qquad (9.16)$$

The roots of this 2×2 determinant give the natural frequencies at different axial load levels P. In the usual way, we have natural frequencies in the absence of axial load given by

$$\omega_1 = 3.533 \left(\frac{EI}{\rho AL^4} \right)^{1/2}, \qquad \omega_2 = 34.81 \left(\frac{EI}{\rho AL^4} \right)^{1/2}, \qquad (9.17)$$

for which the corresponding exact coefficients are 3.516 and 22.03, respectively. As $\omega_1 \to 0$ we obtain the lowest critical load of $P_{cr} = -2.486EI/L^2$ (the exact coefficient is $-\pi^2/4$, or -2.4674). The relation between ω_1^2 and P is linear.

Suppose we have a prismatic beam clamped at both ends and again subject to a constant axial force of magnitude P. We can model this system conveniently by using two elements (of length $l = L/2$). In this case, there is just a SDOF (a lateral deflection at the center). Thus formulating the general eigenvalue problem leads to

$$\left[\frac{24EI}{(l/2)^3} + P\frac{72}{30(l/2)} - \omega^2 \frac{312\rho A(l/2)}{420} \right] W_i = 0. \tag{9.18}$$

We immediately have the expression

$$\omega^2 = \frac{420}{312\rho AL} \left[\frac{192EI}{L^3} + \frac{144P}{30L} \right]. \tag{9.19}$$

Thus, in the absence of an axial load, we obtain the natural-frequency coefficient $\omega_1^2 = 22.7354$ and a critical load of $P_{cr} = -40$. Both of these are relatively close to the exact answers (22.3729 and $-4\pi^2$, or -39.4784). Dividing the beam into more elements improves the accuracy and furnishes higher frequencies and critical loads. Other boundary conditions can be handled very simply, with the allowable nodal deflections determining the size of the system of simultaneous equations to be solved. Because the stiffness method is based on Rayleigh–Ritz, the results generated are upper bounds on the magnitudes of the frequencies and buckling loads.

The preceding formulation does, of course, subsume standard linear elastic analysis. Suppose the clamped beam is subject to a lateral point load F acting at the center (and without the axial load). In this case, the stiffness contribution from each element (clearly we need a minimum of two elements with these boundary conditions) at the center again adds to $192EI/L^3$, which thus results in a lateral deflection of $\delta = FL^3/192EI$ (a familiar result in structural analysis) [6]. Uniformly distributed loads can be handled by the derivation of equivalent nodal forces, and the stiffness method is now widely used in the analysis of frames, including three-dimensional (3D) trusses and frameworks [10].

Using this approach, the FE packages ANSYS and ABAQUS were able to quickly reproduce Figures 7.3, 8.3, and 8.5 (in which the latter was modeled with a large number of slightly different constant section elements). This approach becomes particularly useful when the geometry of the structure becomes more complex. When individual members connect at various angles then an important step in the analysis concerns transformation between local and global coordinate systems. For example, let's reconsider the plane frame shown in Fig. 9.1. With an appropriate nondimensionalized values, a standard ANSYS analysis resulted in the following table:

Load (P)	Ω_1^2	Ω_2^2
0	183.7303	302.3893
6	162.7507	264.0078
12	135.3487	215.7064
18	100.0943	159.9573
24	56.4045	99.1286
30	4.9596	34.7882

An independent ANSYS buckling analysis gave the critical elastic load as $P_{cr} = 30.539 (\equiv 2\Lambda)$ (which compares with the analytical result of 30.6). One of the advantages of FEA is the ease with which parameters can be changed and cases rerun. For example, suppose the side support points were pinned rather than fixed. In this case, it is easy to show that the buckling load (in sway) is reduced to $P_{cr} = 8.04$; that is, we have a more flexible frame with less load-carrying capacity and lower natural frequencies. This is also the case, for example, when the second moment of area of the central beam is reduced. Furthermore, Timoshenko beam elements (incorporating shear effects) can easily be used – these have the effect of slightly increasing the natural frequencies, although this is not necessarily the case when axial loads approach the buckling levels.

The effect of axial loading on more practical multistory frames is considered in Ovunc [17], in which the dynamic response for various geometries and loading conditions is computed, including implications for earthquake-resistant structures.

9.3 A Self-Strained Frame Example

The previous section, focusing on simple systems, was concerned primarily with the reduction of the lowest natural frequency and specifically its approach to zero at buckling. For a more realistic frame or structure (in which the end conditions, or joint stiffnesses, depend on the members meeting at that point), the various natural frequencies will all tend to change with applied loading [18]. Some may increase (if the effect of the loading includes a tendency to cause tension in certain members), decrease, or even cross each other as a system parameter is varied. We saw elements of this behavior in the previous chapters, but here we focus on a (statically indeterminate or redundant) plane truss, or frame, and extract information concerning the lowest few natural frequencies (eigenvalues) analytically.

Analytic Solution. Consider the framework shown in Fig. 9.3. This is a doubly symmetric system with members labeled as 1, 2, and 3 as shown, and the boundary conditions are such that the frame is freely suspended. Because of the application of an

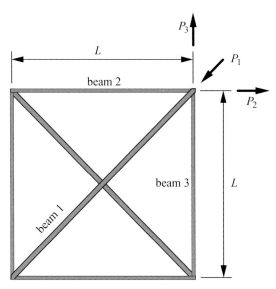

Figure 9.3. A simple truss structure in which loading is applied at a corner.

axial load of magnitude P_1 at the top-right-hand corner, then axial loads $P_2 = P_3 = -P_1 \cos\theta_2 = -P_1(L_2/L_1)$ are also induced. The following nondimensional parameters are introduced: $\bar{x} = x/L_1$, $p = PL_1^2/EI_1$, and $\Omega^2 = (m_1 L_1^4/EI_1)\omega^2$. The frame is square, and it is assumed that all the joints are pinned (to adjacent members rather than to external supports), that the members have the same mass ratio, and that the side members are twice as stiff as the diagonal members. In this case, the frequencies (for the *symmetric* modes only) were computed in Mead [19], and the results are shown in Fig. 9.4. From this figure, we see that the natural frequencies are all positive over the range of load from $p = \pi^2$ to $p = -\sqrt{32}\pi^2 = -55.48$. The positive limit corresponds to the Euler buckling load of the diagonal element, and the negative limit corresponds to the Euler buckling load for a side member. Thus it is interesting to observe that, as the load changes, the lowest frequency may drop but

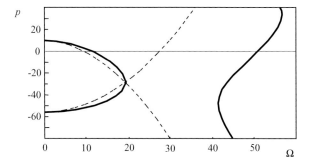

Figure 9.4. Natural frequencies plotted as a function of applied loading for pinned connections. Adapted from Mead [19].

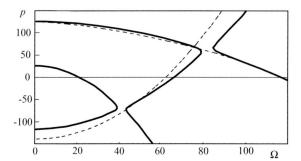

Figure 9.5. Natural frequencies plotted as functions of applied loading for fixed connections. Adapted from Mead [19].

the second lowest natural frequency may increase. Superimposed on this plot are the frequencies of independent pinned beams with the same properties as the frame elements, and indicated by the thin lines.

Now, suppose the corners of the frame are fixed rather than pinned. In this case, the results shown in Fig. 9.5 are obtained. Certain qualitative similarities are apparent between Figs. 9.4 and 9.5, although the critical loads are approximately 2.2 times larger for the latter. The frequency curves pass by each other more closely for the fixed connections. Again some independent beam frequencies are superimposed, although here, the beam boundary conditions were chosen (roughly) to fall midway between pinned and fixed. Mead [19] also plots the mode shapes of the frame at various points along these frequency curves, and some interesting exchanges occur in the relative dominance of the diagonal and side members.

9.4 Modal Analysis

In Chapter 4, we saw how MDOF systems could often be analyzed by standard methods of linear algebra. In a practical environment, modal analysis has become the mainstay of structural testing: A component or structure is excited (by impact, sinusoidal sweep, or broadband forcing, for example), and subsequent dynamic response is interrogated in terms of modal properties. This is achieved with increasingly sophisticated hardware and software capabilities. Models can then be updated with respect to FE modeling [20]. However, a common issue in this process is the extent to which the component is isolated from its surroundings, so that data are not influenced by coupling to the inevitable support mechanism. Isolating the component effectively leads to free–free conditions, and thus axial loads tend to be neglected in this approach. The following section shows the influence of pre-stress on modal testing of a simple frame structure, a structure very similar in form to the one detailed in the previous section [21, 22].

The Simple Truss Again. Consider the pin-jointed truss shown in Fig. 9.6 in which six structural members (pinned at the four corners) make up the frame. This system

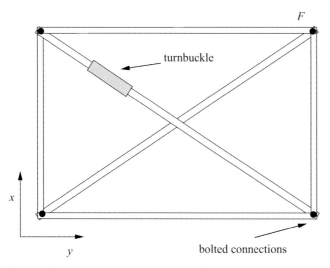

Figure 9.6. A simple truss structure in which prestress is introduced by a turnbuckle.

was intensively studied by Lieven and Greening [21, 22]. Axial loading is introduced into the structure by means of a turnbuckle that alters the length of one of the diagonal members. This system is essentially the same as the one considered in the previous section, but it is now rectangular (rather than square) and is scrutinized by use of experimental modal analysis. The overall dimensions of the truss are 500 mm × 300 mm with each member having a rectangular cross section 15 mm × 6 mm. The frame was excited at the top-right-hand corner (a broadband random input), and the dynamic response was then measured at (24) points throughout the frame. Data were acquired at four different levels of axial load induced by the turnbuckle.

Figure 9.7 shows a frequency-response spectrum taken from the various measuring locations on the frame. The loading condition corresponds to an induced load of 1135 N in the diagonal member. The natural frequencies are identified as the peaks in the spectra. The following table shows the extracted resonant peaks for the nominally unloaded and loaded (1618-N) frames, including the absolute and percentage change over this range. That is, the frequencies have shifted, some up, some down.

Mode	Unloaded (Hz)	Loaded (Hz)	Change (Hz)	Change (%)
1	45.6	36.9	−8.7	−19.1
2	69.0	94.7	25.7	37.2
3	92.2	96.1	3.9	4.3
4	125.3	111.1	−14.2	−11.4
5	129.9	149.9	20	15.4

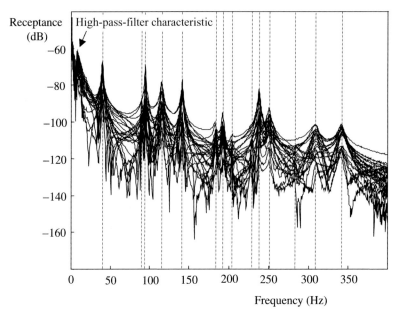

Figure 9.7. Superimposed frequency-response spectra for the frame (point receptance) for a typical loaded case. Reproduced with permission from [21].

These data are also reproduced graphically in Fig. 9.8, including the specific intermediate case relating to Fig. 9.7.

In this study [21], three nominally similar frames were tested to establish the influence of manufacturing tolerance (lack of fit) and hence frame-to-frame variability. It was shown that this effect was small in comparison with the load applied by the turnbuckle. This type of axial loading might be induced by temperature

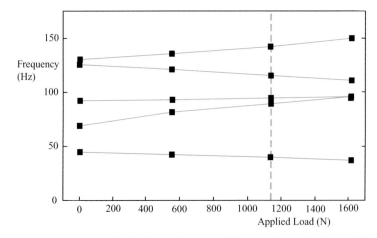

Figure 9.8. Changes in the lowest five natural frequencies as the frame is loaded. After Greening and Lieven [22].

loading, for example. This section shows that care must be taken when modal analysis is conducted on a structure. For example, as the axial load increases and the natural frequencies shift, it may become difficult to identify modes (e.g., modes 2 and 3 appear to coalesce at a certain load condition), especially for higher modes. Only the lowest five modes were included here because a relatively high level of confidence was drawn from their high modal assurance criterion (MAC). This is a technique often used in modal analysis to identify modes and is based on the inner product of the eigenvectors. In this system, the second mode was sometimes difficult to identify in the unloaded frame [21]. The issue of modal analysis will be revisited later in the chapters on nondestructive evaluation and also harmonically forced systems.

9.5 Large-Deflection Analysis

In practical situations, we obtain the dynamic response of axially loaded members as the outcome of a nonlinear analysis, i.e., deflections continuously grow as the frame is loaded and the local dynamic response is computed along the equilibrium path. This is the required approach if solutions relating to relatively large deflections are sought. Hence, buckling is not a distinct event, although it turns out that numerical difficulties are often encountered when the stiffness (matrix) approaches zero (e.g., with a horizontal tangency in the load-deflection curve). This section will again utilize the FEM and will typically follow deflections somewhat greater than the initial postbuckling analyses outlined in Chapter 7. A large-deflection approach based on the classical elastica will be described in a later chapter.

Consider an elastic cantilever consisting of a 10-m-long prismatic member, with a square cross section (2 cm × 2 cm), $E = 209 \times 10^9$ Pa, and $\rho = 7800$ kg/m^3. This corresponds to a very slender geometry based on a slenderness ratio $l/r = 1732$ in which r is the radius of gyration $(= \sqrt{I/a})$. Figure 9.9(a) shows results based on ABAQUS analysis with simple 2D Euler–Bernoulli beam elements and using a Riks analysis, which is a path following technique related to continuation [23–25]. In analyses the beam is also subject to a small lateral load H (to simulate an initial geometric imperfection). As anticipated by theory, there is a (near) linear decay in stiffness (and hence natural frequency squared), with buckling indicated by a complete loss of stiffness. However, because of the postbuckled stiffness of this type of structure, we observe an increase in the lowest natural frequency after the critical load [normalized to occur at unity in part (b)] has been passed. Numerical difficulties were encountered when the axial load reached approximately three times the value of the underlying elastic critical load [which for the input values used gives an elastic critical load of $(\pi^2 EI/(4L^2) = 68.75$ kN] and can be clearly observed as $H \to 0$. Of course, in a practical situation, these levels of deflection might very well violate the assumption of elastic material behavior. At the upper limit of the equilibrium curve the beam is actually deforming much like an elastica—a situation that will be revisited in a later chapter.

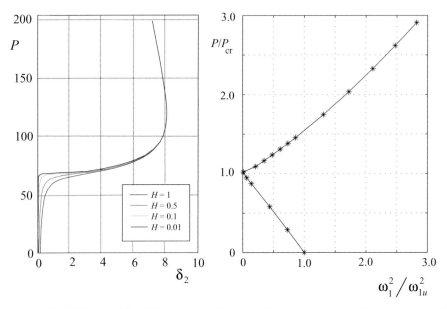

Figure 9.9. (a) The axial load (P) versus tip deflection (δ_2) curve for a cantilever beam and (b) the corresponding nondimensional load-frequency (squared) curve [26].

ABAQUS was also used to solve the follower force problem first encountered in Section 7.6. In this case, it was necessary to use the dynamic simulation capabilities of the software package in order to follow the correct solution. A typical result is shown in Fig. 9.10. As mentioned in Chapter 7, a dynamic analysis must be used to show that buckling occurs for a force of $P_{cr} = 20.05EI/L^2$. With the parameters used as input, this translates to a critical load of about 27 kN. We also see that the instability is of the (oscillatory) flutter type rather than the more familiar (monotonic) divergence type. In some of the computational analyses conducted here, a considerable amount of spatial resolution was needed in order to obtain accurate results, in terms of both static deflections and natural frequencies.

9.6 A Tubular Structure

This chapter has been primarily concerned with plane (i.e., 2D) frames. However, it is easy to see how these analyses can be extended to a third spatial dimension. In this section, we consider a specific form of 3D structure that consists of a hollow, skeletal tube, and for large length-to-diameter ratios approaches the behavior of a beam [26]. An example of such a configuration, sometimes called an *isogrid*, is shown in Fig. 9.11. This system consists of a grid of small composite members arranged in a regular geometry: a combination of circumferential, longitudinal, and diagonal struts. A typical geometry consists of 16 circumferential bays and a helix (which takes approximately two diameters to complete a revolution).

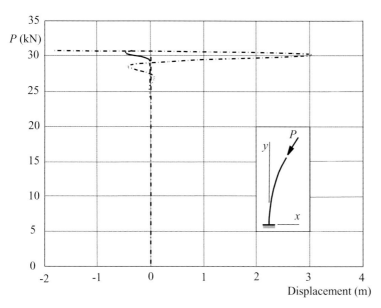

Figure 9.10. The (numerical simulation) response of a cantilever subject to a follower force at its tip [26].

This configuration was then used within ABAQUS to study buckling and dynamic response in terms of natural frequencies and mode shapes by use of an approach similar to that of the previous section but with the simple cantilever replaced with this more complicated structure. For relatively slender structures, it is apparent that the isogrid behaves essentially like a beam, and thus (in terms of rapid reanalysis) should submit to relatively easy parametric studies, at least in this simple configuration. A typical (buckling and vibration) mode shape in bending is shown in Fig. 9.11(b). We also anticipate the issue of a more subtle interpretation of the

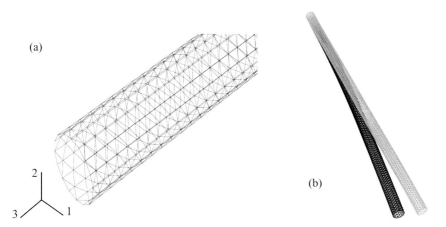

Figure 9.11. (a) A typical isogrid geometry and (b) the first vibration (bending) mode [26].

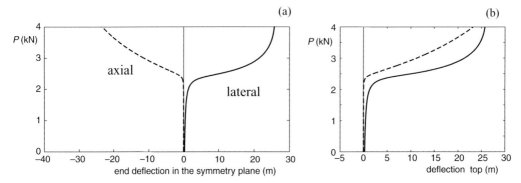

Figure 9.12. The results of a large-deflection static analysis: (a) the complete isogrid and (b) the equivalent beam.

slenderness ratio, because in this shell-like lattice it is likely that local and global behavior may occur.

Consider an isogrid with a length $L = 32$ m, a diameter of $D = 17.78$ cm, and the individual bar's diameter of 5 mm. Figure 9.12(a) shows the results of a large-deflection (Riks) analysis, with the three curves (one lines up with the y axis) indicating the two lateral and one axial direction. In this case the small lateral load was set as 1% of the critical buckling load. The axial and lateral load were concentrated at the free end of the isogrid, with load vectors applied equally at each of the 16 end nodes. The analysis continued until a highly deformed state of the beam is achieved (when the axial load reached about twice the buckling load; this was when ABAQUS started to experience numerical convergence problems). Part (b) shows the results of an equivalent beam analysis in which the cross-sectional properties were effectively averaged and used in a simple (lumped-parameter) analysis.

A complete dynamic (free-vibration) analysis is conducted in which various natural frequencies and their corresponding mode shapes have been computed for various lengths. Figure 9.13 shows the characteristic curve for the lowest natural frequency for this slender isogrid structure based on a Riks analysis. The natural modes of vibration are often similar in form to the buckling modes but also include some interesting breathing and circumferential modes for very high frequencies (but not of much practical interest). With the parameters used in this particular simulation, the critical load occurs at $P_{cr} = 2.47$ kN. We also observe that immediately after buckling the symmetry of the response is broken, and thus two different natural frequencies are apparent in the two orthogonal directions (a small lateral load is again used to promote the onset of deformation). Frequencies about the equivalent beam equilibrium configurations are also shown superimposed in Fig. 9.13.

In Argyris and Mlejnek [27] a number of hanging networks are analyzed by use of FEs. The effect of weight is included, and various time-marching strategies are examined in terms of transient response. The FEM is especially powerful for continua, and the next chapter will make use of this approach for considering the dynamics of axially loaded plates.

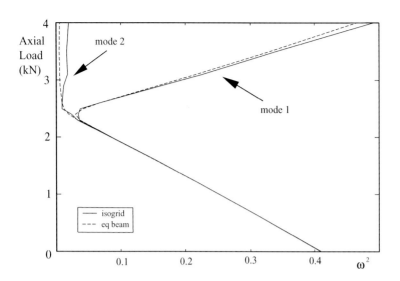

Figure 9.13. The frequency–axial-load relation for an isogrid and its equivalent beam model [26].

References

[1] R.K. Livesley and D.B. Chandler. *Stability Functions for Structural Frameworks*. Manchester University Press, 1956.

[2] M.R. Horne and W. Merchant. *The Stability of Frames*. Pergamon, 1965.

[3] M.D. Capron and F.W. Williams. Exact dynamic stiffnesses for an axially loaded uniform Timoshenko member embedded in an elastic medium. *Journal of Sound and Vibration*, 124:453–66, 1988.

[4] J. Roorda and A.M. Chilver. Frame buckling: An illustration of the perturbation technique. *International Journal of Non-linear Mechanics*, 5:235–46, 1970.

[5] S.P. Timoshenko and D.H. Young. *Elements of Strength of Materials*. Van Nostrand Reinhold, 1977.

[6] R.C. Hibbeler. *Structural Analysis,* 5th ed. Prentice Hall, 2002.

[7] W. Nowacki. *Dynamics of Elastic Systems*. Chapman & Hall, 1963.

[8] R.H. Gallagher. *Finite Element Analysis: Fundamentals*. Prentice Hall, 1975.

[9] O.C. Zienkiewicz and R.L. Taylor. *The Finite Element Method*. McGraw-Hill, 1989.

[10] A. Kassimali. *Matrix Analysis of Structures*. Brooks/Cole, 1999.

[11] R.D. Cook. *Concepts and Applications of Finite Element Analysis*. Wiley, 1981.

[12] K.J. Bathe. *Finite Element Procedures*. Prentice Hall, 1995.

[13] B.A. Coulter and R.E. Miller. Vibration and buckling of beam columns subjected to non-uniform axial loads. *International Journal of Numerical Methods*, 23:1739–55, 1986.

[14] B.J. Hartz. Matrix formulation of structural stability problems. *Journal of the Structural Division, ASCE*, 91:141–57, 1965.

[15] R.W. Clough and J. Penzien. *Dynamics of Structures*. McGraw-Hill, 1982.

[16] J.F. Doyle. *Nonlinear Analysis of Thin-Walled Structures*. Springer, 2001.

[17] B.A. Ovunc. Effect of axial force on frameworks dynamics. *Computers and Structures*, 11:389–95, 1980.

[18] J. Przybylski, L. Tomski, and M. Golebiowska-Rozanow. Free vibration of an axially loaded prestressed planar frame. *Journal of Sound and Vibration*, 189:609–24, 1996.

[19] D.J. Mead. Free vibrations of self-strained assemblies of beams. *Journal of Sound and Vibration*, 249:101–27, 2002.

[20] M.I. Friswell and J.E. Mottershead. *Finite-Element Model Updating in Structural Dynamics*. Kluwer Academic, 1995.

[21] N.A.J. Lieven and P.D. Greening. Effect of experimental pre-stress and residual stress on modal behavior. *Philosophical Transactions of the Royal Society of London*, 359:97–111, 2001.

[22] P.D. Greening and N.A.J. Lieven. Identification and updating of loading in frameworks using dynamic measurements. *Journal of Sound and Vibration*, 260:101–15, 2003.

[23] E. Riks. The application of Newton's method to the problem of elastic stability. *Journal of Applied Mechanics*, 39:1060–6, 1972.

[24] M.A. Crisfield. A fast incremental/iterative solution procedure that handles snap-through. *Computers and Structures*, 13:55–62, 1981.

[25] R. Seydel. *Practical Bifurcation and Stability Analysis*. Springer, 1994.

[26] I. Stanciulescu, L.N. Virgin, and T.A. Laursen. Finite element analysis of slender solar sail booms. *Journal of Spacecraft and Rockets*, 44:528–37, 2007.

[27] G.H. Argyris and H.-P. Mlejnek. *Dynamics of Structures*. North-Holland, 1991.

10 Plates

10.1 Introduction

Often times, plates and panels are subject to in-plane loads. Their dynamic characteristics are influenced in ways not dissimilar to those of axially loaded beams. However, the modeling of these 2D systems is more challenging, involving for example more boundary conditions. Typical plates also exhibit considerable postbuckled stiffness, even to the extent that buckled plates can fulfill useful design purposes, that is, the elastic critical load and ultimate failure load are quite different.

This chapter introduces some basic concepts from the theory of thin, rectangular plates. The bending of plates has received considerable attention over the years and is well established in the literature [1, 2]. In its simplest context, we might consider a long plate supported on only two opposite sides as analogous to a wide beam. More sophisticated analyses would then incorporate large deflections of relatively thick plates including various shapes and higher-order effects [3–5]. It is assumed that the reader is somewhat familiar with simple plate bending theory, both in terms of the governing differential equations and energy considerations. Thus this chapter will focus on the interaction of in-plane forces, large deflections, and dynamic response [6–11].

10.1.1 Brief Review of the Classical Theory

Consider a flat rectangular plate as shown in Fig. 10.1. The plate has a uniform thickness h, and coordinates x and y describe the middle surface of the plate. The z coordinate is directed vertically upward from this plane. It is assumed that the plate material is elastic and isotropic and that the plate is relatively thin (such that transverse shear strain is negligible) [2]. The normal strain (ϵ_z) is negligible, and the plate is in a state of plane stress, with forces denoted by Q for shear, N for in-plane stress, and with bending moments M.

Although a fundamental linear theory of elastic plates can be developed analogous to (linear) Euler–Bernoulli beam theory, we focus on those specific effects relevant to studying the dynamics of axially loaded plates. For this reason, we develop von Karman's plate theory, which takes into account membrane effects. Once these equations have been developed, then some simple examples are shown before additional effects (e.g., initial imperfections and large deflections are included).

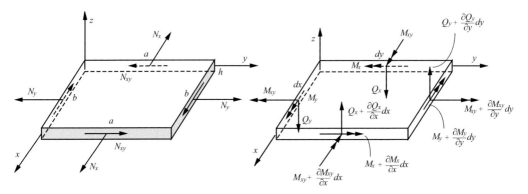

Figure 10.1. Schematic of the compressed rectangular plate.

We start by developing kinematic relations between displacement and strain, then use constitutive laws to relate stress and strain, and then go on to develop the equations of motion by using the standard dynamic equilibrium approach as well as by using energy expressions and Hamilton's principle.

From standard assumptions [12], we write the deflection of the plate in terms of

$$\bar{u} = u - z\frac{\partial w}{\partial x}, \qquad \bar{v} = v - z\frac{\partial w}{\partial y}, \qquad \bar{w} = w, \tag{10.1}$$

where (u, v, w) are measured in the (x, y, z) directions, respectively, and the overbar refers to locations on the midsurface. We next use the strain-displacement relations

$$\epsilon_x = \frac{\partial u}{\partial x} + \frac{1}{2}\left(\frac{\partial w}{\partial x}\right)^2,$$

$$\epsilon_y = \frac{\partial v}{\partial y} + \frac{1}{2}\left(\frac{\partial w}{\partial y}\right)^2,$$

$$\gamma_{xy} = \frac{\partial v}{\partial x} + \frac{\partial u}{\partial y} + \frac{\partial w}{\partial x}\frac{\partial w}{\partial y}, \tag{10.2}$$

in which the last term in each expression is the retained nonlinearity. We thus have

$$\epsilon_x = \bar{\epsilon}_x - z\frac{\partial^2 w}{\partial x^2},$$

$$\epsilon_y = \bar{\epsilon}_y - z\frac{\partial^2 w}{\partial y^2},$$

$$\gamma_{xy} = \bar{\gamma}_{xy} - 2z\frac{\partial^2 w}{\partial x \partial y}, \tag{10.3}$$

where

$$\bar{\epsilon}_x = \frac{\partial \bar{u}}{\partial x} + \frac{1}{2}\left(\frac{\partial w}{\partial x}\right)^2,$$

$$\bar{\epsilon}_y = \frac{\partial \bar{v}}{\partial y} + \frac{1}{2}\left(\frac{\partial w}{\partial y}\right)^2,$$

$$\bar{\gamma}_{xy} = \frac{\partial \bar{u}}{\partial y}\frac{\partial \bar{v}}{\partial x} + \frac{\partial w}{\partial x}\frac{\partial w}{\partial y} \tag{10.4}$$

are the midsurface strains.

For isotropic materials we have

$$\epsilon_x = \frac{1}{E}(\sigma_x - \nu\sigma_y),$$

$$\epsilon_y = \frac{1}{E}(\sigma_y - \nu\sigma_x),$$

$$\gamma_{xy} = \frac{2(1+\nu)}{E}\tau_{xy}, \tag{10.5}$$

where ν is Poisson's ratio, and in terms of the stresses

$$\sigma_x = \frac{E}{1-\nu^2}(\epsilon_x + \nu\epsilon_y),$$

$$\sigma_y = \frac{E}{1-\nu^2}(\epsilon_y + \nu\epsilon_x),$$

$$\tau_{xy} = G\gamma_{xy}, \tag{10.6}$$

in which G is the shear modulus, $G = E/[2(1+\nu)]$.

At this point, we could go on and use these expressions in the strain energy and obtain the governing equations of motion. However, before doing so, we use the alternative approach of equilibrium. Referring to Fig. 10.1 we consider a small element $(dx \times dy)$ of the (constant-thickness) plate, summing forces (shears Q, in-plane forces N) and moments (M), setting them to zero.

Summing forces in the vertical (z) direction we obtain

$$\frac{\partial Q_x}{\partial x} + \frac{\partial Q_y}{\partial y} = \mu\frac{\partial^2 w}{\partial t^2}, \tag{10.7}$$

in which μ is the mass density per unit area. In-plane equilibrium (ignoring in-plane inertia) in the x and y directions gives

$$\frac{\partial N_x}{\partial x} + \frac{\partial N_{xy}}{\partial y} = 0, \tag{10.8}$$

$$\frac{\partial N_{xy}}{\partial x} + \frac{\partial N_y}{\partial y} = 0. \tag{10.9}$$

Taking moments about the x and y axes results in

$$\frac{\partial M_x}{\partial x} + \frac{\partial M_{xy}}{\partial y} = Q_x - N_x\frac{\partial w}{\partial x} - N_{xy}\frac{\partial w}{\partial y}, \tag{10.10}$$

$$\frac{\partial M_{xy}}{\partial x} + \frac{\partial M_y}{\partial y} = Q_y - N_{xy}\frac{\partial w}{\partial x} - N_y\frac{\partial w}{\partial y}. \tag{10.11}$$

Equations (10.7)–(10.11) can be combined to give

$$\frac{\partial^2 M_x}{\partial x^2} + 2\frac{\partial^2 M_{xy}}{\partial x \partial y} + \frac{\partial^2 M_y}{\partial y^2} + \mu\frac{\partial^2 w}{\partial t^2}$$

$$= -\left(N_x\frac{\partial^2 w}{\partial x^2} + 2N_{xy}\frac{\partial^2 w}{\partial x \partial y} + N_y\frac{\partial^2 w}{\partial y^2}\right). \tag{10.12}$$

We obtain the in-plane forces by integrating the stresses over the plate thickness:

$$N_x = \int_{-h/2}^{h/2} \sigma_x dz, \quad N_y = \int_{-h/2}^{h/2} \sigma_y dz, \quad N_{xy} = \int_{-h/2}^{h/2} \tau_{xy} dz, \tag{10.13}$$

and by using Eqs. (10.5), we evaluate them as

$$N_x = \frac{Eh}{(1-\nu^2)}\left(\frac{\partial u}{\partial x} + \nu\frac{\partial v}{\partial y}\right),$$

$$N_y = \frac{Eh}{(1-\nu^2)}\left(\frac{\partial v}{\partial y} + \nu\frac{\partial u}{\partial x}\right),$$

$$N_{xy} = Gh\left(\frac{\partial u}{\partial y} + \frac{\partial v}{\partial x}\right). \tag{10.14}$$

Similarly, the bending moments are given by

$$M_x = \int_{-h/2}^{h/2} \sigma_x z dz, \quad M_y = \int_{-h/2}^{h/2} \sigma_y z dz, \quad M_{xy} = \int_{-h/2}^{h/2} \tau_{xy} z dz, \tag{10.15}$$

which are evaluated as

$$M_x = -D\left(\frac{\partial^2 w}{\partial x^2} + \nu\frac{\partial^2 w}{\partial y^2}\right),$$

$$M_y = -D\left(\frac{\partial^2 w}{\partial y^2} + \nu\frac{\partial^2 w}{\partial x^2}\right),$$

$$M_{xy} = -D(1-\nu)\frac{\partial^2 w}{\partial x \partial y}, \tag{10.16}$$

where $D = Eh^3/[12(1-\nu^2)]$ is the flexural rigidity of the plate (and analogous to EI for the beam).

Substituting Eqs. (10.16) into Eq. (10.12) gives

$$D\left(\frac{\partial^4 w}{\partial x^4} + 2\frac{\partial^4 w}{\partial x^2 \partial y^2} + \frac{\partial^4 w}{\partial y^4}\right) + \mu\frac{\partial^2 w}{\partial t^2}$$

$$= N_x\frac{\partial^2 w}{\partial x^2} + N_y\frac{\partial^2 w}{\partial y^2} + 2N_{xy}\frac{\partial^2 w}{\partial x \partial y}. \tag{10.17}$$

We are primarily interested in how the in-plane (axial) forces N interact with the out-of-plane, (flexural) dynamic behavior described by $w(x, y, t)$. Initial imperfections are not included in the analysis at this stage. Adopting the standard notation (i.e., a double application of the Laplace operator) we write

$$\nabla^4 w = \left(\frac{\partial^4 w}{\partial w^4} + 2\frac{\partial^4 w}{\partial x^2 \partial y^2} + \frac{\partial^4 w}{\partial y^4} \right), \tag{10.18}$$

and we finally have

$$D\nabla^4 w + \mu \frac{\partial^2 w}{\partial t^2} - N_x \frac{\partial^2 w}{\partial x^2} - N_y \frac{\partial^2 w}{\partial y^2} - 2N_{xy}\frac{\partial^2 w}{\partial x \partial y} = 0. \tag{10.19}$$

This is the biharmonic equation with the added effect of constant axial load-ing. It is analogous to beam equation (7.9) and is the basis of much of the linear stability and free-vibration response analysis of plates. It is valid only for small de-flections and represents a differential eigenvalue problem of the kind introduced in Section 4.3.

10.1.2 Strain Energy

As mentioned earlier, it also possible to arrive at Eq. (10.19) by use of an energy approach. We return to the fundamental expression for strain energy of a 3D elastic solid element,

$$U = \int_V \int_{\epsilon_0}^{\epsilon_t} \sigma_{ij}\, d\epsilon_{ij}\, dV, \tag{10.20}$$

which for linear elastic material of the type generally of interest in this book reduces to

$$U = \frac{1}{2} \int_V \sigma_{ij}\, \epsilon_{ij}\, dV, \tag{10.21}$$

where σ_{ij} and ϵ_{ij} are the stress and strain tensors, respectively (the subscripts 0 and t refer to initial and final values of the strain tensor), and V is the volume of the element.

From the assumptions developed earlier, we can write the expression for the strain energy that is due to bending effects:

$$U_b = \int_{-h/2}^{h/2} \int_0^b \int_0^a \frac{1}{2E} \left[\sigma_x^2 + \sigma_y^2 - 2v\sigma_x\sigma_y + 2(1+v)\tau_{xy}^2 \right] dx dy dz, \tag{10.22}$$

which in terms of lateral displacement is given by

$$U_b = \frac{D}{2} \int_0^b \int_0^a \left[\left(\frac{\partial^2 w}{\partial x^2} \right)^2 + \left(\frac{\partial^2 w}{\partial y^2} \right)^2 + 2v\frac{\partial^2 w}{\partial x^2}\frac{\partial^2 w}{\partial y^2} + 2(1-v)\left(\frac{\partial^2 w}{\partial x \partial y} \right)^2 \right] dx dy. \tag{10.23}$$

Similarly, the strain energy that is due to membrane (stretching) effects is given by

$$U_m = \int_0^b \int_0^a (N_x \epsilon_x + N_y \epsilon_y + N_{xy}\epsilon_{xy})\, dx dy, \tag{10.24}$$

in which the simple relations between the tractions N and stresses σ, that is, Eqs. (10.14), have been used. This last expression can alternatively be couched in terms of the work done by the axial loads when deflections are small. In this form, we have implicitly assumed that the tractions are independent of deformations. We will return to the preceding expression later when dealing with (moderately) large deflections.

The kinetic energy is given by

$$T = \int_0^b \int_0^a \frac{1}{2}\mu \left(\frac{\partial w}{\partial t}\right)^2 dxdy, \tag{10.25}$$

in which only translational terms are retained; that is, we neglect rotary inertia effects.

We state Hamilton's principle (from Chapter 2) as

$$\delta \int_{t_1}^{t_2} \{T - (U + V)\}dt = 0, \tag{10.26}$$

where $U = U_b + U_m$ and we assume that there are no externally applied lateral loads (at this point). Placing in the various energy expressions [Eqs. (10.23)–(10.25)] into Eq. (10.26) results in

$$\int_0^t \int_0^b \int_0^a \left\{ \left[\frac{\partial N_x}{\partial x} + \frac{\partial N_{xy}}{\partial y}\right]\delta u + \left[\frac{\partial N_y}{\partial y} + \frac{\partial N_{xy}}{\partial x}\right]\delta v \right.$$

$$+ \left[\left(\frac{\partial N_x}{\partial x} + \frac{\partial N_{xy}}{\partial y}\right)\frac{\partial w}{\partial x} + \left(\frac{\partial N_y}{\partial y} + \frac{\partial N_{xy}}{\partial x}\right)\frac{\partial w}{\partial y}\right.$$

$$+ N_x\frac{\partial^2 w}{\partial x^2} + N_y\frac{\partial^2 w}{\partial y^2} + 2N_{xy}\frac{\partial^2 w}{\partial x \partial y}$$

$$\left. - D\left(\frac{\partial^4 w}{\partial x^4} + 2\frac{\partial^4 w}{\partial x^2 \partial y^2} + \frac{\partial^4 w}{\partial y^4}\right) - \mu\frac{\partial^2 w}{\partial t^2}\right]\delta w \right\} dxdydt = 0. \tag{10.27}$$

For this to be true for all time, we have nontrivial solutions corresponding to the coefficients of δu, δv, and δw being set equal to zero. After some simplifications, these expressions furnish the governing equations in the in-plane and lateral directions as given by Eqs. (10.8), (10.9), and (10.19).

At this point, we simplify things a little by focusing on the underlying linear problem. That is, if we consider cases in which the axial forces are relatively large but the strains are small, then the in-plane and bending problems uncouple.

10.1.3 Boundary and Initial Conditions

Equation (10.19) is fourth order in x and y and second order in t, and hence two boundary conditions are needed along each edge, together with initial displacement and velocity within the plate. For a plate of dimension a units in the x direction the standard boundary conditions at $x = a$ fall into the following categories:

- Clamped edges
  In this case, neither deflection nor slope is permitted, that is,

$$w(a, y) = 0, \tag{10.28}$$

$$\frac{\partial w}{\partial x}(a, y) = 0. \tag{10.29}$$

- Simply supported edges (pinned)
  Neither deflections or bending moments permitted, that is,

$$w(a, y) = 0, \tag{10.30}$$

$$M_x(a, y) = \frac{\partial^2 w}{\partial x^2}(a, y) = 0, \tag{10.31}$$

because $w(x, y) = 0$ causes $\frac{\partial^2 w}{\partial y^2}(a, y) = 0$.

- Free edges
  Because shear deformation has been neglected we can write the following two boundary conditions at a free edge [12]:

$$\left(Q_x + \frac{\partial M_{xy}}{\partial y}\right)(a, y) = 0, \tag{10.32}$$

$$M_x(a, y) = 0. \tag{10.33}$$

It can be shown that these boundary conditions can be alternatively written as

$$\left(\frac{\partial^2 w}{\partial x^2} + v\frac{\partial^2 w}{\partial y^2}\right)\bigg|_{x=a} = 0, \tag{10.34}$$

$$\left(\frac{\partial^3 w}{\partial x^3} + (2 - v)\frac{\partial^3 w}{\partial x \partial y^2}\right)\bigg|_{x=a} = 0. \tag{10.35}$$

We can define similar boundary conditions for the other edges. To these can be added various more subtle boundary conditions, including partial edge restraint, for example, as provided by a supporting beam or stiffener. However, we will have to look more closely at these boundary conditions for the large-deflection problem, because there is coupling between *in-plane* and bending behavior.

The initial conditions are simply

$$w(x, y, 0) = w_0, \tag{10.36}$$

$$\frac{\partial w}{\partial t}(x, y, 0) = \dot{w}_0, \tag{10.37}$$

and we assume, as usual, simple harmonic motion.

Solving the governing equation subject to the appropriate boundary conditions leads to a characteristic equation analogous to those for the beam (but more complicated). Rather than discuss the solutions in general terms we spend the rest of this chapter with specific examples. The general progression of the material will develop from linear free vibration of axially loaded, simply supported plates (for which there is an exact analytical solution), followed by consideration of initial imperfections, approximate analyses, thermal loading (including some experimental results), postbuckling, mode jumping, and finally a brief study will be made of the vibration of an axially loaded cylinder.

10.1.4 The Simplest Case

As usual we start with a relatively simple example. Specifically we seek a relation between the natural frequencies of vibration and axial load for a flat rectangular panel under biaxial compression with simply supported boundary conditions on all four sides. Similarly to the pin-ended beam we find ready access to closed-form solutions in this case. We focus attention on how the flat configuration buckles at a critical value of the axial compression and how the small oscillations behave during this process. We assume a deflection for the plate that satisfies the boundary conditions of zero deflection and bending moment along the edges:

$$w = \frac{\partial^2 w}{\partial x^2} = 0 \quad at \ x = 0, a, \tag{10.38}$$

$$w = \frac{\partial^2 w}{\partial y^2} = 0 \quad at \ y = 0, b. \tag{10.39}$$

It is important to note that for this kind of small-deflection analysis only the transverse (out-of-plane) boundary conditions are required. We shall see later that the in-plane boundary conditions may have a profound effect when larger deflections occur.

For harmonic motion, we have the typical form of response

$$w(x, y, t) = W(x, y) \sin \omega t, \tag{10.40}$$

which is substituted into Eq. (10.19) to give

$$D\nabla^4 W - \mu \omega^2 W = N_x \frac{\partial^2 W}{\partial x^2} + N_y \frac{\partial^2 W}{\partial y^2}. \tag{10.41}$$

We assume a form of spatial response (which satisfies the boundary conditions):

$$W(x, y) = \sum_{m,n=1}^{\infty} A_{mn}(t) \sin \frac{m\pi x}{a} \sin \frac{n\pi y}{b}. \tag{10.42}$$

Substituting this into Eq. (10.41) we get

$$\mu \omega_{mn}^2 = D\left[\left(\frac{m\pi}{a}\right)^2 + \left(\frac{n\pi}{b}\right)^2 \right]^2 + N_x \left(\frac{m\pi}{a}\right)^2 + N_y \left(\frac{n\pi}{b}\right)^2. \tag{10.43}$$

Let's initially consider the special case of uniaxial loading, that is, $N_x = -N$, $N_y = 0$. In this case, we can simplify Eq. (10.43) to

$$\omega_{mn}^2 = \frac{D}{\mu}\left\{ \left[\left(\frac{m\pi}{a}\right)^2 + \left(\frac{n\pi}{b}\right)^2 \right]^2 - \frac{N}{D}\left(\frac{m\pi}{a}\right)^2 \right\}$$

$$= \frac{D\pi^4}{\mu a^4}(m^2 + n^2 k^2)^2 - \frac{N}{\mu}\left(\frac{m\pi}{a}\right)^2, \tag{10.44}$$

where $k = a/b$ is the aspect ratio of the plate. To clarify the relation between the natural frequencies and the axial loading, we nondimensionalize in the following way. When $N = 0$ we have the natural frequencies given by

$$\bar{\omega}_{mn}^2 = \frac{D\pi^4}{\mu a^4}(m^2 + n^2 k^2)^2 \tag{10.45}$$

[note the similarity with Eq. (6.61)], and when the natural frequencies are zero we have the buckling loads

$$N_{\text{cr}} = D\left(\frac{\pi}{ma}\right)^2 (m^2 + n^2 k^2)^2, \tag{10.46}$$

and thus Eq. (10.44) can be written as

$$\left(\frac{\omega_{mn}}{\bar{\omega}_{mn}}\right)^2 = 1 + \frac{N}{N_{\text{cr}}}. \tag{10.47}$$

Again we obtain a linear relation between the square of the natural frequencies and the axial load *for each mode*: We expect to see the lowest natural frequency dropping to zero as the lowest buckling load is approached. The linearity of this relation again depends on the similarity between the vibration and buckling modes. The lowest buckling load depends on the aspect ratio in the following way. We rewrite Eq. (10.46) in the form

$$N_{\text{cr}} = D\left(\frac{\pi}{b}\right)^2 \left(\frac{m}{k} + \frac{n^2 k}{m}\right)^2, \tag{10.48}$$

where the aspect ratio is $k = a/b$. We seek the smallest value of this function to give the critical load. Clearly this occurs when $n = 1$, that is, a half-sine wave in the shorter direction, but plotting Eq. (10.48) for different integer values of m and hence altering the aspect ratio k leads to the plot shown in Fig. 10.2. Therefore, given a certain aspect ratio, this graph will give the critical load together with the number of half-sine waves in the x direction (which is the direction of the loading in this example). A couple of features of this plot are noteworthy: Buckling cannot take place for nondimensional axial loads of less than 4 (for any aspect ratio), and the aspect ratio corresponding to the transition in buckling from mode m to mode $m + 1$ occurs at an aspect ratio of $\sqrt{m(m+1)}$.

The behavior dependence on the aspect ratio also allows us to reinterpret the simple linear relation from Eq. (10.47). This expression is nondimensionalized with respect to buckling and frequency for a given mode. We can gain an alternative and more general view by nondimensionalizing the results with respect to a single (lowest) frequency and load. Figure 10.3(a) shows this relation for an aspect ratio of 2. In this case, the lowest buckling mode (and natural mode of vibration) has a half-sine wave in one direction and a full-sine wave in the other, that is, the (2,1) mode, with a critical load given by $16\pi^2 D/a^2$. This can be seen from Fig. 10.2 along the $(a/b) = 2$ line. At this specific value of the aspect ratio, the (1,1) and (4,1) modes actually have the same buckling load (although of course they are not naturally encountered because of the presence of lower buckling loads). In the absence of axial load, the lowest natural mode of vibration is the (1,1) mode, with a corresponding frequency of $5\pi^2$. The sequence of buckling loads and natural frequencies in the absence of

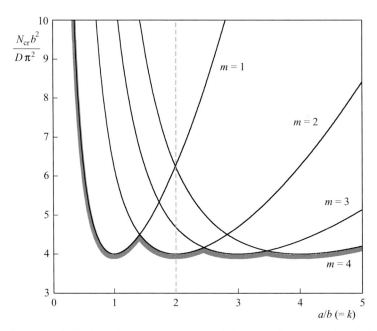

Figure 10.2. Buckling loads versus aspect ratio for a simply supported, uniaxially compressed plate.

axial loading, are indicated by the solid circular and square symbols in Fig. 10.3(a), respectively. The squares of the natural frequencies are plotted in part (b), a result well established in the literature [2, 13, 14]. Thus, the fundamental frequency will always correspond to the mode shape with $n = 1$ but not necessarily $m = 1$. For the plate with an aspect ratio of 2, this changeover occurs at 80% of the buckling load [and indicated by the open circle in part (b)]. It can also be shown that, for a plate

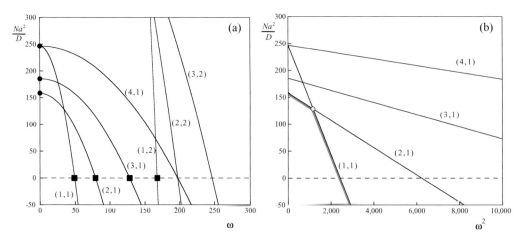

Figure 10.3. The relation between the natural frequencies and axial load for a uniaxially loaded, simply supported plate with an aspect ratio of 2.

with an aspect ratio of 3, the fundamental mode changes from (1,1) to (2,1) at 63% of its critical load and from (2,1) to (3,1) at 85% of its critical load. This is quite different from the beam. We also note that the critical load scales with the cube of the plate thickness and the natural frequency with the 3/2 power of the thickness.

An alternative solution procedure can also be followed by assuming an exponential form for the deflected shape, analogous to the earlier chapter on beams. Also, exact solutions in terms of the Levy solution can be obtained for rectangular plates with two opposite sides simply supported [1, 4].

10.1.5 Initial Imperfections

As for struts, the behavior of axially loaded plates may be profoundly affected by the presence of initial geometric imperfections, especially in the vicinity of critical loading conditions. We can incorporate these into the plate theory by appropriately, including the total deflection, made up of an initial deflection w_0 together with additional deflection w_1 caused by the application of a transverse load, say. The total deflection is then $(w = w_0 + w_1)$, and this is used in in Eq. (10.19). Similarly, we modify the strain energy in Eq. (10.23).

As a simple example, we go back to consider the simply supported rectangular plate under uniaxial compression N. We assume the plate has an initial curvature given by

$$w_0 = Q_0 \sin \frac{\pi x}{a} \sin \frac{\pi y}{b}, \tag{10.49}$$

in which Q_0 is understood to be small. The plate deflection is $w = w_0 + w_1$ and its maximum value (at the center point) will then be given by

$$w_{\max} = \frac{Q_0}{1 - \alpha}, \tag{10.50}$$

where

$$\alpha = \frac{N}{\left(\frac{\pi^2 D}{a^2}\right)\left[1 + \left(\frac{a^2}{b^2}\right)\right]^2}. \tag{10.51}$$

For example, for a square plate we have $a = b$ and a critical load given by $N_{cr} = 4\pi^2 D/a^2$, and thus $N \to N_{cr}$, we have $\alpha \to 1$ and $w_{\max} \to \infty$, and we see that the effect of the initial imperfection is to magnify subsequent lateral deflection especially in the vicinity of buckling for the initially perfect plate, a situation similar to the strut.

10.2 The Ritz and Finite-Element Approaches

Adopting a similar approach to the approximate analysis of axially loaded beams, we can also attack problems involving the dynamics of plates by assuming the response, in the usual way, comprises a finite number of degrees of freedom,

$$w = \sum_{i=1}^{N_i} W_i(x, y) q_i(t), \tag{10.52}$$

where the $q_i(t)$ are the generalized coordinates (assumed to be harmonically vary-ing in time) and $W_i(x, y)$ are the shape functions satisfying the geometric boundary conditions. In this section, it will be convenient to use matrix notation: $w = \{W\}^T\{q\}$.

We now return to the expression for strain energy [Eq. (10.22)] and write the internal virtual work in terms of bending moments and the plate curvatures:

$$\delta W_{int} = -\int_A [M_x \delta K_x + M_y \delta K_y + 2M_{xy} \delta K_{xy}] \, dA, \tag{10.53}$$

in which $K_x = -\partial^2 w/\partial x^2$, and so on. The integrand in Eq. (10.53) can also be written in matrix notation, that is, $\{\delta K\}^T\{M\}$. Use is now made of the moment–curvature relations [Eqs. (10.16)]: $\{M\} = [D]\{K\}$, where

$$[D] = D \begin{bmatrix} 1 & v & 0 \\ v & 1 & 0 \\ 0 & 0 & (1-v)/2 \end{bmatrix}. \tag{10.54}$$

The curvatures are then related to the assumed displacements by $\{K\} = [B]\{q\}$, where

$$[B] = \begin{bmatrix} -\{\partial^2 W/\partial x^2\}^T \\ -\{\partial^2 W/\partial y^2\}^T \\ -2\{\partial^2 W/\partial x \partial y\}^T \end{bmatrix}. \tag{10.55}$$

Therefore, the internal work that is due to bending can be written as

$$\delta W_{int} = -\int_A \{\delta q\}^T [B]^T [D] [B] \{q\} \, dA \tag{10.56}$$

$$= -\{\delta q\}^T [K] \{q\}, \tag{10.57}$$

and this provides the definition of the stiffness matrix: $[K] = \int_A [B]^T [D] [B] \, dA$.

The virtual work that is due to stretching [Eq. (10.24)] can also be rewritten as

$$\delta W_{int} = -\int_A \{\delta \theta\}^T [N] \{\theta\} \, dA, \tag{10.58}$$

where the slope vector $\theta = \{\partial w/\partial x, \partial w/\partial y\}$ can also be related to the assumed dis-placements by $\{\theta\} = [G]\{q\}$, where

$$[G] = \begin{bmatrix} -\{\partial W/\partial x\}^T \\ -\{\partial W/\partial y\}^T \end{bmatrix}. \tag{10.59}$$

Thus we can write the internal work that is due to in-plane stretching as

$$\delta W_{int} = -\int_A \{\delta q\}^T [G]^T [N] [G] \{q\} \, dA, \tag{10.60}$$

$$= -\{\delta q\}^T [K_G] \{q\}, \tag{10.61}$$

and this provides the definition of the *geometric* stiffness matrix: $[K_G] = \int_A [G]^T[N][G]dA$; see Kapur and Hartz [15].

A consistent mass-matrix formulation can be developed by use of virtual work and D'Alembert's principle for the inertia loads to give

$$[M] = \int_A \mu [W]^T[W]dA. \tag{10.62}$$

Alternatively, the mass can be lumped at discrete locations on the plate.

The choice of shape functions, W_i in Eq. (10.52), is a crucial step in this whole process. For a standard Ritz analysis, we choose functions satisfying the geometric boundary conditions; for FEs we often choose local polynomials that are then assembled for the whole structure. In either case, there is a wide choice, with FE packages offering a variety of element types. In terms of matrix notation, we have

$$[M]\{\ddot{q}\} + \{[K] - \lambda[K_G]\}\{q\} = 0, \tag{10.63}$$

and assuming harmonic motion, that is, $\{q\} = \{q_0\}e^{i\omega t}$, we arrive at

$$(\{[K] - \lambda[K_G]\} - \omega^2[M])\{q_0\} = 0. \tag{10.64}$$

For nontrivial solutions, we set the determinant equal to zero (in practice, for large systems, an iterative solution approach is preferred). This is a linear eigenvalue problem that subsumes a linear-elastic buckling analysis (with $\omega = 0$) and a linear-vibration analysis (with $\lambda = 0$). Again ANSYS, with an external load, and ABAQUS were able to reproduce the results shown in Fig. 10.3 with minimal effort.

An early FEA of axially loaded plates was reported by Anderson et al. [16]. They showed the reduction of natural frequencies as a function of axial load, and the following table shows a subset of their results for a simply supported square plate under uniaxial loading.

Load (% of critical)	Ω_1	Ω_1^2
−200	34.150	1166.2
−100	27.811	773.5
−50	24.048	578.3
0	19.596	384.0
40	15.139	229.2
80	8.661	75.0
90	6.046	36.6
95	3.705	13.7
99	1.410	2.0

This analysis used 128 triangular (noncomforming) elements. The coefficient of the lowest natural frequency at zero axial load is 19.596, and the critical load (at zero frequency) is 3.97, both of which compare well with the analytical values of $2\pi^2 =$

19.739 and 4 from Eq. (10.44), respectively. Note that Rayleigh–Ritz and Galerkin would give values no lower than the exact.

10.3 A Fully Clamped Plate

A flat plate fixed against deflection and rotation on all four sides represents an important class of problem. Because closed-form analytical solutions are not available in this case, we use the approximate method developed in the previous section to obtain a relation between the dynamics and stability for the lowest mode. We assume that the plate is square (of length a) and is again loaded in a uniaxial fashion (in the x direction) so that $N_x = -N$.

Again we enter the analysis at an intermediate point by simply indicating that the strain energy in bending (consistent with the level of approximation in the previous section) for a square ($a \times a$) plate is given by Eq. (10.23). The potential energy associated with the external axial load is given by

$$V = \int_0^a \frac{N}{2} \int_0^a \left(\frac{\partial w}{\partial x}\right)^2 dxdy. \tag{10.65}$$

The kinetic energy is

$$T = \frac{1}{2}m \int_0^a \int_0^a \left(\frac{\partial w}{\partial t}\right)^2 dxdy. \tag{10.66}$$

The fixed nature of the boundary conditions means that

$$w = \frac{\partial w}{\partial x} = 0 \quad at \ x = 0, a, \tag{10.67}$$

$$w = \frac{\partial w}{\partial y} = 0 \quad at \ y = 0, a, \tag{10.68}$$

and a relatively simple function satisfying these boundary conditions is

$$w = A(t) \left(1 - \cos \frac{2\pi x}{a}\right)\left(1 - \cos \frac{2\pi y}{a}\right). \tag{10.69}$$

Evaluating the energy terms by use of Eq. (10.69) leads to

$$U = \frac{16D\pi^4 A^2}{a^2}, \tag{10.70}$$

$$V = \frac{3N\pi^2 A^2}{2}, \tag{10.71}$$

$$T = \frac{9ma^2 \dot{A}^2}{8}. \tag{10.72}$$

We can again use our standard theory to immediately write

$$\omega^2 - \frac{4\pi^2}{9ma^4}(32D\pi^2 - 3Na^2). \tag{10.73}$$

As the frequency drops to zero we have the critical load $N_{cr} = 10.67D\pi^2/a^2$ and without axial load we get the natural frequency $\omega = 37.22\sqrt{D/ma^4}$ (compared with "exact" coefficients of 10.07 [5] and 35.99 [2], respectively). Note that by clamping the edges of the plate the critical load has increased by a factor of about 2.5 over the critical load for simply supported boundary conditons (for a strut it is a factor of 4).

In an earlier chapter, we used a Duncan polynomial to arrive at an approximate solution for a clamped beam. For example, conducting a similar analysis for the plate by assuming the first Duncan polynomial in each direction leads to a critical load factor of 10.94. Higher modes can be handled by use of this approach, but with more involved algebra.

Thermal Loading. Under the influence of an applied thermal loading condition, a plate will experience compression (in both directions) because of its constrained edges [17, 18]. It can be shown that, for a thermal load, uniformly distributed through the plate thickness, the following stress resultants are generated:

$$N_x = N_y = -\frac{Eh\alpha\Delta T}{1-\nu}, \tag{10.74}$$

in which ΔT is the applied thermal gradient, h is the plate thickness, and α is the (material-dependent) coefficient of thermal expansion.

Consider a fully clamped panel and use the Ritz approximation

$$W_i(x, y) = \sum_{i=1}^{N_i}\sum_{j=1}^{N_j} q_{ij}(t)X_i(x)Y_j(y), \tag{10.75}$$

in which

$$X_i(x) = \left(\frac{x}{a}\right)^{i+1} - 2\left(\frac{x}{a}\right)^{i+2} + \left(\frac{x}{a}\right)^{i+3}, \tag{10.76}$$

$$Y_i(y) = \left(\frac{y}{a}\right)^{i+1} - 2\left(\frac{y}{a}\right)^{i+2} + \left(\frac{y}{a}\right)^{i+3}, \tag{10.77}$$

with six modes taken in each direction, that is, $i = 1, 2, \ldots, 6$ and $j = 1, 2, \ldots, 6$. The resulting eigenvalue problem is solved to show the relation between the applied thermal load and the squares of the natural frequencies as shown in Fig. 10.4 for a plate with an aspect ratio of 2.7. In this analysis, the natural frequencies and thermal loading have been normalized in the following way:

$$\bar{N} = \frac{N_0}{N_{\text{ref}}}, \tag{10.78}$$

in which

$$N_0 = \frac{\alpha\Delta T Eh}{1-\nu}, \qquad N_{\text{ref}} = \frac{D\pi^2}{b^2}, \tag{10.79}$$

and

$$\bar{\omega} = \frac{\omega}{\omega_{\text{ref}}}, \tag{10.80}$$

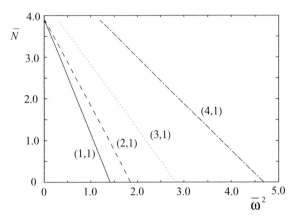

Figure 10.4. The relation between the natural frequencies and axial load for a thermally loaded, clamped plate with an aspect ratio of 2.7.

in which

$$\omega_{\text{ref}} = \left(\frac{2\pi^2}{b^2}\right)\sqrt{\frac{D}{\mu h}}. \tag{10.81}$$

In these nondimensional terms, the result given in Leissa [2] in the absence of thermal loading would correspond to a natural frequency (squared) of 1.4 in Fig. 10.4. Similarly, the nondimensional critical buckling temperature (i.e., where the the lowest natural frequency drops to zero) should occur in the vicinity of 3.9 based on detailed FEA [16]. It is interesting to note that the critical temperature for a clamped plate is a little over three times that of a corresponding plate with simply supported boundary conditions.

10.4 Moderately Large Deflections

It has already been mentioned that plates often have significant postbuckled strength [19]. This suggests that dynamic behavior in the postbuckled regime may also be of interest. Furthermore, plates may exhibit large deflections that result in a coupling between the stretching and bending deformations in much the same way as for the beam considered in Section 7.5. Figure 10.5 shows a schematic of a plate undergoing relatively large deformation. The standard plate theory incorporating finite (but not small) deflections [including the nonlinear terms in strain-displacement equations (10.2)] again results in the von Karman nonlinear partial differential equation developed earlier:

$$D\nabla^4 w - \left[\frac{\partial^2 F}{\partial y^2}\frac{\partial^2 w}{\partial x^2} + \frac{\partial^2 F}{\partial x^2}\frac{\partial^2 w}{\partial y^2} - 2\frac{\partial^2 F}{\partial x\partial y}\frac{\partial^2 w}{\partial x\partial y}\right] + \mu\frac{\partial^2 w}{\partial t^2} = 0, \tag{10.82}$$

where w is the lateral displacement of the panel. In the small-deflection theory the in-plane forces were simply given. Now they are unknown functions of x and y but

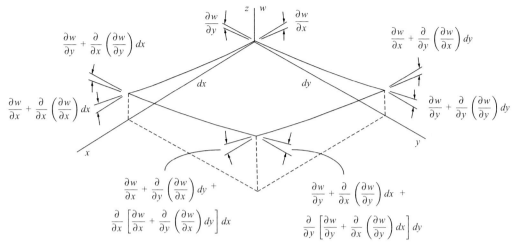

Figure 10.5. Large deformations of a rectangular plate [2].

can be specified in terms of a function $F(x, y)$, called the Airy function:

$$N_x = \frac{\partial^2 F}{\partial y^2},$$ (10.83)

$$N_y = \frac{\partial^2 F}{\partial x^2},$$ (10.84)

$$N_{xy} = -\frac{\partial^2 F}{\partial x \partial y},$$ (10.85)

which, because of compatibility between the in-plane loads and the lateral displacements, leads to a second equation:

$$\nabla^4 F = E\left[\left(\frac{\partial^2 w}{\partial x \partial y}\right)^2 - \frac{\partial^2 w}{\partial x^2}\frac{\partial^2 w}{\partial y^2} \right].$$ (10.86)

The two nonlinear fourth-order partial differential equations (10.82) and (10.86) are solved together. In-plane boundary conditions must also be carefully considered [20, 21].

10.5 Postbuckling

In this section, we again make use of large-deflection modeling but this time the axial-load effects will be produced by thermal loading [22–24]. As we have seen repeatedly throughout this book, we would expect axial loading to reduce the effective lateral stiffness and therefore natural frequencies. For an initially flat plate, a critical temperature will cause buckling, and we might expect the plate to bifurcate into either of the (symmetric) postbuckled equilibrium configurations available. However, we have also seen how structural systems of this kind will tend to have some kind of initial imperfection (often an initial geometric deflection, axial-load offset, or small lateral force) such that the symmetry is broken and the plate continuously deflects with an increase in the rate of deflection in the vicinity of the critical temperature for the underlying *perfect* geometry: The plate has a preferred direction of buckling displacement. In this case, because there is no distinct bifurcation, the lowest natural frequency does not drop to zero but rather tends to a minimum before increasing as the postbuckled stiffness takes over.

Because of its practical significance we shall take a closer look at the dynamics associated with thermal buckling. We shall look initially at a fully clamped plate and follow frequencies into the postbuckled regime, and then consider a simply supported plate that is heated sufficiently beyond initial buckling so that *mode jumping* occurs [25–27]. Extensive use will be made of Galerkin's method, and a FE study will be conducted for verification purposes [28, 29].

We again make use of von Karman's equations including the effects of a small initial geometric imperfection (w_0), a little damping (β), and a lateral load (Δp), that is, by modifying Eqs. (10.82) and (10.86) we have [20]

$$\bar{\nabla}^4 w - \left(\frac{a_x}{b_y}\right)^2 \left[\frac{\partial^2 F}{\partial \eta^2}\left(\frac{\partial^2 w}{\partial \xi^2} + \frac{\partial^2 w_0}{\partial \xi^2}\right) + \frac{\partial^2 F}{\partial \xi^2}\left(\frac{\partial^2 w}{\partial \eta^2} + \frac{\partial^2 w_0}{\partial \eta^2}\right)\right.$$

$$\left. - 2\frac{\partial^2 F}{\partial \xi \partial \eta}\left(\frac{\partial^2 w}{\partial \xi \partial \eta} + \frac{\partial^2 w_0}{\partial \xi \partial \eta}\right)\right] + \beta\frac{\partial w}{\partial \tau} + \frac{\partial^2 w}{\partial \tau^2} + \Delta p = 0, \tag{10.87}$$

$$\bar{\nabla}^4 F = 12(1-v^2)\left(\frac{a_x}{b_y}\right)^2 \left[\left(\frac{\partial^2 w}{\partial \xi \partial \eta}\right)^2 + 2\left(\frac{\partial^2 w}{\partial \xi \partial \eta}\right)\left(\frac{\partial^2 w_0}{\partial \xi \partial \eta}\right)\right.$$

$$\left. - \frac{\partial^2 w}{\partial \xi^2}\frac{\partial^2 w}{\partial \eta^2} - \frac{\partial^2 w}{\partial \xi^2}\frac{\partial^2 w_0}{\partial \eta^2} - \frac{\partial^2 w}{\partial \eta^2}\frac{\partial^2 w_0}{\partial \xi^2}\right], \tag{10.88}$$

where scaled length ($\xi = x/a$, $\eta = y/b$) and time [$\tau = t\sqrt{D/(\mu h a^4)}$] dimensions have been introduced and are reflected in the overbars on the Laplace operator.

Suppose we are primarily interested in the lower modes (say, nine total, involving three in each direction); then we can apply Galerkin's method to this problem by assuming mode shapes of the form

$$w(\xi, \eta, \tau) = \sum_{i=1}^{3}\sum_{j=1}^{3} a_{ij}(\tau)\Psi_i(\xi)\Phi_j(\eta), \tag{10.89}$$

where $\Psi_i(\xi)$ and $\Phi_j(\eta)$ are spatial beam mode shapes that satisfy the zero-deflection and zero-slope boundary conditions for a fully clamped panel, for example,

$$\Psi_i(\xi) = \cos([i-1]\pi\xi) - \cos([i+1]\pi\xi), \tag{10.90}$$

$$\Phi_j(\eta) = \cos([j-1]\pi\eta) - \cos([j+1]\pi\eta). \tag{10.91}$$

The initial imperfection w_0 is assumed to be given by the (1,1) mode.

The thermal effects enter the Airy stress function according to

$$\left(\frac{D}{b^2}\right)\frac{\partial^2 F}{\partial \eta^2} = N_\xi - \frac{Eh}{1-\nu}\alpha\Delta T, \tag{10.92}$$

$$\left(\frac{D}{a^2}\right)\frac{\partial^2 F}{\partial \nu^2} = N_\nu - \frac{Eh}{1-\nu}\alpha\Delta T. \tag{10.93}$$

Further details of this modeling can be found in Murphy et al. [24, 30, 31]. In these references, discussion is also included on allowing a certain amount of in-plane deformation along the edges of the plate (because this is what may happen in an experimental situation), and the use of the Southwell plot.

The solution procedure then follows the basic Galerkin steps of substituting the assumed modes (including the initial shape) and Airy stress function into the partial differential equation of motion, multiplying by the modes, and integrating over the domain (orthogonality). This, of course, leaves a set of coupled (nonlinear) ordinary differential equations. They can be linearized at given temperature levels to yield local natural frequencies of vibration.

Before moving on to consider detailed solutions we summarize the anticipated behavior conceptually as shown in Fig. 10.6. If heating begins from ambient (room) temperature, we expect buckling to take place. A cooling temperature would lead to stretching. Initially the potential energy is locally quadratic, and small oscillations would occur about the flat configuration. Beyond the critical thermal buckling temperature, the potential energy typically has a *double-well* characteristic, for example, at the temperature indicated by V_3 (in analogy to the postbuckling dynamics of the strut); see Fig. 3.7(a). Again, with an initial imperfection there is no distinct buckling, but rather an increase in the rate of lateral deflection. The only distinct instability is a saddle-node at temperature point B where the secondary solution changes its stability property. However, we know that this solution path is not typically obtained for smooth changes in the thermal loading. However, a *decrease*

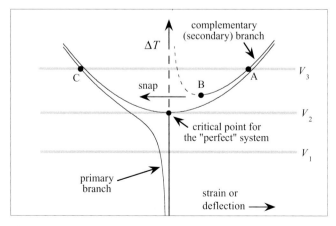

Figure 10.6. Schematic of the static equilibrium behavior as a function of temperature rise. The dashed curve starting from point B is an unstable equilibrium path.

in temperature from high values (for example if we could get to point A) passing through point B will result in a dynamic snap as the system inevitably jumps to the primary equilibrium path [32]. This type of behavior will occur regardless of specific boundary conditions.

At relatively low, or prebuckling, temperatures (e.g., V_1), the response is effectively linear (provided the lateral deflection does not induce significant membrane effects). As the temperature reaches its critical value, the underlying potential energy ceases to be a minimum (at least, in the initially perfect geometry). With the presence of an initial geometric imperfection the two postbuckled equilibrium configurations are slightly asymmetric [see Fig. 3.8(a)].

The natural frequencies of the system will change according to the local stiffness (for a given temperature); that is, we linearize about equilibrium in either potential-energy well. For low temperatures, we anticipate that the lowest natural frequency will initially decrease. Linearization leads to our standard eigenvalue problem:

$$[M]\ddot{\mathbf{x}} = -[J]\mathbf{x}, \qquad (10.94)$$

where $[J]$ is the Jacobian representing the linearized stiffness (see Section 4.2). The natural frequencies can be obtained numerically as functions of temperature.

Given an aspect ratio of 1.25 we obtain results for the lowest mode (1,1) for the relation between the natural frequency and temperature for both the perfect and imperfect geometries, as shown in Fig. 10.7(a). These results are also shown nondimensionalized in part (b), in which the frequency squared is plotted. On

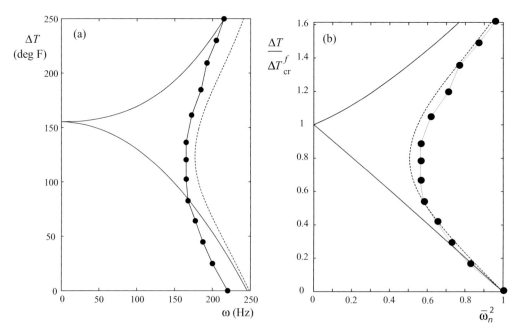

Figure 10.7. Thermal load versus frequency for a clamped plate. Solid circles are experimental data, continuous lines are theoretical, and dashed curves include the effect of an initial imperfection. (a) Dimensional results and (b) normalized results and with frequency squared [24].

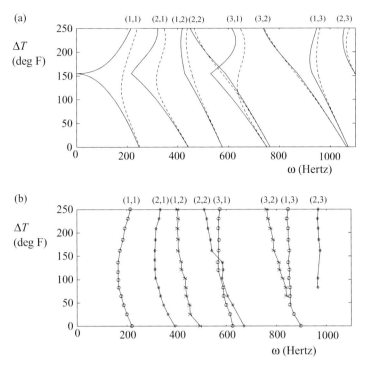

Figure 10.8. The natural frequencies for a thermally loaded panel with an aspect ratio of 1.25: (a) analytic results and (b) experimental results [24].

subsequent heating the plate stiffens in its postbuckled configuration and the frequency increases. This behavior is similar to the axially loaded strut, and a relation anticipated by the abstract form of Fig. 3.8(b).

The lowest nine natural frequencies for the numerical model are shown in Fig. 10.8(a) with dimensional units [20]. There is a mode switch as the temperature changes, a feature we have seen a number of times before.

Experimental Description

An experimental verification of the preceding analysis is described in Murphy et al. [24], in which a panel was subjected to thermal loads. Rather than impact hammer being used to conduct a transient dynamic test to extract the natural frequencies, the panel is subject to narrowband acoustic excitation that excited the natural frequencies resonantly. In Fig. 10.9, a photographic image of a thin panel with dimensions 0.381 m × 0.305 m × 3.175 mm and made of AISI stainless steel is shown. The deformation (both static and dynamic) of the panel was measured by the placement of high-temperature strain gauges near the lower left corner of the plate. Because of a little in-plane slippage in the boundary clamping, a Southwell plot approach was used to account for this effect [33]. The lowest natural frequency was extracted as a function of applied temperature, and nondimensionalized results are shown together with theory in Fig. 10.7. Note that the theoretical relation between the natural

Figure 10.9. Photograph of the panel in the Thermal Acoustic Fatigue Apparatus at NASA Langley.

frequency (squared) and the axial load before and after buckling resembles the ratio given by Eq. (7.53) for the strut. A similar study for all the modes considered is summarized in Fig. 10.8(b). An example of two modes shapes (obtained with a laser scanning velocity vibrometer) is shown in Fig. 10.10 [34]. The panel was excited at different forcing frequencies to excite resonance, with typical (3,2) and (2,1) mode shapes illustrated in part (a) and (b), respectively.

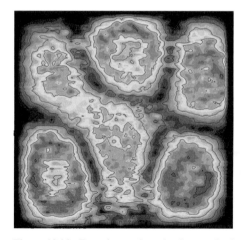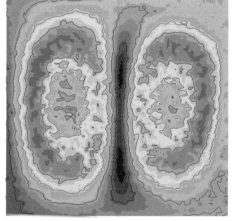

Figure 10.10. Experimental mode shapes obtained with a scanning laser vibrometer: (a) the (3,2) mode and (b) the (2,1) mode [34].

10.6 Mode Jumping

10.6.1 Introduction

One of the most interesting aspects of panels is that they may exhibit finite jumps in their postbuckled equilibrium behavior when subject to high levels of compressive axial load [25, 26]. This form of secondary bifurcation is associated with subtle interplay between buckling modes and is often referred to as mode jumping [27, 35, 36] and has received considerable attention in the literature [37, 38]. Tracking the movement of natural frequencies before, and during, this phenomenon can shed light on this interesting aspect of the dynamics of axially loaded structures.

Interest in mode jumping was initiated by the classic study of Stein [25]. His system consisted of a rectangular panel, stiffened at regular intervals by ribs. Hence a single bay of the panel could be considered to have simply supported edges along its long sides (i.e., restrained against deflection but not rotation). The two short edges (the mechanically loaded sides) consisted of clamped boundary conditions. Under increasing loading (in the long direction) Stein noticed that in the postbuckled regime the plate would suddenly jump to a different buckled equilibrium configuration: mode jumping. This typically corresponds to an often sudden change of the wavenumber of the buckled form. It may also happen that on subsequent unloading the system may not follow the original path: Hysteresis occurs [39]. Secondary bifurcation was also observed in Augusti's model (see Section 5.7), although in that case there was no adjacent stable equilibrium for the system to jump to, and hence the system effectively collapsed.

We finish this section by considering a thin elastic plate that is pinned (lateral deflection is prevented in the in-plane direction as well) on all four sides and with in-plane stress caused again by thermal loading. It has a small initial imperfection and is subject to elevated temperatures such that mode jumping in buckled patterns occurs [40]. We monitor the change in natural frequencies during this process by using both a Galerkin approach (incorporating AUTO [41, 42]) and a FEA [28, 29]. Again a numerical simulation during a parameter sweep reveals an interesting sequence of (instability) events: events manifested in the behavior of local vibrations.

10.6.2 The Analytic Approach

We return to Fig. 10.1 and make use of our earlier theory appropriate to initial imperfections, large deformations, and thermal loading [Eqs. (10.87) and (10.88)]. Because we will subsequently compare with numerical results by using FEs, it is convenient to identify a representative location on the central line with respect to the plate width (and indicated in the figures) to monitor deflection and dynamic response.

Because the out-of-plane boundary conditions are simply supported it is appropriate to assume the following functions for the deflected shape:

$$w(x, y) = \sum_{k,l} A_{kl} \sin(k\pi x) \sin\left(\frac{l\pi y}{r}\right),$$ (10.95)

$$w_0(x, y) = \sum_{k,l} A_{0kl} \sin(k\pi x) \sin\left(\frac{l\pi y}{r}\right),$$ (10.96)

where A_{kl} represents the amplitude of the buckling mode with k and l half-waves over the long and short directions, respectively and r is the aspect ratio. Odd values of the subscripts k and l denote the symmetrical modes, and even values denote the antisymmetrical ones. Galerkin's method is then used to generate the equations of motion, which are solved by use of the continuation package AUTO. With the temperature used as the control parameter, AUTO will follow the equilibrium path and track the bifurcation points. We can calculate the natural frequencies and vibration mode shapes (expressed in terms of the assumed modal amplitudes) of the plate by locally linearizing the system in the vicinity of equilibrium and solving the associated eigenvalue problem.

Equilibrium Paths

Consider a specific plate with $a = 762$ mm, $b = 282.2$ mm (and thus an aspect ratio of 0.37), and thickness $t = 1.9844$ mm. Young's modulus is $E = 70$ GPa, Poisson's ratio $v = 0.33$, thermal coefficient of linear expansion $\alpha = 23 \times 10^{-6}$, and mass density $\mu = 2.7143$ kg/m^3. The nondimensional damping ratio $\zeta = 0.05$. Also, the amplitude of the initial imperfection of the plate is assumed to be 30% of the plate thickness.

The bifurcation diagram generated by AUTO is shown in Fig. 10.11. The plate initially buckles (with a very small increase in temperature) in the vicinity of the critical load for the perfect geometry near $\Delta T = 1.5\,^\circ$C. Again, this really should not be thought of as buckling, because deflections increase from the start of the application of the axial loading. A secondary bifurcation occurs near $\Delta T = 69\,^\circ$C that results in a sudden dynamic jump to one of the two available postcritical branches. A more detailed study has revealed that it is important to use a considerable number of assumed modes in order to accurately capture this highly nonlinear behavior. In the computation of Fig. 10.11, $\{m\} = \{1, 2, 3, \ldots, 9\}$ and $\{n\} = \{1, 2, 3\}$. It is interesting to note that the primary postbuckled branch consists of only purely symmetric modes. The square data points in Fig. 10.11 show where AUTO detects an instability. This is associated with the behavior of the underlying eigenvalues of the system and reflects various bifurcation types (see Chapter 3). The dashed lines correspond to unstable paths, and it should be borne in mind that this figure is a bifurcation diagram based on a projection of a relatively high-order dynamical system. After the mode jump, there is a significant asymmetric component. This can be observed in a contour plot of the buckled shapes, also shown in Fig. 10.11.

An interesting aspect of this behavior is hysteresis. Suppose, after mode jumping has occurred, we start to *reduce* the applied temperature. In this case the plate

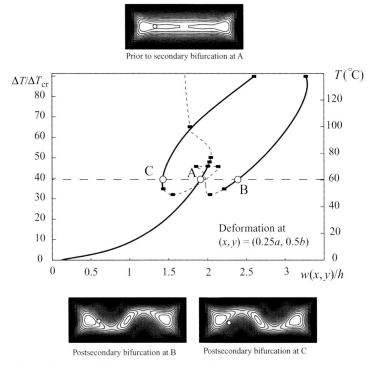

Figure 10.11. Equilibrium path of a specific point on the plate including critical temperatures.

would mode jump back onto the primary postbuckled path but at a temperature different from the original secondary bifurcation temperature. Thus we see that over a finite range of temperatures (of approximately 20 °C) there are in fact three coexisting stable equilibria. It is possible to perturb the system from one state to another, a feature we first observed in the behavior of discrete-link models [43].

Free Vibration

A plot of the three lowest natural frequencies as functions of temperature is shown in Fig. 10.12. Immediately after initial buckling the lowest mode of vibration is characterized by the (1,1) mode. This persists until the temperature change has reached a few °C when the (2,1) mode becomes the lowest. A number of subsequent vibration mode switches occur at higher temperatures. This type of mode switching was also observed in Section 10.1 with mechanical axial loading. The lowest natural frequency drops to zero at the critical secondary buckling temperature. This behavior was also observed for the Augusti model considered in Chapter 5. The hysteretic behavior is also confirmed because the lowest three natural frequencies associated with the post-mode-jumped equilibrium path extend to lower temperatures than those of the secondary bifurcation values. It appears that these paths also lose their stability at a nondimensional temperature close to $\Delta T/\Delta T_{\mathrm{cr}} = 87$. Some representative vibration mode shapes corresponding to specific temperatures are also shown in Fig. 10.12.

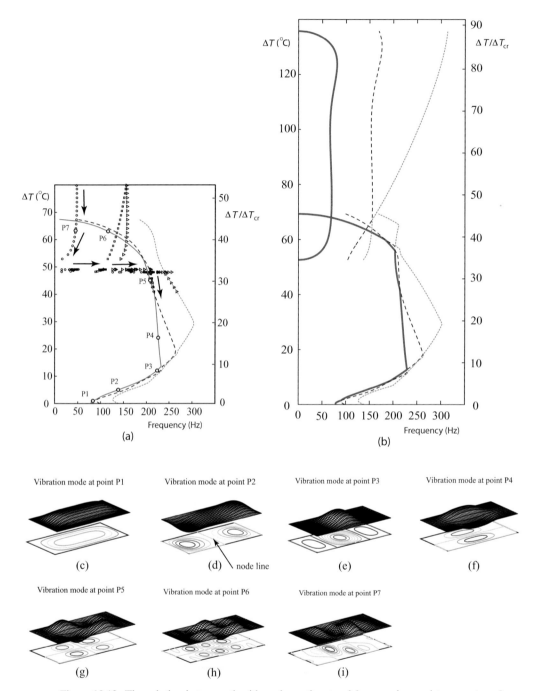

Figure 10.12. The relation between the (three lowest) natural frequencies, and temperature for a simply supported plate with an aspect ratio of 0.37: (a) finite-element results, (b) semi-analytical (AUTO) and (c)–(i) snapshots of the lowest vibration mode at specific temperature levels.

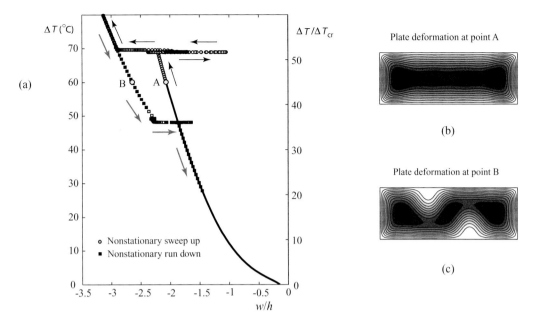

Figure 10.13. Transient response [at location $w(0.3a, 0.2b)$] obtained with a rate of increase of temperature of $\Delta T/\Delta t = 10\,^{\circ}\text{C/s}$ [29].

10.6.3 Finite-Element Transient Results

As done elsewhere in this book, it is illustrative to evolve the loading (in this case temperature) with a linear sweep. Figure 10.13 shows this temperature ramp starting from ambient, through initial buckling, then secondary buckling, and followed by a reverse sweep that reveals the hysteresis. We note that the rate of sweep may affect which of the two symmetric postcritical equilibria is followed. This is a subtle aspect of nonlinear dynamic behavior that has received recent interest as an example of *indeterminate* bifurcation [44]. The arrows indicate the path followed as the temperature is changed, with parts (b) and (c) showing the buckled patterns [45]. Another post-critical equilibrium configuration is also present but is not naturally revealed by a simple sweep of the control parameter [28, 29].

10.7 Cylindrical Shells

The structural (static and dynamic) analysis of shells is a vast subject [1, 46, 47] and has important practical applications. We finish this chapter by briefly looking at the vibrations of thin cylindrical shells subject to axial loads [48]. We also note that axial effects may also be induced by relatively large displacements (akin to struts and plates); this situation will be revisited in the final chapter of this book on large deflections.

   Consider the schematic of a thin cylindrical shell shown in Fig. 10.14. The three important spatial dimensions are the length L, the radius R, and the thickness h. We might anticipate this cylinder behaving in a way somewhat analogous to a beam if

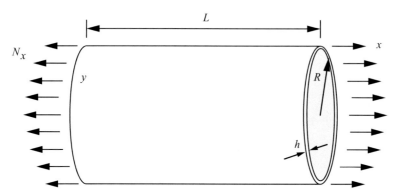

Figure 10.14. Schematic of a long cylindrical shell with axial loading.

L/R is large. Furthermore, we might expect axial and torsional modes to be present depending on the geometry. We focus attention on an intermediate class of problem for which Donnell and Flügge theory [49] has been shown to be appropriate. Under certain circumstances the stress–strain-deformation relations can be simplified in a manner similar to that of Section 10.1 on plates. The resulting governing equation of motion then takes the form

$$DV^8 w + \mu \frac{\partial^2 w}{\partial t^2} + \frac{Eh}{R^2} \frac{\partial^4 w}{\partial x^4} - N_x \nabla^4 \frac{\partial^2 w}{\partial x^2} = 0, \tag{10.97}$$

for uniform axial loading, that is, N_x constant, where the longitudinal and circumferential coordinates are (x, y). For ease of analysis it is assumed that simply supported (shear diaphragm) boundary conditions are present such that the midsurface lateral deformation of the cylinder w can be written as

$$w = A_0 \sin \frac{m\pi x}{L} \sin \frac{n\pi y}{\pi R} \cos \omega t, \tag{10.98}$$

where m is the number of half-waves in the longitudinal u direction and n is the number of half-waves in the circumferential v direction.

Equation (10.98) is placed in equation of motion (10.97), and the conventional solution procedure is followed. In the absence of axial loading, the frequencies can be shown to be given by

$$f_{nm} = \frac{\lambda_{nm}}{2\pi R} \sqrt{\frac{E}{\mu(1 - v^2)}}, \tag{10.99}$$

where

$$\lambda_{nm} = \frac{\sqrt{(1 - v^2)(m\pi R/L)^4 + (h^2/12R^2)\left[n^2 + (m\pi R/L)^2\right]^4}}{n^2 + (m\pi R/L)^2}. \tag{10.100}$$

Figure 10.15 shows the frequencies plotted as a function of the aspect ratio L/R for a thin shell with $R/h = 500$ and $v = 0.3$ [50, 51].

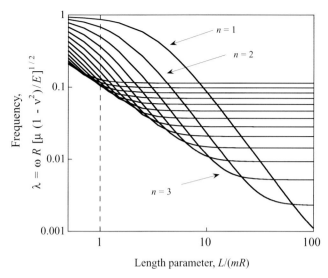

Figure 10.15. Natural frequencies for cylinders for different geometric parameters [50, 51].

As a specific example consider the case of a shell with $L/R = 1$. The mode corresponding to the lowest frequency is actually given by $n = 11$. Suppose the shell thickness is $h = 0.1$ cm, Young's modulus $E = 2 \times 10^6$ dyn/cm^2, and density $\mu = 8 \times 10^{-6}$ dyn/s^2/cm^4. The fundamental natural frequency [Eq. (10.99)] is $f = 174$ Hz.

Now suppose the cylinder is subject to an applied axial load of uniform intensity N_x. It can be shown that the effective natural frequency is altered according to

$$\omega_{nm}^2\big|_{N_x} = \omega_{nm}^2\big|_{N_x=0} + \frac{N_x n^2 \pi^2}{\mu h L^2}. \tag{10.101}$$

Given the same shell geometry as previously, we use the frequency dropping to zero to obtain the elastic buckling load $N_x/Eh = -12.12 \times 10^{-4}$ [13]. From static considerations, Batdorf [52] showed that the buckling stress ($\sigma = N_x/h$) for this type of cylinder was $\sigma_{cr} = -0.6Eh/R$, thus confirming this result.

For more slender geometries, that is, $L/R > 10$, the beamlike bending modes start to dominate, and in this region Flügge shell theory provides more accurate solutions [46]. In the other extreme, the behavior tends to be more platelike. Other studies have included the effects of initial imperfections [53] and composites [54, 55].

Equation (10.101) thus gives the familiar-looking result shown in Fig. 10.16 in which both the frequency and axial load are nondimensionalized. The linearity in the axial-force–frequency (squared) relation was encountered earlier in a variety of systems for which the buckling and vibration modes were similar. It is the relative simplicity of this relation that provides compelling motivation for nondestructive testing purposes, that is, using frequencies to predict buckling, and this is the subject of the next chapter.

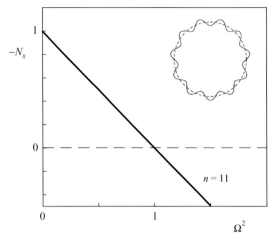

Figure 10.16. Fundamental natural frequency for a specific cylinder as a function of axial load.

Other Issues

The stiffening effect that is due to a spinning circular plate [56] is a problem of practical interest and is related to the spinning beam analysis in Chapter 7. Certain plate problems are conveniently solved by the methods of Lagrange multipliers [57] and finite differences [58]. Later chapters will also revisit shells in terms of step loading and parametric excitation.

References

[1]   S.P. Timoshenko and S. Woinowsky-Krieger. *Theory of Plates and Shells*, 2nd ed. McGraw-Hill, 1968.

[2]   A.W. Leissa. Vibration of plates. Technical Report SP–160, NASA, 1969.

[3]   T. von Karman, E.E. Sechler, and L.H. Donnell. The strength of thin plates in compression. *Transactions of ASME*, 54:53–7, 1932.

[4]   S. Levy. Bending of rectangular plates with large deflections. Technical Report 737, NACA, 1942.

[5]   S. Levy. Buckling of rectangular plates with built-in edges. *Journal of Applied Mechanics*, 9:171–4, 1942.

[6]   H.-N. Chu and G. Herrmann. Influence of large amplitudes on free flexural vibrations of rectangular elastic plates. *Journal of Applied Mechanics*, 23:532–40, 1956.

[7]   S.F. Bassily and S.M. Dickinson. Buckling and lateral vibration of rectangular plates subject to in-plane loads—a Ritz approach. *Journal of Sound and Vibration*, 24:219–39, 1972.

[8]   S.M. Dickinson. The buckling and frequency of flexural vibration of rectangular isotropic and orthotropic plates using Rayleigh's method. *Journal of Sound and Vibration*, 61:1–8, 1978.

[9]   C.F. Ng and R.G. White. Dynamic behavior of postbuckled isotropic plates under in-plane compression. *Journal of Sound and Vibration*, 120:1–18, 1988.

[10] G.H. Bryan. On the stability of a plane plate under thrusts in its own plane with applications to the buckling of the sides of a ship. *Proceedings of the London Mathematical Society*, 22:54–67, 1891.

[11] S. Ilanko. Vibration and post-buckling of in-plane loaded rectangular plates using a multiterm Galerkin's method. *Journal of Applied Mechanics*, 69:589–92, 2002.

[12] A.C. Ugural. *Stresses in Plates and Shells*. McGraw-Hill, 1999.

[13] G. Herrmann and J. Shaw. Vibration of thin shells under initial stress. *Journal of Engineering Mechanics*, 91:37–59, 1965.

[14] J.F. Doyle. *Nonlinear Analysis of Thin-Walled Structures*. Springer, 2001.

[15] K.K. Kapur and B.J. Hartz. Stability of plates using the finite element method. *Journal of Engineering Mechanics*, 92:177–95, 1966.

[16] R.G. Anderson, B.M. Irons, and O.C. Zienkiewicz. Vibration and stability of plates using finite elements. *International Journal of Solids and Structures*, 4:1031–55, 1968.

[17] B.A. Boley and J.H. Weiner. *Theory of Thermal Stresses*. Wiley, 1960.

[18] D.J. Johns. *Thermal Stress Analysis*. Pergamon, 1965.

[19] J. Marcinowski. Postbuckling behaviour of rectangular plates in axial compression. *Archives of Civil Engineering*, 45:275–88, 1999.

[20] K.D. Murphy. *Theoretical and experimental studies in nonlinear dynamics and stability of elastic structures*. Ph.D. dissertation, Duke University, 1994.

[21] R.E. Kielb and L.S. Han. Vibration and buckling of rectangular plates under in-plane hydrostatic loading. *Journal of Sound and Vibration*, 70:543–55, 1980.

[22] R.E. Kielb. Thermal buckling of uniform rectangular plates. Technical Report, U.S. Air Force Wright-Patterson, ASD-TR-75-37, 1976.

[23] T.R. Tauchert. Thermally induced flexure, buckling, and vibration of plates. *Applied Mechanics Reviews*, 44:347–60, 1991.

[24] K.D. Murphy, L.N. Virgin, and S.A. Rizzi. The effect of thermal prestress on the free vibration characteristics of clamped rectangular plates: Theory and experiment. *Journal of Vibration and Acoustics*, 119:243–9, 1997.

[25] M. Stein. Loads and deformation of buckled rectangular plates. Technical Report R–40, NASA, 1959.

[26] D.G. Schaeffer and M. Golubitsky. Boundary conditions and mode jumping in the buckling of rectangular plates. *Communications in Mathematics and Physics*, 69:209–36, 1979.

[27] R. Maaskant and J. Roorda. Mode jumping in biaxially compressed plates. *International Journal of Solids and Structures*, 29:1209–19, 1991.

[28] H. Chen and L.N. Virgin. Finite element analysis of postbuckling dynamics in plates: Part I: An asymptotic approach. *International Journal of Solids and Structures*, 43:3983–4007, 2006.

[29] H. Chen and L.N. Virgin. Finite element analysis of postbuckling dynamics in plates: Part II: A nonstationary analysis. *International Journal of Solids and Structures*, 43:4008–27, 2006.

[30] K.D. Murphy, L.N. Virgin, and S.A. Rizzi. Characterizing the dynamic response of a thermally loaded, acoustically excited plate. *Journal of Sound and Vibration*, 196:635–58, 1996.

[31] K.D. Murphy, L.N. Virgin, and S.A. Rizzi. Experimental snap-through boundaries for acoustically excited, thermally buckled plates. *Experimental Mechanics*, 36:312–7, 1996.

[32] L.N. Virgin. Parametric studies of the dynamic evolution through a fold. *Journal of Sound and Vibration*, 110:99–109, 1986.

[33] R.V. Southwell. On the analysis of experimental observations in problems of elastic stability. *Proceedings of the Royal Society of London*, 135A:601–16, 1932.

[34] K.D. Murphy, L.N. Virgin, and S.A. Rizzi. Free vibration of thermally loaded panels including initial imperfections and post-buckling effects. Technical Memorandum 109097, NASA, 1994.

[35] M. Uemura and O. Byon. Secondary buckling of a flat plate under uniaxial compression – Part 1: Theoretical analysis of simply supported flat plate. *International Journal of Non-Linear Mechanics*, 12:355–70, 1977.

[36] E. Riks, C.C. Rankin, and F.A. Brogan. On the solution of mode jumping phenomena in thin-walled shell structures. *Computer Methods in Applied Mechanics and Engineering*, 36:59–92, 1996.

[37] H. Troger and A. Steindl. *Nonlinear Stability and Bifurcation Theory: An Introduction for Engineers and Applied Scientists*. Springer-Verlag, 1991.

[38] P.R. Everall and G.W. Hunt. Mode jumping in the buckling of struts and plates: A comparative study. *International Journal of Non-Linear Mechanics*, 35:1067–79, 2000.

[39] G.W. Hunt and P.R. Everall. Arnold tongues and mode-jumping in the supercritical post-buckling of an archetypal elastic structure. *Proceedings of the Royal Society of London A*, 445:125–40, 1999.

[40] T. Nakamura and K. Uetani. The secondary buckling and post-buckling behaviors of rectangular plates. *International Journal of Mechanical Sciences*, 21:265–86, 1979.

[41] E.J. Doedel. *AUTO – Software for continuation and bifurcation problems in ordinary differential equations*. California Institute of Technology, 1986.

[42] E.J. Doedel, A.R. Champneys, T.F. Fairgrieve, Y.A. Kuznetsov, B. Sandstede, and X.J. Wang. Auto97: Continuation and bifurcation software for ordinary differential equations. Technical Report, Department of Computer Science, Concordia University, Montreal, Canada, 1997 (available by FTP from ftp.cs.concordia.ca in directory pub/doedel/auto).

[43] W.J. Supple. On the change in buckle pattern in elastic structures. *International Journal of Mechanical Sciences*, 10:737–45, 1968.

[44] J.M.T. Thompson and H.B. Stewart. *Nonlinear Dynamics and Chaos*, 2nd ed. Wiley, 2002.

[45] H. Chen. *Nonlinear analysis of post-buckling dynamics and higher order instabilities of flexible structures*. Ph.D. dissertation, Duke University, 2004.

[46] A.W. Leissa. Vibration of shells. Technical Report SP-288, NASA, 1973.

[47] E.H. Dowell. *Aeroelasticity of Plates and Shells*. Noordhoff, 1975.

[48] T. von Karman and H.S. Tsien. The buckling of thin cylindrical shells under axial compression. *Journal of the Aeronautical Sciences*, 8:303–12, 1941.

[49] L.H. Donnell. A new theory for the buckling of thin cylinders under axial compression and bending. *Transactions of ASME*, 56:796–806, 1934.

[50] R.D. Blevins. *Formulas for Natural Frequencies and Mode Shapes*. Van Nostrand Rheinhold, 1979.

[51] K. Forsberg. A review of analytical methods used to determine the modal characteristics of cylindrical shells. NASA Report CR-613, Lockheed Aircraft Company, CA, September 1966.

[52] S.B. Batdorf. A simplified method of elastic-stability analysis for thin cylindrical shells. Technical Report 874, NACA, 1947.

[53] A.E. Armenakas. Influence of initial stress on the vibrations of simply supported circular cylindrical shells. *AIAA Journal*, 2:1607–12, 1964.

[54] H.S. Shen. Thermomechanical post buckling analysis of imperfect laminated plates using a higher-order shear-deformation theory. *Computers and Structures*, 66:395–409, 1998.

[55] C.-S. Chen and C.-P. Fung. Non-linear vibration of initially stressed hybrid composite plates. *Journal of Sound and Vibration*, 274:1013–29, 2004.

[56] R.G. Parker and C.D. Mote. Tuning of the natural frequency spectrum of a circular plate by in-plane stress. *Journal of Sound and Vibration*, 145:95–110, 1991.

[57] J.H. Ginsberg. *Advanced Engineering Dynamics*. Cambridge University Press, 1995.

[58] F. Bleich. *Buckling Strength of Metal Structures*. McGraw-Hill, 1952.

11 Nondestructive Testing

11.1 Introduction

Previous chapters have repeatedly illustrated the often well-defined relation between axial load and natural frequency. In some cases, for example, a simply supported strut, the equivalence of the buckling and vibration modes results in an exactly *linear* relation between the axial load (providing it is less than critical) and square of the effective natural frequency:

$$\frac{\omega^2}{\omega_0^2} = 1 - \frac{P}{P_{cr}}. \tag{11.1}$$

In other cases, this relation is very nearly linear. For example, consider a simple cantilever. The fundamental frequencies in bending have the mode shapes

$$W(x) = \cosh\left(\frac{\lambda_i x}{L}\right) - \cos\left(\frac{\lambda_i x}{L}\right) - \sigma_i\left[\sinh\left(\frac{\lambda_i x}{L}\right) - \sin\left(\frac{\lambda_i x}{L}\right)\right], \tag{11.2}$$

with $\sigma_1 = 0.7341$, $\lambda_1 = 1.8751$ for the lowest mode and corresponding frequency [i.e., $\omega_1 = 3.516\sqrt{EI/(mL^4)}$]. The buckling mode for a cantilever with an end load is given by

$$W(x) = 1 - \cos\left(\frac{\pi x}{2L}\right), \tag{11.3}$$

with a critical load of $P_{cr} = \pi^2 EI/(4L^2)$. These normalized shapes are plotted in Fig. 11.1. They are close, but unlike the pinned–pinned (and some sliding boundary conditions) case, they are not equal. However, also plotted in this figure (as the dashed curve) is the buckling mode shape corresponding to the cantilever subject to self-weight. This shape is computed numerically and the critical parameter was established as $h_{cr} = 1.986$ in Section 7.9 (and equivalent to $\alpha = -7.837$). This is much closer to the fundamental mode of vibration, and in fact, the difference between them is never more than 1%.

The equivalence of the vibration and buckling mode shapes results in the linear relation between axial load and frequency, that is, the extent to which the vibration mode shape is changed by axial loading. Thus we have a frequency (squared) versus load relation that is closer to linearity for the vibrations of a cantilever subject to self-weight than an end load. However, even for the end-loaded cantilever case a simple use of ABAQUS shows that when $(P/P_{cr}) = 0.51875$ we obtain a lowest natural

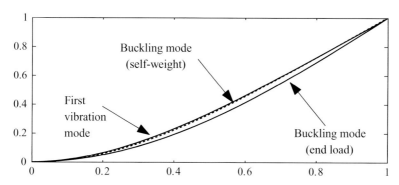

Figure 11.1. Comparison of vibration and buckling mode shapes for a uniform cantilever.

frequency of $(\omega/\omega_0)^2 = 0.43804$ compared with an estimate of $(\omega/\omega_0)^2 = 0.41825$ suggested by Eq. (11.1).

Thus Eq. (11.1) brings the possibility of using dynamics as a means of assessing axial-load effects, including the prediction of buckling [1–3]. In static buckling tests it is often unavoidable that specimens are destroyed during the experimental procedure (often the result of plastic deformation during large deflections). The Southwell plot is a related static approach that also exploits a linear extrapolation to predict buckling nondestructively [4]. Correlation studies between dynamic response and stiffness are also used to determine the actual boundary conditions as well [5, 6]. The simplicity of this relation can be used to nondestructively test axially loaded slender structural elements through monitoring of dynamic response [7–12].

11.1.1 The Southwell Plot

In Eq. (7.57) we saw how a small initial geometric imperfection tended to amplify the lateral deflections of a strut, especially as the buckling load is approached. Suppose we have a simply supported beam as shown in Fig. 11.2(a). We can measure the lateral deflections from the initially bent configuration, w_0, but here we measure the total lateral deflection, w, from the straight configuration. If we assume the initial deflection is in the form of a half-sine wave of amplitude Q_0, we can plot the amplification effect [Eq. (7.57)] as shown in Fig. 11.2(b). This, of course, assumes small deflections. However, in an experimental context what we would actually measure would typically be the lateral deflection over and above the initial deflection, which we can call $\delta = w - w_0$ and thus (at the midpoint of the strut)

$$\delta = \frac{Q_0}{1 - P/P_E} - Q_0 = Q_0 \frac{P/P_E}{1 - P/P_E}. \tag{11.4}$$

Equation (11.4) can be arranged in the form

$$\frac{\delta}{P} = \frac{\delta}{P_E} + \frac{Q_0}{P_E}, \tag{11.5}$$

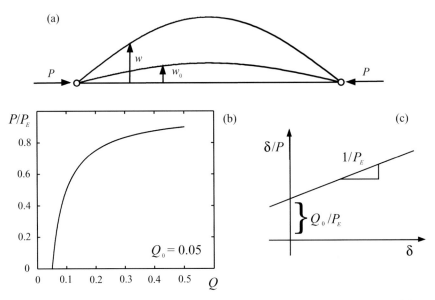

Figure 11.2. The Southwell plot: (a) strut geometry with an initial imperfection, (b) axial-load–lateral-deflection relation, and (c) Southwell plot.

so that if we plot δ/P as a function of δ we get a straight line in which the intercept is given by Q_0/P_E and the slope is given by $1/P_E$. Southwell [4] recognized the usefulness of this approach to determine both the critical load and initial imperfection, and this is shown schematically in Fig. 11.2(c).

A Southwell plot based on experimental data is shown in Fig. 11.3 [13]. Here, the data suggest a critical load (slope) in the vicinity of 87 N and an imperfection of $\epsilon \approx 0.05$ or about 3 deg; values not unreasonable when compared with the data presented in Fig. 5.10(c). Although there are limitations to this approach, the key utility here is that the linear relation allows for extrapolation. This provides some

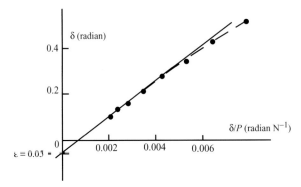

Figure 11.3. The Southwell plot obtained with experimental data taken from the system shown in Fig. 5.9. Adapted from Croll and Walker [13].

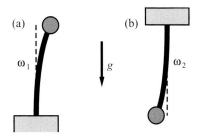

Figure 11.4. The effect of axial-load direction on natural frequencies.

motivation for exploring related concepts in the dynamic testing of structures, in which vibration testing *in situ* is a well-established procedure (e.g., in structural health monitoring [14]).

11.1.2 Examples

The relation between axial load and lateral vibrations and its potential use for nondestructive evaluation purposes goes back to Sommerfeld [15]. He made the simple observation that the fundamental natural frequencies of the two systems shown in Fig. 11.4 were quite different (with $\omega_2 > \omega_1$). He concluded that the greater the compressive stress, the lower the natural frequency of lateral vibration. With tensile stress, an increase in natural frequency was observed. Furthermore, in the former case it was noted that the frequency dropped to zero as the compressive load approached its critical value.

We can conduct a simple analysis of this system by using the methods developed earlier in this book. Suppose the strut has a length l, end mass m, flexural rigidity EI, and oscillates in gravity g. Introducing the nondimensional parameter $\alpha = \sqrt{mg/EI}$, we can readily show that when the strut has the mass placed at its top [Fig. 11.4(a)], the natural frequency is given by [16]

$$\omega = \sqrt{g\alpha/(\tan \alpha l - \alpha l)}, \qquad (11.6)$$

which remains positive until buckling occurs at $m_c = \pi^2 EI/4gl^2$. When the strut is turned upside down [Fig. 11.4(b)] the natural frequency becomes

$$\omega = \sqrt{g\alpha/(\alpha l - \tanh \alpha l)}. \qquad (11.7)$$

Thus, suppose we have a mass that corresponds to about the half the critical load, that is, $m = (\pi^2 EI)/(8gl^2)$; then $\alpha = \pi/(2\sqrt{2}l)$ and the natural frequency of the system in part (a) would be $1.106\sqrt{g/l}$, as opposed to $1.904\sqrt{g/l}$ for the system in part (b). In fact, even a mass that causes buckling in part (a) would result in oscillations of frequency $1.55\sqrt{g/l}$ in the inverted system [part (b)].

It is quite easy to demonstrate this experimentally [16]. Consider a simple polycarbonate cantilever strip, as shown in the inset to Fig. 11.5. If we consider the beam mass as being negligible compared with the concentrated mass added to its free end, then the theoretical results given by Eqs. (11.6) and (11.7) apply. For the specific case of a strip with $L = 0.181$ m, a second moment of area $I = 1.903 \times 10^{-12}$ m^4, and

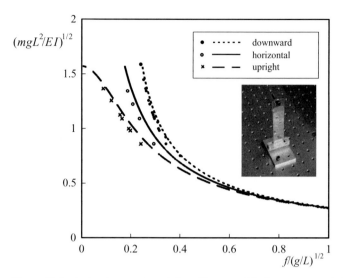

Figure 11.5. A simple experimental cantilever and its frequency variation with end load for different orientations.

Young's modulus $E = 2.4 \times 10^9$ N/m^2, we add increments of end mass and measure the resulting fundamental frequency. The nondimensionalized results are shown in Fig. 11.5 together with the theoretical results. Also included is the simple horizontal cantilever result based on the lumped stiffness approximation, that is, $\omega = \sqrt{K_e/m}$ in which $K_e = 3EI/L^3$ [17]. The effect of the mass of the beam itself can be included either in this lumped analysis or by use of a more sophisticated approach (adding a little distributed mass would tend to shift the data points up slightly), but the trend describing the effect of gravity (and hence axial loading) is clear. We also note at this point that experiments on cantilevers with self-weight loading are relatively easy to set up. The results from tests with other boundary and loading conditions, for example, in a testing machine, need more careful interpretation, as discussed in Section 7.2.

As a reference point the typical amount of end mass the strut was able to withstand before appreciably starting to droop to one side was about 27 g. The Euler load for a cantilever is $EI\pi^2/(4L^2)$, which gives a value of $m_c = 35$ g, and given the inevitable initial imperfections in the system this magnitude is not unreasonable. Furthermore, when no end mass was added the strut vibrated with a measured natural frequency of a little over 6 Hz (in fact 6.075, 6.2375, and 6.4 in its upright, horizontal, and downward orientations). The theory of continuous elastic beams covered in Chapter 7 listed a fundamental natural frequency for a cantilever of $\omega = 3.52\sqrt{EI/mL^4}$ and with the total mass of the strip measured at $mL = 5.94 \times 10^{-3}$ kg this corresponds to a predicted frequency of 6.38 Hz.

We can also reinterpret Fig. 7.18 at this point. Recall that this plot referred to a slender continuous strut subject to self-weight loading (rather than a concentrated end mass). Plotting the original "weight" parameter $|\alpha|$ as a function of frequency *squared* gives the results shown in Fig. 11.6(a). In the absence of gravity we would

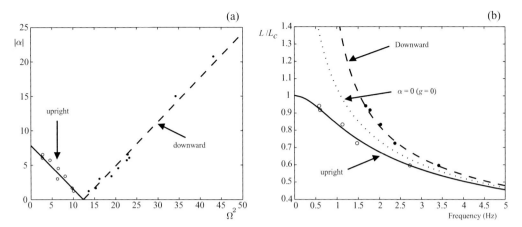

Figure 11.6. The frequencies of a simple but heavy experimental cantilever. The solid line represents the upright case, and the dashed line represents the hanging-down orientation: (a) $|\alpha|$ versus the fundamental frequency squared and (b) alternative plot of the same results.

expect the frequency to be proportional to the inverse of the length squared, and this case is shown too. However, in the upright configuration, as the critical length is approached the stiffness is diminished such that the frequency drops to zero at the critical length. If we plot the dimensional frequency versus the length (normalized by the critical length), we get the results shown in Fig. 11.6(b). However, not all the data from part (a) are included because of different thicknesses. The near linear relationship is, of course, a suitable form for extrapolation. Thus we might measure the fundamental natural frequency for a number of different α values (specifically changing the length L) and fitting a straight line to this data [Fig. 11.6(b)] we would *predict* buckling in the vicinity of $\alpha \approx 7.8$. Recall that in this plot the "weight" α is a nondimensional parameter given by $\alpha = mgL^3/(EI)$, and hence with mass per unit length of 0.0147 kg/m, cross-sectional dimensions of 25.4 mm × 0.508 mm, and Young's modulus of 2.4 GPa we get the actual length at buckling of about 0.33 m. The cantilever that hangs down never buckles as the length increases of course.

We expect the natural frequency of a system to reduce if more mass is added to it. But what the preceding setup shows is that if the mass acts through gravity then it may reduce the *stiffness* of the system, and it is this tendency that can be exploited in terms of nondestructive (stability) testing. In other words, if we refer back to Fig. 11.5 and consider a fixed value of the end mass toward higher values (where gravity has more effect) then the effect of orientation (and whether the effective axial load is compressive or tensile) is apparent. A good deal of the earlier material in this book has highlighted ways in which this trend may be relatively simple [18].

If the trend is linear then it also provides the possibility of predicting the elastic buckling not only from, in principle, measurement of the lower natural frequency at two distinct axial-loading conditions but even when one or more of these loads is tensile. In a practical (experimental) context, the boundary conditions may not be

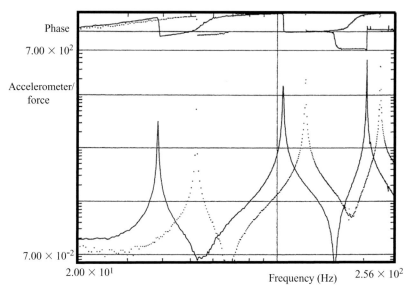

Figure 11.7. Frequency content of a prismatic beam showing the shift in resonant frequencies under the application of axial loading. Reproduced with permission from Elsevier [19].

known. Consider the results shown in Fig. 11.7, which were described in Livingston et al. [19]. This frequency spectrum was obtained from a prismatic beam by experimental modal analysis as part of a larger study in the context of system identification and parameter estimation. Over this frequency range the lowest three frequencies are quite distinct. The solid line corresponds to (practically) zero axial loading, with the dotted line showing the shift to higher frequencies when the beam is subject to a tensile axial load (approximately of a similar magnitude to that of the Euler buckling load with boundary conditions somewhat intermediate between clamped and pinned) [20, 21].

11.2 Some Background

As mentioned in the introduction to this chapter, the idea of using dynamic behavior to predict buckling has received attention from a number of researchers over the years. However, major contributions were made by Massonet [1], who considered a variety of structural systems from a theoretical standpoint, and Lurie [2], who showed the utility of this approach including experiments. Even when the mode of vibration and buckling mode are not identical the load–frequency (-squared) relation may be almost linear. Lurie used an energy approach to show that an upper limit for the frequency of axially loaded thin beams resulted in a relation

$$1 \geq \frac{m\omega^2}{EI} \frac{\int_0^l w^2 dx}{\int_0^l \left(\frac{d^2 w}{dx^2}\right)^2 dx} + \frac{P}{EI} \frac{\int_0^l \left(\frac{dw}{dx}\right)^2 dx}{\int_0^l \left(\frac{d^2 w}{dx^2}\right)^2 dx}. \tag{11.8}$$

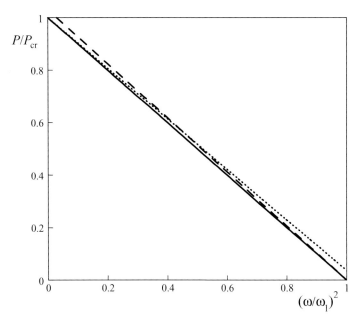

Figure 11.8. Upper and lower bounds of the frequency–load relation for a clamped–clamped beam.

For example, consider a clamped–clamped beam for which we know that $P_{cr} = 4\pi^2 EI/l^2$ and $\omega_n(P=0) = 22.373\sqrt{EI/(ml^4)}$, and using the buckling mode shape

$$w = A\left[1 - \cos\frac{2\pi x}{l}\right] \tag{11.9}$$

results in an expression

$$1 \geq 0.9635\left(\frac{\omega}{\omega_n}\right)^2 + \frac{P}{P_{cr}}. \tag{11.10}$$

Using the lowest mode of vibration [22] results in

$$1 \geq \left(\frac{\omega}{\omega_n}\right)^2 + 0.9704\frac{P}{P_{cr}}. \tag{11.11}$$

Relations (11.10) and (11.11) are plotted as inequalities in Fig. 11.8 together with the linear relation [Eq. (11.1)]. Thus we see the possibility of exploiting the linear relation between the square of the lowest natural frequency and the level of axial loading to extrapolate critical conditions [23].

Underlying General Theory. We have repeatedly looked at systems with a stiffness that tended to be diminished by the presence of (compressive) axial loading. In terms of potential energy, we can write this as

$$V = U(q_i) - \eta_k E^k(q)_i, \tag{11.12}$$

with the quadratic approximation in the form of the inner product

$$V = \frac{1}{2} < q, (U - \eta_k E^k)q >, \tag{11.13}$$

where the η_k $(k = 1, 2, \ldots, m)$ are independent parameters and U is the strain energy (symmetric and positive-definite). In terms of the equations of motion, we use Lagrange's equation to obtain

$$M\ddot{q} + (U - \eta_k E^k)q = 0, \tag{11.14}$$

and assuming harmonic motion in the usual way, $q = ue^{\lambda t}$, we obtain the characteristic equation

$$|M\lambda^2 + U - \eta_k E^k| = 0. \tag{11.15}$$

For conservative systems, we have $\lambda = i\omega$ with λ^2 identified as the negative of the square of the natural frequencies (see Chapter 4). We are, of course, primarily interested in systems for which $q = 0$ represents a stable system but may become unstable (at buckling), and this occurs when one of the eigenvalues vanishes. Although instability may occur by means of a complex pair of eigenvalues in nonconservative systems (flutter, e.g., Beck's problem, Section 7.8), in this book we remain primarily focused on the conditions under which a real eigenvalue vanishes at the *divergence* boundary.

The relation between ω^2 and η_k constitutes the characteristic curve (or surface, when more than one parameter is present). It has been proven [24, 25] that, for conservative systems with a trivial equilibrium state, any number of degrees of freedom, and equations of motion that are linear in the parameters η_k, the surface involving the fundamental frequency cannot have convexity toward the origin. Furthermore, it also follows that the fundamental surface is a plane (or straight line for a system with a single parameter) if the matrices M, U, and E^k can be reduced to a diagonal form simultaneously.

A useful implication of this convexity property (and of obvious usefulness in the context of the present chapter) can be concluded. The divergence boundary is contained in the characteristic curve, and we obtain it by setting $\omega^2 = 0$. If a single parameter ξ (load) is acting on the system, then it is possible to obtain an (upper bound) estimate of the critical value. From Fig. 11.9, we see that if we know the frequencies at two values of the loading parameter ξ, $\omega_{11}^2(\xi_1)$ and $\omega_{12}^2(\xi_2)$, we can gain an estimate of frequencies at other loading values. Of course, if the characteristic curve is a straight line (e.g., if the equations uncouple) then this estimate will be exact. By extrapolating a line joining them, we obtain an *upper bound* on the critical value of ξ from the intersection with the ξ axis. Clearly, the accuracy of the estimate also depends on the location of the two reference points. This is an issue that will be discussed later.

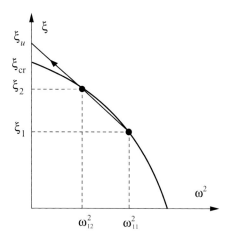

Figure 11.9. Convexity of the characteristic curve and its implication for providing a lower bound.

11.3 Snap-Through Revisited

In the previous section, we considered systems for which the fundamental equilibrium path was the trivial one. In snap-through buckling, we might still expect to monitor the lowest natural frequency to predict instability, but the nonlinearity of the underlying equilibrium curve can also have an influence. To quantify this, we go back to one of our standard forms from Chapter 3 in which we consider the dynamics of a system in the vicinity of a saddle-node bifurcation:

$$\ddot{X} - X^2 - \lambda = 0, \tag{11.16}$$

where both the deflection X and the load parameter λ are measured from the origin. Now suppose we have an equilibrium position (X_e), and we wish to study the behavior of small oscillations about it. We can expand Eq. (11.16) in the usual way by replacing X with $X_e + x$ that leads to

$$\ddot{x} - X_e^2 - 2X_e x - x^2 - \lambda = 0. \tag{11.17}$$

The x^2 term can be dropped because it is small, and because of equilibrium we also have $-X_e^2 - \lambda = 0$, and thus we are left with

$$\ddot{x} - 2X_e x = 0. \tag{11.18}$$

This system has the natural frequency

$$\omega = \sqrt{-2X_e} = \sqrt{+2(-\lambda)^{1/2}}, \tag{11.19}$$

and thus, for large negative λ say, we observe a linear relation between the loading parameter and the *fourth* power of the natural frequency [26].

Effect of Damping. So far we have concentrated mainly on undamped systems. In most of the mechanical systems of interest, there is usually a little energy

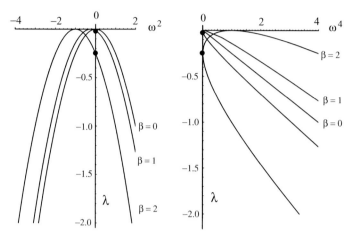

Figure 11.10. The effect of damping on the frequency–load-parameter relation.

dissipation, and we shall assume that this takes the form of a linear-viscous damping (see Section 3.5). Thus we consider

$$\ddot{X} + \beta\dot{X} - X^2 - \lambda = 0. \tag{11.20}$$

Conducting an analysis similar to that of the previous section we arrive at relationships between the natural frequency (the harmonic factor in the decaying, oscillating motion) and load parameter of

$$\omega^2 = \pm 2(-\lambda)^{1/2} - (\beta/2)^2,$$
$$\omega^4 = 4(-\lambda) \pm 4(-\lambda)^{1/2}(\beta/2)^2 + (\beta/2)^4. \tag{11.21}$$

These expressions are plotted in Fig. 11.10 for three values of damping including the undamped case. We see that damping has the effect of causing the natural frequency to diminish to zero *prior* to buckling. For example, with a damping level of $\beta = 2$ (and assuming the damping coefficient is constant), we observe that oscillations will cease when the load reaches a value of about $\lambda = -0.25$. One way of thinking about this is to recall the standard expression for a damping ratio: $\zeta = c/(2m\omega_n)$, but now the stiffness is reducing and thus, although the damping coefficient is constant, the damping ratio increases such that damping effectively becomes critically damped (to use the definitions introduced in Section 3.1) just prior to the stiffness dropping to zero.

We can again integrate the equation of motion while slowly sweeping through the load parameter (as was done in Chapter 3). For example, Fig. 11.11 shows nine trajectories generated for system equation (11.20) and with a constant value of the initial total energy with $\beta = 0.5$, λ evolved at the rate $30t$, and the initial conditions prescribed by $\dot{x}(0) = 0.0$, $x(0) = -\sqrt{300} + A$, where A varied between -8 and 8 in increments of 2. A number of interesting features can be seen in this figure. Damping does indeed appear to make the oscillations die out prior to instability, although this *drifting* system is never, of course, quite in equilibrium. We also see that

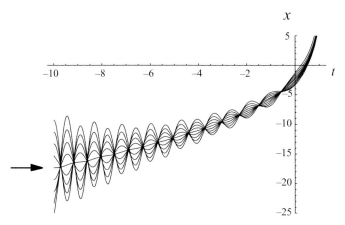

Figure 11.11. Some trajectories plotted as time series as the system is swept toward the saddle-node bifurcation.

the larger-amplitude oscillations display a degree of asymmetry—this feature is not unexpected because for a given amplitude the motion evolves toward the instability as the potential-energy well shrinks, from one side.

A simple experimental verification of this situation is shown in Fig. 11.12. Here, a flexible rod with an end mass in a heavily postbuckled configuration was subject to base rotation such that a saddle-node bifurcation was encountered. The base rotation can be thought of as *control* in our standard system of gravity acting on the mass. The system follows its complementary equilibrium path during which time the frequency of natural (superimposed) oscillations are measured. The jump at the saddle-node bifurcation is represented schematically as A–B in Fig. 11.12(b). The measured frequencies and their relation to the control parameter are shown plotted in Fig. 11.13. Part (b) shows some times series in which the parameter r is a measure of the rate at which the base is rotated. We see that raising the frequency to the fourth power provides a more linear relationship than for the second power with which to predict criticality [26].

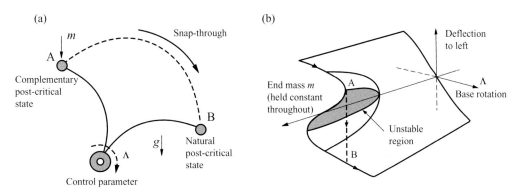

Figure 11.12. (a) A flexible strut with an end mass and (b) control surface showing a transition through bifurcation [26].

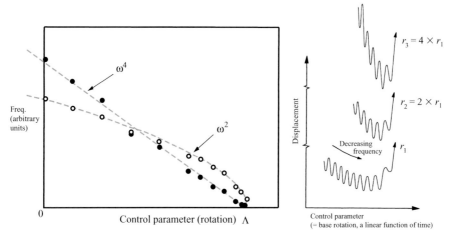

Figure 11.13. (a) Measured frequencies for the flexible strut rotated through a saddle-node bifurcation and (b) some sample time series with different rates of rotation [26].

The concept of using the reduction in frequency as a stability predictor has also been used in the context of secondary buckling [27]. Referring back to Fig. 5.21, it is apparent that once the Augusti model has buckled into its primary mode, it is the frequency of the *second mode* that can be used to infer the approach of the secondary bifurcation as this frequency tends toward zero.

11.4 Range of Prediction

The previous section indicated that rather than there being a universal relation between frequency and load, which would make extrapolation fairly straightforward, we see that damping, changing boundary conditions, initial imperfections, and type of instability, all conspire to make predictions more difficult. For example, Lurie [2] pointed out the increasingly important role played by initial imperfections as the critical load is approached (especially for plates). An experimental example of this effect [28] is shown in Fig. 11.14(a). Part (b) shows the results from a laminated composite column in which Chailleux et al. [28] identified three distinct regions, with region II providing the most useful (linear) relation for prediction purposes. Also in part (b), the authors noted that with very low load levels they experienced some clearance in the boundary conditions. As pointed out by Lurie [2], it may be possible to test specimens in tension, in which case initial imperfections will have minimal influence. In both of these sets of results, we observe that the linear relation breaks down near buckling (a feature first observed in Fig. 7.6).

Given that raising the frequency to either the second or fourth power might be a more appropriate predictor, it seems reasonable to raise the frequency to various powers in order to see how the subsequent curve might change from concave to convex, which then has clear implications for lower bound estimates. For example, the frequency–load relations for an elastic arch were given in Eqs. (8.76) and (8.77) for a shallow, and less shallow arch, respectively, and shown in Fig. 8.11. Plaut and Virgin [29] studied the effect of extrapolating frequency raised to various powers

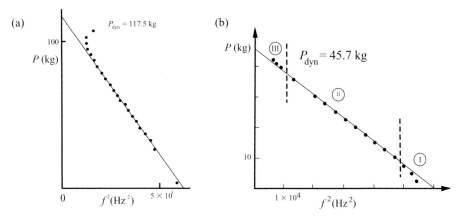

Figure 11.14. The load–frequency (-squared) relation for (a) rectangular duraluminum plate and (b) laminated composite column. Adapted from Chailleux et al. [28].

with a special reference for the range over which data were measured. That is, by raising the frequency to various powers a value is reached whereby the relation shown in Fig. 8.11 changes from convex to concave, with the transition point providing a close-to-linear relationship.

For example, Plaut and Virgin [29] used the criterion suggested by Singer et al. [12] using (numerical) data from the simple elastic arch considered earlier. Over a broad range of loading conditions, i.e, not necessarily close to buckling, we typically do *not* observe a (near) linear-frequency-squared–load relation. However, we can fit data to a relation of the form

$$p = C - D\Omega^r, \tag{11.22}$$

in which p is the load and Ω is the frequency ω_f [from Eq. (8.76) or (8.77)] nondimensionalized by the frequency under zero load. They showed that the values of r varied according to the range of load levels considered but that a change in curvature of the (frequency)^r versus load relation could be extracted to provide reasonable upper and lower bound estimates of the buckling load. That is, a value for r is sought such that it results in the most linear relation between p and (frequency)^r. The following table shows an example for the case $\lambda = 3$ (i.e., the less shallow arch).

N	C	D	r
4	11.04	11.04	2.15
5	9.30	9.32	2.69
6	9.25	9.27	2.71
7	9.35	9.37	2.69
8	9.38	9.40	2.65
9	9.46	9.47	2.61
10	9.48	9.48	2.60
11	9.53	9.52	2.57
12	9.59	9.57	2.52

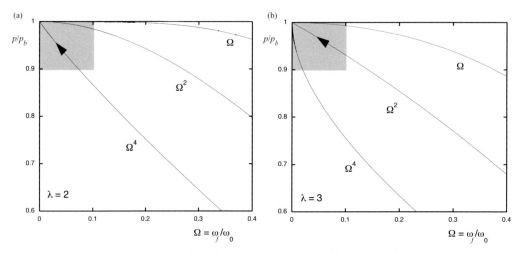

Figure 11.15. Frequency–load relations for an elastic arch: (a) $\lambda = 2$ and (b) $\lambda = 3$.

In this table, N represents the number of data points used in the fit and spreads roughly evenly from zero to critical loading, and the last column gives the exponent. In this nondimensionalization, the critical load (which corresponds to a bifurcation rather than to a limit point) occurs at $p = 9.71$. The data were subject to a nonlinear-least-squares fit and we observe a range of optimal exponents depending on the number of data points used.

The shallow arch is a useful device for illustrating frequency–load effects and applicability in terms of predicting instability because the λ parameter (i.e., the rise of the arch), can change the nature of the critical point [30]. For example, in Fig. 8.11, when $\lambda = 2$, the lowest frequency drops to zero when the arch experiences a saddle-node bifurcation and the arch snaps through to its inverted position. For a less shallow arch (e.g., when $\lambda = 3$), there is bifurcation and the arch experiences an asymmetric buckling (with a full-sine mode). However, the relations between frequencies and the type of instability (for example, as described in Section 3.4), and the material earlier in this section, were based on *local* generic behavior. If we take a closer look at Fig. 8.11 and focus in on a close proximity to buckling we get the results shown in Fig. 11.15. Here, the load has been normalized such that buckling corresponds to $p/p_b = 1$. Thus in the vicinity of the instability it is the frequency raised to the fourth power that is more useful for predicting the saddle-node [part (a)], and frequency squared for the branching bifurcation [part (b)].

11.5 A Box Column

A notable piece of work on the dynamic nondestructive evaluation of structures can be found in Jubb et al. [31]. The authors conducted some tests on box columns, which provided a clever way of studying plates by incorporating simply supported edges. One of their main goals was to establish a means of assessing the effects of residual stresses on the stiffness, dynamics, and stability of a typical structure. They suggested using the following variation on the frequency (f)–load (σ) theme:

$$k\frac{\sigma_r}{\sigma_{\text{cr}}} + \frac{\sigma_a}{\sigma_{\text{cr}}} + \left(\frac{f}{f_0}\right)^2 = 1. \qquad (11.23)$$

In this expression, the residual stress, σ_r, is added to the external stress, and k is a constant (less than unity) that takes account of stress distribution. We have already encountered how the (ordering of) modes of vibration of a plate depend on axial load (see Fig. 10.3), and in the experiments of Jubb et al. [31] they chose an aspect ratio of 4. In the absence of axial load, if the frequency corresponding to four half-sine waves in the longitudinal direction ($m = 4$) is denoted by 1.0, the lower modes turn out to have relative frequencies of 0.610 ($m = 3$), 0.391 ($m = 2$), and 0.282 ($m = 1$). As the axial force increases, these frequencies decrease (linearly with frequency squared) but at different rates such that it is the $m = 4$ mode that drops to zero at buckling (i.e., $P/P_{\text{cr}} = 1$).

The experimental results are shown in Fig. 11.16. The lowest four natural frequencies are plotted as functions of axial load. Two important points can be extracted from the results. First, the welded corners induce a degree of residual stress such that the initiation of an applied axial load does not correspond to zero axial load in the frequency–load relation. Second, a degree of postbuckling stiffening is apparent in each of the modes—this is a feature anticipated by the analysis of Section 10.4. Therefore it may be important to monitor the first few lowest natural frequencies in order to capture the appropriate buckling mode.

11.6 Plates and Shells

A thorough body of work on nondestructive testing of cylindrical shells under axial loading by use of dynamic (vibration) characteristics is due to Singer and his colleagues [11, 12]. An example of this type of research is shown in Fig. 11.17. Here a series of tests was conducted on cylindrical shells to see if the lowest frequency of lateral vibrations could be useful in predicting buckling. The cylinder included rib stiffeners, which had the effect of reducing some of the imperfection sensitivity typically encountered in axially loaded shells. Figure 11.17(a) shows a conventional frequency-squared versus load plot. Part (b) shows the same data with the frequency raised to the 2.9th power [12].

Figure 11.18 shows a plot in which (lower-load-level) frequencies are raised to various powers and then extrapolated as suggested in Plaut and Virgin [29].

The simplicity of the frequency–load relationship (or a variation thereof) can also be exploited when more complex structures are studied. For example, consider the panel (representative of a typical aeronautical configuration) shown in the lower part of Fig. 11.19. An eigenanalysis was conducted by Williams et al. [32] using the computer program VICONOPT. This structure, which might be a typical component in an aircraft fuselage, has ends that are simply supported such that sinusoidal modes result in the longitudinal direction. These are characterized by half-wavelengths λ,

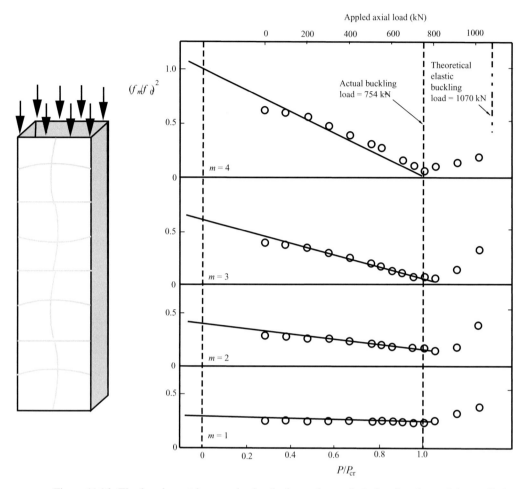

Figure 11.16. The four lowest frequencies for the box column plotted as functions of the applied axial load. Adapted from Jubb et al. [31].

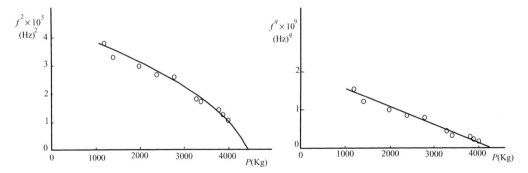

Figure 11.17. The lowest vibration frequency of an axially loaded cylinder: (a) frequency squared and (b) frequency raised to the power $q = 2.9$. Adapted from Singer et al. [12].

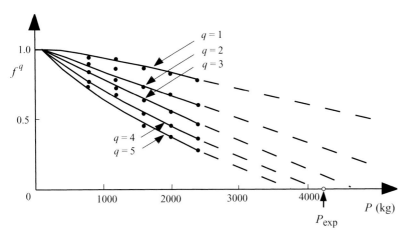

Figure 11.18. The same frequencies as in Fig. 11.17 but raised to various powers and subsequently extrapolated (linearly) to the buckling load. Adapted from Singer et al. [12].

which are integral fractions, l/i, of the length of the structure, l. A set of eigenvalues (natural frequencies) can be computed for values of λ.

Suppose we have a structure that is isotropic (but need not necessarily be so) and with the geometric properties shown in the lower part of the figure. The notation $\alpha = 1000/E$ and $\beta = \sqrt{(4000\rho/E)}$ are introduced, and Poisson's rato is $\nu = 0.3$. Suppose we wish to extract the natural frequencies below $6/\beta l$ when $l = 6b$. The computed characteristic curves shown in the upper part of Fig. 11.19 indicate natural frequencies where the vertical dashed lines intercept the curves, with the small markers indicating a frequency of $6/\beta l$. Thus we observe the frequencies listed in the top rows of the following table (the entries are in terms of the reciprocal of βl):

n_j	1.37(1)	1.66(1)	2.02(1)	2.29(1)	4.43(2)	4.82(2)
n_{oj}	0.82(1)	1.25(1)	1.70(1)	2.01(1)	3.85(2)	4.29(2)
n_j (cont'd)		5.25(1)	5.40(2)	5.99(2)	5.99(1)	–
n_{oj} (cont'd)		5.13(1)	4.94(2)	5.57(2)	5.89(1)	5.76(3)

Now, if the panel is subjected to a compressive load $\sigma = 0.0012E$ we can exploit the equivalence of the vibration and buckling modes to compute the reduced natural frequencies that are due to the presence of the axial load. That is, we make use of the relation [of the form of Eq. (11.1)]:

$$n_{oj} = \left[n_j^2 - \frac{\alpha\sigma}{\beta^2\lambda^2} \right]^{1/2}. \qquad (11.24)$$

Thus $(\alpha\sigma)^{1/2} \approx 1.1$ and for the various λ we get the altered natural frequencies as shown in the lower rows of the preceding table. We note how the order has changed, the equal values have separated, and a new natural frequency has fallen into the range of interest because of the presence of the axial load.

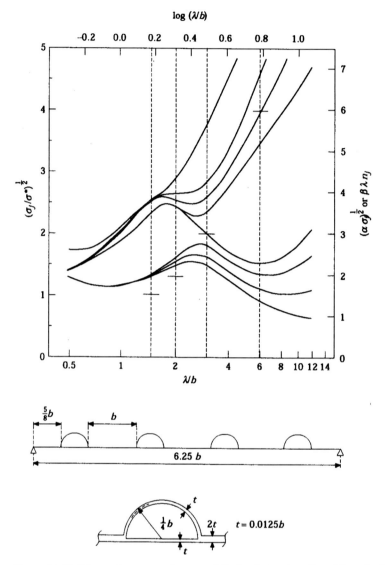

Figure 11.19. Eigenvalue curves for an axially loaded, prismatic, stiffened, plated structure. Reproduced with permission from Elsevier [32].

We note here also that the linearity in the axial load versus frequency-squared relation can also be used to infer the percentage error in neglecting shear effects in the modeling [33]. Another important practical application related to this approach is determining the level of stresses in pressure vessels [34].

References

[1]   C. Massonnet. Les relations entre les modes normaux de vibration et la stabilité des systemes élastiques. Technical Report, Bulletin des cours et des laboratoires d'essais des constructions du genie civil et d'hydraulique fluviale, Brussels, Belgium, I (1,2), 1–353, 1940.

[2] H. Lurie. Lateral vibrations as related to structural stability. *Journal of Applied Mechanics*, 19:195–204, 1952.

[3] C. Sundararajan. Frequency analysis of axially loaded structures. *AIAA Journal*, 30:1139–41, 1992.

[4] R.V. Southwell. On the analysis of experimental observations in problems of elastic stability. *Proceedings of the Royal Society of London*, 135A:601–16, 1932.

[5] J. Ari-Gur, T. Weller, and J. Singer. Experimental and theoretical studies of columns under axial loading. *International Journal of Solids and Structures*, 18:619–41, 1982.

[6] M.A. Souza. The effects of initial imperfection and changing support conditions on the vibration of structural elements liable to buckling. *Thin-Walled Structures*, 5:411–23, 1987.

[7] R.G. White. Evaluation of the dynamic characteristics of structures by transient testing. *Journal of Sound and Vibration*, 15:147–61, 1971.

[8] A. Segall and M. Baruch. A nondestructive dynamic method for the determination of the critical load of elastic columns. *Experimental Mechanics*, 20:285–8, 1980.

[9] P.M. Mujumdar and S. Suryanarayan. Nondestructive techniques for prediction of buckling loads – a review. *Journal of the Aeronautical Society of India*, 41:205–23, 1989.

[10] M.A. Souza and L.M.B. Assaid. A new technique for the prediction of buckling loads from nondestructive vibration tests. *Experimental Mechanics*, 31:93–7, 1991.

[11] J. Singer, J. Arbocz, and T. Weller. *Buckling Experiments, Vol. 1*. Wiley, 1998.

[12] J. Singer, J. Arbocz, and T. Weller. *Buckling Experiments, Vol. 2*. Wiley, 2002.

[13] J.G.A. Croll and A.C. Walker. *Elements of Structural Stability*. Wiley, 1972.

[14] C.R. Farrar, S.W. Doebling, and D.A. Nix. Vibration-based structural damage identification. *Philosophical Transactions of the Royal Society*, A359:131–49, 2001.

[15] A. Sommerfeld. Eine einfache Vorrichtung zur veranschaulichung des Knickungsvorganges. *Zeitschrift des Verein deutscher Ingenieure*, pp. 1320–3, 1905.

[16] A.B. Pippard. *Response and Stability*. Cambridge University Press, 1985.

[17] T.D. Burton. *Introduction to Dynamic Systems Analysis*. McGraw-Hill, 1994.

[18] A. Segall and G.S. Springer. A dynamic method for measuring the critical loads of elastic flat panels. *Experimental Mechanics*, 26:354–9, 1986.

[19] T. Livingston, J.G. Béliveau, and D.R. Huston. Estimation of axial load in prismatic members using flexural vibrations. *Journal of Sound and Vibration*, 179:899–908, 1995.

[20] A.L. Sweet and J. Genin. Identification of a model for predicting elastic buckling. *Journal of Sound and Vibration*, 14:317–24, 1971.

[21] A.L. Sweet, J. Genin, and P.F. Mlakar. Determination of column-buckling criteria using vibratory data. *Experimental Mechanics*, 17:385–91, 1977.

[22] D.J. Inman. *Engineering Vibration*. Prentice Hall, 2000.

[23] P.-Y. Shih and H.L. Schreyer. Lower bounds to fundamental frequencies and buckling loads of columns and plates. *International Journal of Solids and Structures*, 14:1013–26, 1978.

[24] K. Huseyin and J. Roorda. The loading–frequency relationship in multiple eigenvalue problems. *Journal of Applied Mechanics*, 38:1007–11, 1971.

[25] K. Huseyin. *Multiple Parameter Stability Theory and Its Applications*. Oxford University Press, 1986.

[26] L.N. Virgin. Parametric studies of the dynamic evolution through a fold. *Journal of Sound and Vibration*, 110:99–109, 1986.

[27] L.N. Virgin and R.H. Plaut. Use of frequency data to predict secondary bifurcation. *Journal of Sound and Vibration*, 251:919–26, 2002.

[28] A. Chailleux, Y. Hans, and G. Verchery. Experimental study of the buckling of laminated composite columns and plates. *International Journal of Mechanical Sciences*, 17:489–98, 1975.

[29] R.H. Plaut and L.N. Virgin. Use of frequency data to predict buckling. *Journal of Engineering Mechanics*, 116:2330–5, 1990.

[30] R.H. Plaut and E.R. Johnson. The effect of initial thrust and elastic foundation on the vibration frequencies of a shallow arch. *Journal of Sound and Vibration*, 78:565–71, 1981.

[31] J.E.M. Jubb, I.G. Phillips, and H. Becker. Interrelation of structural stability, stiffness, residual stress and natural frequency. *Journal of Sound and Vibration*, 39:121–34, 1975.

[32] F.W. Williams, P.N. Bennett, and D. Kennedy. Curves for natural frequencies of axially compressed prismatic plate assemblies. *Journal of Sound and Vibration*, 194:13–24, 1996.

[33] P.N. Bennett and F.W. Williams. Insight into the sensitivity to axial compressive load of the natural frequencies of structures which include shear deformation. *Journal of Sound and Vibration*, 209:707–22, 1998.

[34] R.R. Archer. On the influence of uniform stress states on the natural frequencies of spherical shells. *Journal of Applied Mechanics*, pp. 502–5, September 1962.

12 Highly Deformed Structures

In most practical applications, Euler–Bernoulli beam theory is often sufficient to provide useful information about the relation between axial loading and free vibrations. However, there are a number of instances in which the axial loading, or some related effect, results in relatively highly deflected states of the system, especially when the structure under consideration is very slender. For example, a pipeline or cable is characterized by having one of its dimensions very much greater than the other two, and the loads to which it is subject may often result in large deflections, even in cases in which self-weight is the only appreciable loading [1, 2]. Elastic bending stiffness does not necessarily dominate the effects of gravity, for example. In these cases, a more sophisticated description of the geometry is needed, and it is these types of flexible structures that form the basis for this chapter. In Chapter 7, we saw how *initial postbuckling* could be handled by retaining extra terms in the various energy expressions. But now, we allow (static) deflections to become large by using an arc-length description of the geometry and then consider small-amplitude oscillations about these nonlinear equilibrium configurations. In the final section, a FE solution is also shown for a specific case (essentially with the same approach as used toward the end of Chapter 9). It also turns out that experimental verification is relatively easy, especially if thermoplastics like polycarbonate are used.

12.1 Introduction to the Elastica

We start, in the usual way, by considering the behavior of an initially straight, inextensible, prismatic, thin elastic beam. The curvature of such a system is given by

$$\frac{1}{\rho} = \frac{d\theta}{dS}, \tag{12.1}$$

in which the deformed geometry of the system is described in terms of the arc-length coordinates S and θ, as shown for an axially loaded clamped beam in Fig. 12.1. In terms of Cartesian coordinates, we can write the angle as

$$\theta = \tan^{-1} \frac{dY}{dX}, \tag{12.2}$$

and, using

$$\frac{dS}{dX} = \sqrt{1 + (dY/dX)^2}, \tag{12.3}$$

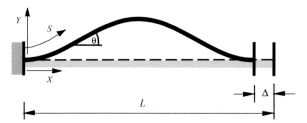

Figure 12.1. A thin elastic beam with clamped end conditions subjected to end shortening.

we obtain the familiar expression for curvature

$$\frac{1}{\rho} = \frac{d\theta}{dS} = \frac{d\theta}{dX}\frac{dX}{dS} = \frac{d^2Y/dX^2}{[1 + (dY/dX)^2]^{3/2}}. \tag{12.4}$$

Expanding the right-hand side as a Taylor series leads to

$$\frac{1}{\rho} = (d^2Y/dX^2)[1 + (3/2)(dY/dX)^2 + \cdots +]^{-1}. \tag{12.5}$$

Thus, if dY/dX is small, then we obtain the simple expression for curvature familiar from Euler–Bernoulli beam theory and generally used as an analytical basis in Chapters 7 and 8. In Eq. (7.40) the curvature was developed in terms of Lagrangian coordinates, and the next term in the expansion was retained for moderately large slopes. Both descriptions appear in the literature [3].

However, we now consider the fully (but still elastic) nonlinear system with no restriction on deflections. Returning to the example in Fig. 12.1 (and assuming no gravitational effects just yet) we can write the governing (elastica) equation in the relatively simple form

$$\frac{d^2\theta}{dS^2} = -\frac{P}{EI}\sin\theta, \tag{12.6}$$

in which P is an axial load associated with the end shortening Δ. This form is restricted to prismatic members with no forcing acting along its length but is a form familiar from the swings of a pendulum [4]. The boundary conditions must then be specified to obtain a solution to a specific problem. However, this equation is not easy to solve. Analytical solutions are available through elliptic integrals [5], and this form also allows mildly nonlinear solutions to be obtained with perturbation methods (see Naschie [3]), but we shall adopt numerical methods to solve this type of problem, that is, to obtain the deflected configuration under various axial loads and subsequent vibration properties about equilibrium configurations (which may be highly nontrivial) [6–8]. Four case studies will be described in which the primary difference between the cases concerns the boundary conditions. Experimental verification is included for each.

12.2 The Governing Equations

If we return to a prismatic strip but now inclined to the horizontal by an angle β, and characterized by flexural rigidity EI, length L, and weight (per unit length) W, we can describe the geometry in terms of coordinates $X(S, T)$ and $Y(S, T)$, and rotation $\theta(S, T)$ with respect to the X axis, where S is the arc length and T is time.

The internal forces in the strip are denoted $P(S, T)$ and $Q(S, T)$ parallel to the X and Y axes, respectively, and the bending moment is $M(S, T)$. The governing equations can be written as [9]

$$\partial X/\partial S = \cos\theta, \qquad \partial Y/\partial S = \sin\theta,$$
$$\partial\theta/\partial S = M/EI, \qquad \partial M/\partial S = Q\cos\theta - P\sin\theta, \tag{12.7}$$

and, in addition, we have dynamic equilibrium,

$$\partial P/\partial S = -(W/g)\partial^2 X/\partial T^2 - W\sin\beta,$$
$$\partial Q/\partial S = -W\cos\beta - (W/g)\partial^2 Y/\partial T^2. \tag{12.8}$$

Damping can be added at this point, but relative to the experimental results to be discussed later the damping is very light and is neglected in the analytical description.

We introduce convenient nondimensional quantities that are especially useful in the context of this kind of nonlinear formulation

$$w = WL^3/EI, \quad x = X/L, \quad y = Y/L, \quad s = S/L, \quad p = PL^2/EI,$$
$$q = QL^2/EI, \quad m = ML/EI, \quad t = (T/L^2)\sqrt{EIg/W}, \quad \Omega = \omega L^2\sqrt{W/EIg}, \tag{12.9}$$

where ω is a dimensional vibration frequency. In nondimensional terms, equilibrium equations (12.8) thus become

$$\partial x/\partial s = \cos\theta, \qquad \partial y/\partial s = \sin\theta,$$
$$\partial\theta/\partial s = m, \qquad \partial m/\partial s = q\cos\theta - p\sin\theta, \tag{12.10}$$

and for linear vibrations [Eqs. (12.8)]

$$\partial p/\partial s = -w\sin\beta - \partial^2 x/\partial t^2, \qquad \partial q/\partial s = -w\cos\beta - \partial^2 y/\partial t^2. \tag{12.11}$$

Assuming harmonic motion appropriate to small-amplitude vibration in the usual way, we can write the variables in the form

$$x(s, t) = x_e(s) + x_d(s)\sin\Omega t, \qquad y(s, t) = y_e(s) + y_d(s)\sin\Omega t,$$
$$\theta(s, t) = \theta_e(s) + \theta_d(s)\sin\Omega t, \qquad m(s, t) = m_e(s) + m_d(s)\sin\Omega t, \tag{12.12}$$
$$p(s, t) = p_e(s) + p_d(s)\sin\Omega t, \qquad q(s, t) = q_e(s) + q_d(s)\sin\Omega t,$$

where subscripts e and d denote equilibrium and dynamic quantities, respectively. At equilibrium, the equations are now given by

$$x'_e = \cos\theta_e, \quad y'_e = \sin\theta_e,$$
$$\theta'_e = m_e, \quad m'_e = q_e \cos\theta_e - p_e \sin\theta_e, \tag{12.13}$$

where the prime is used to denote the derivative with respect to s and where the internal forces can be written as

$$p_e(s) = p_0 - sw\sin\beta, \quad q_e(s) = q_0 - sw\cos\beta, \tag{12.14}$$

where p_0 and q_0 are constants representing values at $s = 0$.

We can determine equilibrium shapes by solving Eqs. (12.14), and then we obtain small vibrations about these equilibrium solutions by solving the resulting linear equations in the dynamic variables:

$$x'_d = -\theta_d \sin\theta_e, \quad y'_d = \theta_d \cos\theta_e,$$
$$\theta'_d = m_d, \quad m'_d = (q_d - p_e\theta_d)\cos\theta_e - (p_d + q_e\theta_d)\sin\theta_e, \tag{12.15}$$
$$p'_d = \Omega^2 x_d, \quad q'_d = \Omega^2 y_d.$$

The general approach used here for solving these types of nonlinear boundary-value problems is based on the shooting method [9, 10]. With this approach, the known boundary conditions at one end are used together with educated guesses of the unknown boundary conditions, and the nonlinear ordinary differential equations are solved numerically. However, the boundary conditions at the far end will not typically be satisfied, and the *error* is then used to iteratively re-solve the system until a tolerance has been achieved [11]. This is basically a root-finding approach and can be significantly simplified by use of some of the built-in capabilities of MATLAB or Mathematica, for example. Continuation is an alternative solution procedure [12].

12.3 Case Study A: Self-Weight Loading Revisited

In Section 7.9, we considered the effect of self-weight on the dynamics of an upright cantilever by using a Rayleigh–Ritz approach and included some simple experimental results. In the experimental results shown in Fig. 7.15(a) a degree of postbuckled stiffness can be observed for moderately large deflections. We take another look at this system but now using the elastica approach outlined in the previous section (as originally shown in Fig. 7.13). Figure 12.2 shows the arc-length coordinates together with some typical equilibrium configurations. These, of course, go well beyond the range of validity of Euler–Bernoulli theory.

Before some further results are presented, the appropriate nondimensionalization is mentioned. For the examples to be considered later in this chapter, it is natural to normalize the "weight" according to the first element in Eqs. (12.9) with the distance between the ends providing a natural control parameter. For the upright

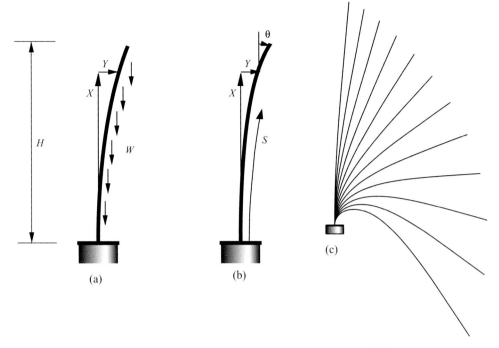

Figure 12.2. The slender column subject to self-weight: (a) basic geometry, (b) arc-length coordinate system, and (c) some typical deflected shapes.

column (and also the pinched-loop configuration to be considered later), it is more convenient (especially for subsequent comparisons with experimental results) to use $a = (EI/W)^{1/3}$ as the key nondimensional parameter. This is somewhat different from the scheme presented in Eqs. (12.9), which are used later. We use $h = H/a$ as the control, where H is the height of the column (and equivalent to L), such that increasing the length of the system, in the presence of gravity, increases the effective axial loading and leads to instability.

12.3.1 Numerical Results

For a vertical upright column (i.e., with the clamped end at the bottom), we can set $\beta = \pi/2$, and the gravity acts in the negative x direction. The boundary conditions are fully clamped at the base, that is, $x_e = y_e = \theta_e = 0$ when $s = 0$, and free at the tip, that is, $m_e = 0$ when $s = h$, and the shooting method is used to determine the unknowns at the tip to within a prescribed accuracy. A summary of the equilibrium solutions for the heavy column are shown in Fig. 12.3(a) as a bifurcation diagram, that is, a measure of the nondimensional (lateral tip) deflection, $y(h)$ (see Fig. 12.2) plotted against the control parameter (nondimensional column height h). A typical break in symmetry will be provided by some initial geometric imperfection, but this effect is not included here [9]. The trivial solution is $y_e(x) = 0$. The first bifurcation point occurs at the critical height $h_{cr} = 1.986$ (compare with the analyses described

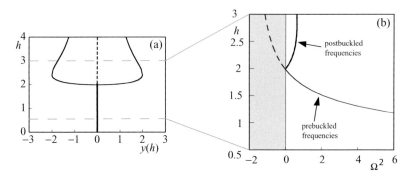

Figure 12.3. (a) Equilibrium paths for the column including gravity and (b) corresponding frequencies of small-amplitude vibrations.

in Section 7.9 [13]), and the trivial solution is unstable for larger values of h. The nature of the bifurcation point is supercritical [i.e., stable symmetric; see Fig. 3.7(a)], and the column smoothly begins to droop as the height is increased past its critical value. Again, this is behavior somewhat familiar from the approximate analytical solution. The stability of these equilibria can be obtained from a linear-vibration analysis.

Fundamental vibration frequencies are plotted in Fig. 12.3(b). Negative values of Ω^2 (shaded gray) are associated with unstable equilibrium states and with motions that grow exponentially (see Section 3.1). The fundamental frequency is zero at the critical height (at least for the geometrically perfect, undamped case). The curve is convex toward the origin, unlike typical characteristic curves in which a loading parameter (rather than the height) is plotted versus the frequency squared [14]—see Section 11.3. Fundamental frequencies for vibrations about the stable postbuckled equilibrium path for the perfect column are plotted also.

12.3.2 Experiments

We refer back to Section 7.9.2 that showed some experimental results for a (circular-cross-section) cantilever [9]. The rod was placed in an upright position and the length was incrementally increased. The results from these experiments are shown in Fig. 7.15, which suggests buckling between $H = 15$ and 20 cm. In this case, an estimate of the flexural rigidity was obtained from the (gravitional) droop of a horizontal cantilever and using the theoretical critical length, $1.986(EI/W)^{1/3}$ resulted in an estimated critical height very close to 20 cm. This is close to the height associated with the minimum value of the fundamental frequency, shown in Fig. 7.15(b). The column clearly exhibits a supercritical bifurcation.

A flat, slender strip of polycarbonate was used to take some additional data (natural frequencies) in which the geometry of the cross section ensured unambiguous deflection in a plane. The free end of the strut was subjected to a small perturbation and subsequent oscillations were monitored by a laser vibrometer. The fundamental frequency content was then extracted, with the results shown in

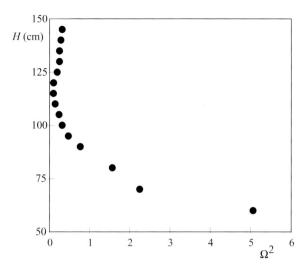

Figure 12.4. Fundamental frequency for a polycarbonate strip.

Fig. 12.4. The reduction in the lowest natural frequency can clearly be seen as buckling is approached (the theoretical critical length is approximately 110 cm in this case) together with an increase in the postbuckling frequencies. The form of the length–frequency relation follows the theoretical curves depicted in Fig. 12.3(b) quite closely, especially if an initial imperfection had been incorporated into the analysis. Also, the inevitable presence of a little damping has a minor effect on the frequencies, but again, this is not considered in the analysis here.

12.4 Case Study B: A Heavy Beam

Suppose the column is now rotated back to its horizontal configuration ($\beta = 0$) and both ends are constrained against lateral deflection and rotation. Rather than the column height (length), it is more convenient to use the axial (imposed) end shortening of the strip Δ as the control parameter (see Fig. 12.1), and this can also be placed into nondimensional terms by use of $\delta = \Delta/L$ [15–17]. Now we use the nondimensionalization used in Eqs. (12.9) rather than a from the previous section. In Chapter 7, we obtained the critical buckling load for a clamped–clamped beam of $4\pi^2 EI/L^2$ and lowest natural frequency of $22.37\sqrt{EI/(\rho AL^4)}$ for the straight beam (of length L and flexural rigidity EI) without including the effect of gravity (acting laterally on the beam and giving a constant weight W per unit length).

Now the boundary conditions when $s = 0$ are $x_e = y_e = \theta_e = 0$, and the quantities $m_e(0)$, p_0, and q_0 are determined by the shooting method, based on satisfying the conditions $x_e = 1 - \delta$ and $y_e = \theta_e = 0$ at $s = 1$ with sufficient accuracy.

A Note on Lift-Off and Self-Contact. For a very long strip, subject to end shortening, only a central part of the beam will tend to lift-off [18]. As the ends of the strip are pushed toward each other, the length of the proportion of the beam that does not lift off diminishes until the whole strip is characterized by a nonzero lateral deflection (other than immediately at the clamped ends). For convenience, the strip is called "short" if it does not touch the foundation between its ends. If there is a flat section resting on the foundation at both ends, the strip is called "long." The extent to which the beam is long or short depends to a large extent on its weight/stiffness ratio. A practical aspect of this is that for the long beam a zero-frequency traveling wave is observed [18].

Another interesting feature of this type of system is that, for sufficiently high values of end shortening, *self-contact* may occur, in which two points on the strip contact each other and the segment between has a teardrop (or pinched-loop) shape [19]. A geometry related to this specific case will be considered in a later section of this chapter.

12.4.1 Numerical Results

Equilibrium paths are depicted in Fig. 12.5 in the plane of end shortening δ versus axial load p_0, along with some corresponding equilibrium shapes, for weight parameters $w = 0, 25, 125, 250,$ and 343. As a reference point, for $w = 0$ (no weight), the fixed–fixed strip buckles at $p_0 = 4\pi^2$, and shapes at points A, B, and C along the postbuckling path are shown. This result was essentially obtained in Chapter 7 in which the increase in deflection after buckling (indicating a degree of postbuckled stiffness) was first observed. Self-contact occurs at C, when $\delta = 0.849$ and $p_0 = 72.18$ [18], with midpoint deflection $y(0.5) = 0.403$. Under increasing δ, the near-horizontal path to the right of C is followed, with large increases in p_0 associated with small increases in δ.

For heavy strips ($w > 0$), as δ is increased from zero, the strip is initially long but then becomes short. For $w = 25$, symmetric self-contact occurs when $\delta = 0.845$,

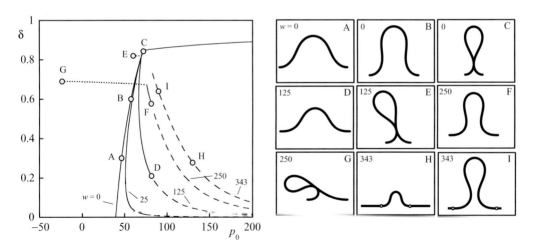

Figure 12.5. Equilibrium shapes and end shortening as functions of axial load for horizontal strip with weights (from left to right) $w = 0, 25, 125, 250,$ and 343.

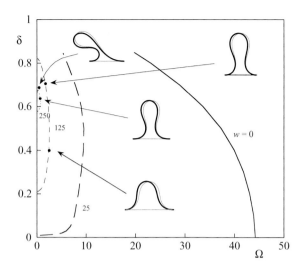

Figure 12.6. End shortening as a function of fundamental frequency for short strips with $w = 0, 25,$ 125, and 250, and mode shapes; equilibrium shapes are shown in gray.

$p_0 = 71.65$, and $y_e(0.5) = 0.401$. For $w = 125$, the symmetric shape becomes unstable when $\delta = 0.820$ and $p_0 = 69.74$, and the stable equilibrium associated with tilted (asymmetric) shapes bifurcates leftward in Fig. 12.5 toward point E, where self-contact occurs with $\delta = 0.820$ and $p_0 = 60.33$. These equilibrium shapes correspond only to positive lateral deflection (lift-off). In the *absence* of a foundation a variety of other equilibrium configurations are possible (although most of these are unstable) [20].

Small vibrations about equilibrium are shown in Fig. 12.6 in terms of the fundamental frequency for $w = 0, 25, 125$, and 250, along with the mode shapes for four specific cases. As the end shortening δ is increased, the fundamental frequency is zero at the transition from the long to short equilibrium shape, and then is zero again when the symmetric shape becomes unstable (as seen for weights w of 125 and 250).

For $w = 0$, the frequency at $\delta = 0$ is $\Omega = 44.36$, corresponding to the second mode of a fixed–fixed column subjected to an axial load $p_0 = 4\pi^2$. When $\Omega = 19.81$ for $w = 0$, and also when $\Omega = 5.35$ for $w = 25$, symmetric self-contact occurs, and these two curves in Fig. 12.6 are ended. For $w = 125$, as δ is increased beyond the value 0.820 where the symmetric shape becomes unstable, the strip tilts and the frequency increases until self-contact occurs when $\Omega = 0.155$ (and $\delta = 0.820$ still). An asymmetric mode along the path for $w = 250$ is shown in the top-left part of Fig. 12.6.

12.4.2 Experiments

It turns out that although the types of structures described in this section exhibit very large deflections it is relatively straightforward to confirm much of this behavior experimentally. Throughout this chapter some experimental studies in which thin polycarbonate strips are used are described. The general approach to measuring frequencies and mode shapes under different levels of axial loading is similar for each of the systems in this chapter. For this very flexible elastic material, the specific weight was measured at 11.2 kN/m^3 and Young's modulus was 2.4 GPa. The strips were 7.62 mm wide. Two thicknesses (0.508 and 1.016 mm) and two lengths (0.532 and 0.832 m) were used in combinations to yield three different values for the nondimensional weight w.

The strip was clamped such that one end could move toward the other in 6.3-mm increments to create the end shortening (see Fig. 12.1). This is the main control parameter in this example, that is, the actual (arc) length of the beam is held constant. An alternative means of changing the system would be to feed additional material in from one side, which was effectively done for the system shown in Fig. 3.9. The strip was deflected beyond the transition from long to short equilibrium before any measurements were taken for both the static and dynamic experiments.

A point-to-point laser vibrometer was used to measure the velocity at a user-prescribed point on the strip (avoiding any obvious node). For modal measurements, the velocity at multiple points was taken. For δ values ranging from 0.021 to 0.917, the first four frequencies were obtained by excitation of the strip by an impact hammer at different locations along the strip. Frequency measurements were also independently confirmed by measurement of the beam response to forced excitation, that is, use of a sine sweep (from 0.008 to 50 Hz) applied to the baseplate by an electromagnetic shaker.

For the modal analysis, the strip was again excited with a modal impact hammer. Data were acquired by the vibrometer (utilizing Bruel and Kjaer PULSE signal-processing software) and analyzed with ME'scope VES to generate a frequency-response function for each measurement point. This same approach was then used to measure the response at 30 different points along the strip. Vibration modes associated with the first few frequencies were constructed.

In Fig. 12.7, experimental and analytical results are compared, with frequencies given in hertz. The open circles are associated with tests on a strip of length 0.532 m and thickness 1.016 mm, giving $w = 8.145$. The solid circles in Fig. 12.7 correspond to tests on a strip of length 0.532 m again, but a thickness of 0.508 mm, so that $w = 32.56$. Several data points at high values of end shortening are associated with strips having self-contact. For the leftmost results, the open triangles were obtained experimentally for a longer strip, with length 0.832 m and thickness 0.508 mm (corresponding to $w = 124.7$).

In terms of higher frequencies, Fig. 12.8 shows the first four frequencies (in hertz) for the specific case $w = 32.56$. Black circles correspond to data acquired from an impact test; open circles denote results obtained by forced vibration.

The fundamental frequency is zero when the strip becomes short at $\delta = 0.017$ (with $p_0 = 80.86$). This can be inferred from Fig. 12.5. The frequency increases and then decreases. Self-contact occurs when $\delta = 0.844$ and $p_0 = 71.47$. Vibration mode shapes are shown in Fig. 12.9 for $\delta = 0.117$, with the analytical shapes on the left and the experimental shapes on the right. Further results (including cases for $\beta \neq 0$) can be found in Santillan et al. [18].

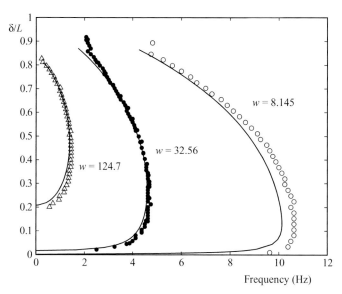

Figure 12.7. End shortening as a function of fundamental frequency for horizontal strips. Solid curves correspond to $w = 124.7$, 32.56, and 8.145; $\triangle$, $\bullet$, and $\circ$, experiment.

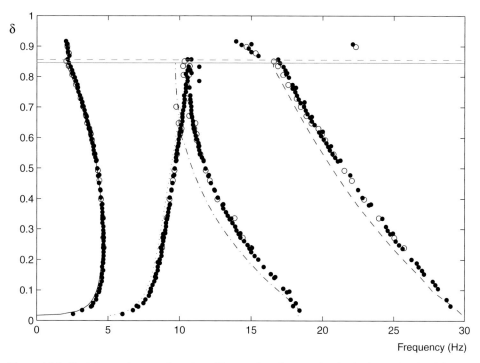

Figure 12.8. End shortening as a function of lowest four frequencies for horizontal strip with $w = 32.56$. experiment: $\circ$ (forced) and $\bullet$ (free); dashed curves, theory. Self-contact is indicated by the horizontal gray lines: continuous, theory; dashed, experiment.

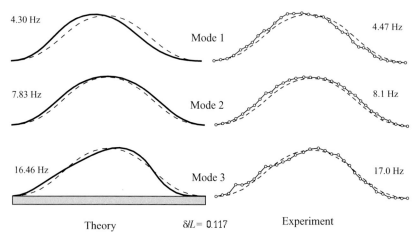

Figure 12.9. First three vibration modes from analysis and experiment for horizontal strip with $w = 32.56$ and $\delta = 0.117$; equilibrium shape is dashed.

A relatively highly deflected (or prestressed) elastic beam with pinned ends can also be considered to be an arch. Such a structure was analyzed by Perkins [21] in which a slightly different analytical approach was used to determine the natural frequencies of small-amplitude vibrations about highly nonlinear equilibria. He used a variational formulation to obtain the results shown in Fig. 12.10, in which the four lowest natural frequencies are plotted as a function of the nondimensional end load n. Some mode shapes at specific values of the end load are superimposed (for $n = 5, 15, 21.55$). The lowest natural frequency dropping to zero at the Euler buckling load (corresponding to $n = \pi^2$) is observed together with the subsequent jump to an asymmetric mode. He also conducted some experiments, and a comparison between theoretical and measured frequencies is summarized in the table following (in which H is the separation distance between the pinned ends, and L is arc length):

H/L	n	ω_1			ω_2			ω_3		
		Th.	Exp.	%	Th.	Exp.	%	Th.	Exp.	%
0.8	11	26.9	25.7	4.8	77.3	75.9	1.9	146	140	4.7
0.35	15	13.6	13.3	2.1	62.8	61.7	1.9	133.7	135	1.7
0.06	20	4.57	4.7	2.7	53.1	52.6	0.9	122.8	113.7	8.0

12.5 Case Study C: A Pinched Loop

An interesting extension to the analysis of the first case study can be made if the ends of the clamped beam are rotated such that they are pressed flat together, as shown in Fig. 12.11. Also shown in this figure (part c) is a snapshot of a highly deflected shape corresponding to a length beyond the buckling point for the upright equilibrium configuration. Again the parameter a is used for the nondimensionalization and the behavior is investigated through the evolution of the nondimensional

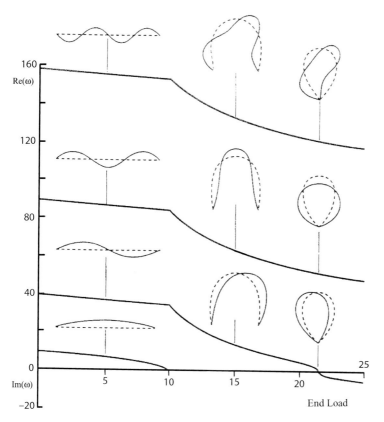

Figure 12.10. Natural frequencies and mode shapes of an elastic arch as functions of end load n. Reproduced with permission from ASME [21].

control parameter $l = L/a$. The boundary conditions are basically the same as for the clamped beam, but now we have $\theta_e(l) = -\pi$, and although it is easy to incline the pinched support [18], we focus attention on the upright system, that is, $\beta = \pi/2$ in the governing equations (and again include gravitational effects).

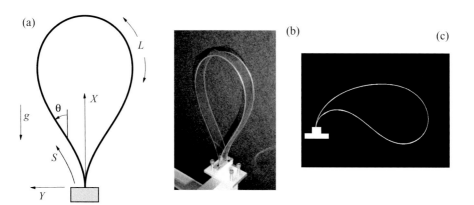

Figure 12.11. (a) Geometry of pinched loop, (b) photographic image of experimental setup, and (c) a highly deflected configuration.

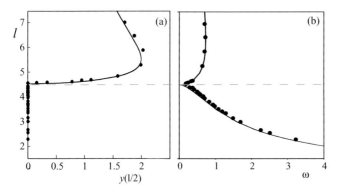

Figure 12.12. Horizontal midpoint deflection of upright loop as a function of length. Solid curve, numerical; •, experimental.

Because the theoretical approach is now well established, both theoretical and experimental results are plotted together in this section. The experimental results were based on a strip of cross-sectional dimensions 25.4 mm × 0.508 mm, which corresponds to the reference length a of 0.167 m, and the nondimensional vibration frequency ω of 0.130 times the dimensional frequency (where $\omega = \Omega\sqrt{a/g}$) that is due to the scaling of time: $t = T\sqrt{g/a}$—again note the difference with Eqs. (12.9). In this section, most of the results are presented in terms of nondimensional quantities of frequency ω and the coordinates x and y of the midpoint (where $s = 0.5$).

Equilibrium results for the loop are depicted in Fig. 12.12(a). The horizontal deflection of the midpoint is plotted as a function of the length of the loop, with the appearance of a critical length of $l = 4.50$ (signifying the onset of a supercritical pitchfork bifurcation).

A typical experimental frequency spectrum is depicted in Fig. 12.13, obtained with the laser vibrometer discussed in the previous section. The strip from which

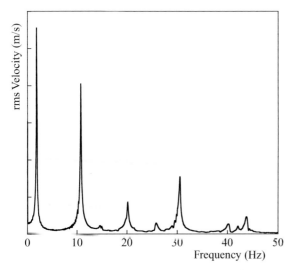

Figure 12.13. Typical frequency spectrum for upright loop with $l = 3.06$; main peaks at 1.81, 10.73, 20.16, 30.53 Hz.

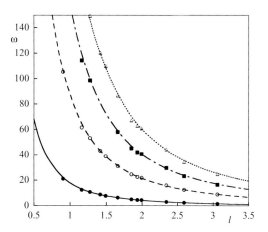

Figure 12.14. Lowest four natural frequencies of upright loop as functions of length.

these data were collected is 0.5096 m long ($l = 3.06$). The results were averaged over four time series. The four lowest-dimensional (in-plane) frequencies are distinct at 1.81, 10.73, 20.16, and 30.53 Hz. This procedure was repeated for several thicknesses and a number of lengths. The four lowest measured frequencies are denoted by circles, squares, and triangles in Fig. 12.14 over a range of nondimensional lengths (these vibrations are all about the trivial equilibrium configuration for this range of l). Solid curves represent the analytical results, and in general we see the anticipated reduction in the natural frequency as the loop length increases, thus observing the *softening* effect of gravity (for the upright orientation).

The mode shapes were also determined experimentally and analytically for some cases. The first four mode shapes for the upright case with $l = 3.48$ and a strip thickness of 0.508 mm are depicted in Fig. 12.15, along with the equilibrium shape. The corresponding measured dimensional frequencies are 1.3, 8.26, 15.8, and 23.9 Hz. As expected, the second and fourth modes are symmetric with respect to the vertical axis.

Finally, the fundamental vibration frequency is plotted in Fig. 12.12(b) as a function of the nondimensional length for relatively long loop lengths (including postcritical drooping). The experimental data points were obtained as the average values from multiple-frequency spectra. As the length of the loop is increased, the fundamental frequency decreases until it is effectively zero at the critical length $l \approx 4.5$. Extrapolation of measured fundamental frequencies at smaller lengths to the length at zero frequency could be used to predict the critical length (see Chapter 11). As the length is increased further and the loop droops (postbuckling deformation), the fundamental frequency increases.

12.6 Case Study D: A Beam Loaded by a Cable

In the final section of this chapter, we again consider the free vibrations of an elastic structural system characterized by large deflection that is due primarily to axial-load

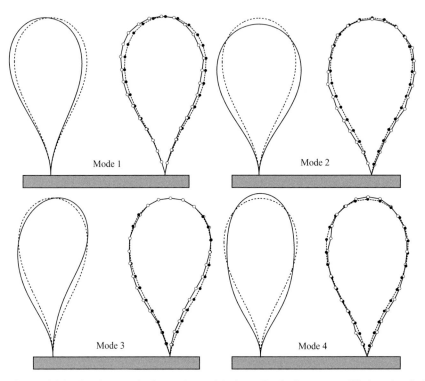

Figure 12.15. First four mode shapes for upright loop. Dashed curve, equilibrium (analytical); solid curve, mode (analytical); ●, equilibrium (experimental); ○, mode (experimental).

effects. A prismatic cantilever beam has a cable attached to its free end that is then pulled in a direction depending on the location of its far-end attachment point [22]. This can be considered as somewhat intermediate between the cantilever subject to an increasing end load that maintains its direction and Beck's problem (Chapter 7), although it can be shown that this is a conservative system. If the distant end of the cable is attached to the beam on its axis, then Timoshenko and Gere [5] showed that the critical load is equal to the Euler load (but for a pinned–pinned beam). If the far-end attachment point is offset, then symmetry is broken and the increase in load results in the nonlinear deflection of the cantilever in much the same way that initial geometric imperfections influence the behavior of axially loaded systems in general. The cable loading has a considerably greater effect than gravity and hence weight is not included in this analysis.

A schematic of the system is shown in Fig. 12.16. The primary method of analysis is again based on a shooting method solution of the boundary-value problem. Again some experimental verification is presented and some FE solutions are also included. Here, the tension in the (axially very stiff) cable is the principal control parameter (as the far end is pulled through the attachment point), with natural frequencies and mode shapes again characterizing the vibration of small-amplitude motion about even highly deflected equilibria.

We still basically have the same form of the governing static equations but with the addition of the horizontal component of the end force that complicates the

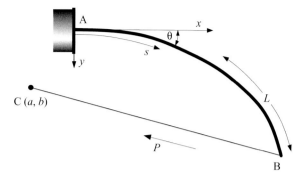

Figure 12.16. Geometry of a slender cantilever column loaded by a cable passing from the free end to a point near the base.

boundary conditions. Given the x and y offset coordinates a and b, respectively, the boundary condition (moment) at the ($s = 0$) clamped end is

$$M_e(0) = P_{ve}(L)b - P_{he}(L)a, \qquad (12.16)$$

and the load is adjusted (by use of Newton's method) until $M_e(L) = 0$ is satisfied.

In the results to be presented, some comparisons will be made with FE results obtained with ABAQUS. Beam elements (B31), suitable for large deflections, were used in which 1000 elements were employed to ensure spatial convergence. A truss element was used for modeling the cable, and an extreme negative thermal load was applied to drastically reduce the length of the cable and thus pull on the end of the cantilever. A path-following algorithm based on Riks method was employed [23, 24], that is, the same approach as used for the highly deflected cantilever in Chapter 9.

Figure 12.17 shows an experimental setup for the cable–beam system. A thin polycarbonate beam of dimensions 0.762 m long with a rectangular cross section of 25.4 mm × 4.8 mm is configured as a cantilever. The elastic modulus and density were given earlier in this chapter. The cable was made of high-strength woven steel wire with a stiffness of 11.67 kN/m. The cable was then connected to the base

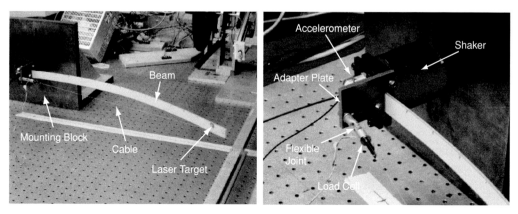

Figure 12.17. The experimental system of cantilever beam and end-loading cable.

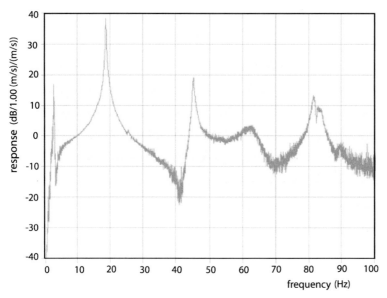

Figure 12.18. A typical frequency-response spectrum from the cable–beam system.

by a flexible joint with the other end attached to the tip of the cantilever. A load cell was incorporated into the wire connection in order to monitor the tensile force in the system. In some cases, the wire was replaced with shorter lengths to facilitate accurate tension measurements. The base of the cantilever was mounted to an electromagnetic shaker, with the excitation measured by an accelerometer. The response of the beam was again measured by a laser velocity vibrometer in the usual way.

Standard data acquisition and signal-processing data were used (the Bruel and Kjaer PULSE system), and a typical frequency response is shown in Fig. 12.18. This particular spectrum was taken from a system in which the nondimensional offset was $a/L = 0.0375, b/L = 0.0167$, and the applied load was $P/P_{cr} = 1.12$, in which P_{cr} corresponds to the elastic critical (Euler) load for the underlying case with no offset. The excitation employed here was a pseudorandom input signal over the range 0 to 100 Hz and with a sampling rate of $\Delta f = 0.03125$ Hz. Appropriate windowing and averaging were used to improve the quality of the data [25]. Vibration mode shapes were then extracted by the standard approach at a specific level of the cable tension.

Figure 12.19 shows the static and dynamic response for the beam with $a/L = 0.0375, b/L = 0.0167$. Part (a) shows both shooting results and ABAQUS together with some experimental data points for the lateral deflection of the tip. The inset shows a couple of deflected shapes. Part (b) shows the corresponding fundamental frequency, which also changes as a function of the tension in the cable. The frequency does not exhibit the type of monotonic decay we observed in earlier sections of this book. The frequency in part (b) is nondimensionalized by the fundamental frequency for a cantilever beam with no cable attached ($\omega = 3.516\sqrt{EI/mL^4}$), and it is interesting to see the effect the cable has (for both FE and shooting results) as the

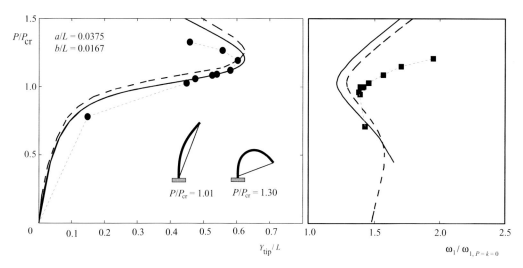

Figure 12.19. (a) Load-deflection characteristic with $b/L = 0.0167$ and (b) the corresponding fundamental natural frequency. The solid curves represent the FEA solution, dashed for analytical results obtained with the shooting method.

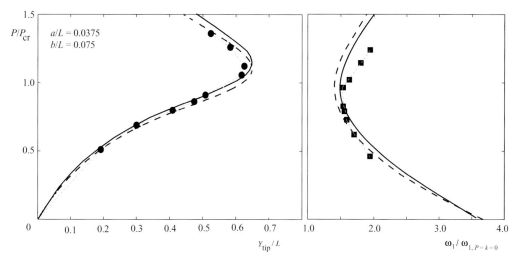

Figure 12.20. (a) Load-deflection characteristic with $b/L = 0.075$ and (b) the corresponding fundamental natural frequency. The solid curves represent the FEA solution, dashed for analytical results obtained with the shooting method.

tension tends to zero, i.e., even at zero tension the cable provides some constraint at the "free" end. Figure 12.20 shows similar results but with a larger static offset: $a/L = 0.0375, b/L = 0.075$. The behavior is seen to be quite sensitive to the magnitude of the cable offset.

A similar procedure was followed to obtain the higher frequencies and the first four, for the smaller offset, are shown in Fig. 12.21. The gray dashed curve at $P/P_{cr} = 1.12$ indicates the level at which the frequency spectrum illustrated in Fig. 12.18 was taken. Although the frequencies are separated in a somewhat typical

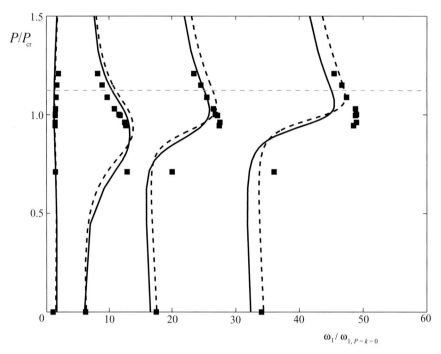

Figure 12.21. The four lowest natural frequencies plotted as functions of the tension in the cable with $b/L = 0.0167$. The solid curves represent the FEA solution, dashed for analytical results obtained with the shooting method.

spread for beam vibrations the effect of the tension loading depends on the specific frequency. Some vibration modes shapes are shown in Fig. 12.22 for the larger offset value. There is good agreement between the results from a shooting analysis (eigenvectors), ABAQUS, and experimentally determined mode shapes. There is an arbitrary phase in some of these plots. Finally, one of the potential applications of this type of system is solar sails (see also Section 9.6). In these innovative structural systems, the idea is to use the Sun's photons for propulsion based on a very lightweight but large surface area, rather like a kite. However, to keep such a membrane taut, it would need to be attached to relatively stiff but inevitably slender booms [26] that might typically lead to offset loading of the type considered in this section. There might also be some advantage to using tapered booms (see Section 8.3 and Holland et al. [27]).

12.7 The Softening Loop Revisited

Before we leave this chapter, a brief result is given that uses the elastica to solve the generating example described in Section 3.5, again with the shooting method. By incorporating a geometrically nonlinear (softening) moment–curvature relation from experiments, we obtain the subcritical pitchfork bifurcation result shown in Fig. 12.23. The dashed curve is the result when the effect of a small initial geometric imperfection is included in the analysis (i.e., the symmetry is broken). The lower part

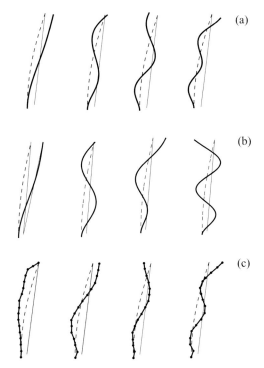

Figure 12.22. The lowest four vibration modes with $b/L = 0.075$: (a) shooting, (b) ABAQUS, and (c) experimental.

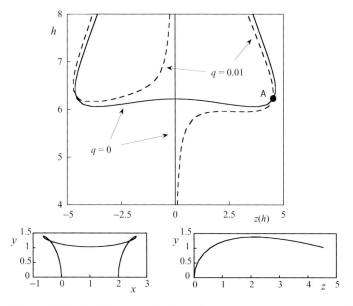

Figure 12.23. Equilibrium paths for softening loop, where q is an initial imperfection, $q = 0$ or $q = 0.01$. Two views of the deflected configuration corresponding to point A are also shown [10].

of this figure shows two views of the drooped configuration immediately after the initial instability. The transition between the upright and this drooped configuration is dynamic because there is no locally adjacent stable equilibrium, thus confirming the *subcritical* qualitative behavior from Figs. 3.10(c) and 3.10(d). Further details of the solution technique can be found in Plaut and Virgin [10].

In experiments on a loop with axisymmetric section properties, the natural frequencies consist of out-of-plane, in-plane, and twisting modes, and the lowest mode for each case is shown for oscillations about the (prebuckled) upright position in Fig. 12.24. Here h is a nondimensional parameter associated with one-half the length of the loop. Appropriately nondimensionalized, this also confirms part of the qualitative picture from Figs. 3.10(b) and 3.10(d), as well as the experimental data in Fig. 3.14, thus again confirming the trend of the lowest natural frequency toward zero at buckling.

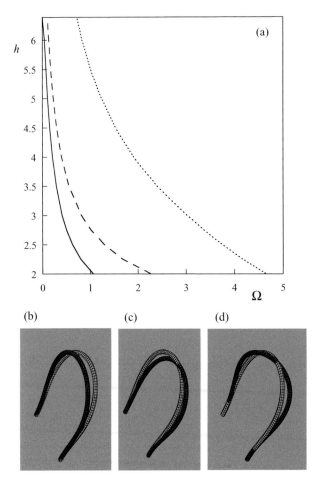

Figure 12.24. (a) Half-length versus lowest frequencies of each type, and corresponding modes of vibration, (b) out-of-plane (———); (c) in-plane (- - - - -); (d) twist (........).

References

[1] J.F. Wilson. *Dynamics of Offshore Structures*. Wiley, 2002.

[2] Y. Bai and Q. Bai. *Subsea Pipelines and Risers*. Elsevier, 2005.

[3] M.S. El Naschie. *Stress, Stability and Chaos in Structural Engineering: An Energy Approach*. McGraw-Hill, 1990.

[4] D.W. Jordan and P. Smith. *Nonlinear Ordinary Differential Equations*. Oxford University Press, 1999.

[5] S.P. Timoshenko and J.M. Gere. *Theory of Elastic Stability,* 2nd ed. McGraw-Hill, 1961.

[6] J.P. Cusumano. *Low-Dimensional, Chaotic, Nonplanar Motions of the Elastica: Experiment and Theory*. Ph.D. dissertation, Cornell University, 1990.

[7] C. Gatti-Bono and N. C. Perkins. Dynamic analysis of loop formation in cables under compression. *International Journal of Offshore and Polar Engineering*, 12:217–22, 2002.

[8] D. Addessi, W. Lacarbonara, and A. Paolone. On the linear normal modes of planar pre-stressed curved beams. *Journal of Sound and Vibration*, 284:1075–97, 2005.

[9] L.N. Virgin and R.H. Plaut. Postbuckling and vibrations of linearly elastic and softening columns under self-weight. *International Journal of Solids and Structures*, 41:4989–5001, 2004.

[10] R.H. Plaut and L.N. Virgin. Three-dimensional postbuckling and vibration of vertical half-loop under self-weight. *International Journal of Solids and Structures*, 41:4975–88, 2004.

[11] C.J. Goh and C.M. Wang. Generalized shooting method for elastic stability analysis and optimization of structural members. *Computers and Structures*, 38:73–81, 1990.

[12] M.A. Crisfield. *Nonlinear Finite Element Analysis of Solids and Structures, Vol. 2: Advanced Topics*. Wiley, 1997.

[13] A.G. Greenhill. Determination of the greatest height consistent with stability that a vertical pole or mast can be made, and of the greatest height to which a tree of given proportions can grow. *Proceedings of the Cambridge Philosophical Society*, 4:65–73, 1881.

[14] K. Huseyin. *Multiple Parameter Stability Theory and Its Applications*. Oxford University Press, 1986.

[15] R. Schmidt and D.A. DaDeppo. Large deflection of heavy cantilever beams and columns. *Quarterly Journal of Applied Mathematics*, 28:441–4, 1970.

[16] C.Y. Wang. A critical review of the heavy elastica. *International Journal of Mechanical Sciences*, 28:549–59, 1986.

[17] S.-B. Hsu and S.-F. Hwang. Analysis of large deformation of a heavy cantilever. *SIAM Journal on Mathematical Analysis*, 19:854–66, 1988.

[18] S. Santillan, L.N. Virgin, and R.H. Plaut. Post-buckling and vibration of heavy beam on horizontal or inclined rigid foundation. *Journal of Applied Mechanics*, 73:664–71, 2006.

[19] S. Santillan, L.N. Virgin, and R.H. Plaut. Equilibria and vibration of a heavy pinched loop. *Journal of Sound and Vibration*, 288:81–90, 2005.

[20] J.G.A. Croll. Some comments on the mechanics of thermal buckling. *The Structural Engineer*, 83:127–32, 2005.

[21] N.C. Perkins. Planar vibration of an elastica arch: Theory and experiment. *Journal of Vibration and Acoustics*, 112:374–9, 1990.

[22] H.G. McComb. Large deflection of a cantilever beam under arbitrarily directed tip load. Technical Report, Technical Memorandum 86442, NASA Langley Research Center, 1985.

[23] E. Riks. The application of Newton's method to the problem of elastic stability. *Journal of Applied Mechanics*, 39:1060–6, 1972.

[24] D.B. Holland, I. Stanciulescu, L.N. Virgin, and R.H. Plaut. Vibration and large deflection of cantilevered elastica compressed by angled cable. *AIAA Journal*, 44:1468–76, 2006.

[25] T.G. Beckwith, R.D. Marangoni, and J.H. Lienhard. *Mechanical Measurements*. Addison-Wesley, 1993.

[26] I. Stanciulescu, L.N. Virgin, and T.A. Laursen. Finite element analysis of slender solar sail booms. *Journal of Spacecraft and Rockets*, 44:528–37, 2007.

[27] D.B. Holland, L.N. Virgin, and R.H. Plaut. Large deflections and vibration of a tapered cantilever pulled at its tip by a cable. *Journal of Sound and Vibration*, 2007, to appear.

13 Suddenly Applied Loads

13.1 Load Classification

The early parts of this book focused attention on the dynamics of structures in which axial loading was increased *quasi-statically*. Thus dynamic response was typically considered under effectively *set* axial-loading conditions and observed in terms of free vibrations. However, it is just as common for either the axial load to be applied dynamically or an axially loaded structure to be subjected to dynamic lateral loading as well [1]. In these cases, it is not uncommon for the maximum response to occur during transient motion. In this chapter, we look at a number of different scenarios in which a structure with a constant axial load is then subject to various types of (dynamic) disturbance forces. These will range from a slow, but nonnegligible, increase in axial loading, to suddenly applied loading (e.g., an impulse or step input [2–5]).

We have already seen (e.g., Fig. 5.7) that the free response of an undamped nonlinear system is strongly influenced by the magnitude of the initial conditions. In some, a large initial velocity can be considered as an impulse. Before moving on to consider some specific structural systems, we return to the simple (underdamped) *linear* oscillator from Chapter 3 (Fig. 3.1). However, we now add a step input applied directly to the mass such that the governing equation of motion is given by

$$m\ddot{x} + c\dot{x} + kx = F_0, \tag{13.1}$$

in which we assume a system initially at rest, that is, $x(0) = \dot{x}(0) = 0$. It is easy to show (by use of Laplace transforms for instance) that the response of this system is given by

$$\frac{x(t)}{F_0/k} = \bar{x}(t) = 1 - \frac{\omega_n}{\omega_d}e^{-\zeta\omega_n t}\sin(\omega_d t + \phi_1), \tag{13.2}$$

where $\phi_1 = \cos^{-1}\zeta$, $\omega_n^2 = k/m$, $\omega_d = \omega_n\sqrt{1 - \zeta^2}$, $\zeta = c/(2m\omega_n)$.

A typical response (normalized with respect to the magnitude of the input) is shown in Fig. 13.1 with $\omega_n = 1$, $\zeta = 0.1$. This type of step response can be characterized by a number of metrics, for example, how long the transient takes to die out or by how much the response overshoots the final resting position initially. Clearly, damping has a key role to play here. It can be shown that the maximum percentage overshoot (OS, i.e., $\bar{x}$ over and above unity) is

$$OS = 100e^{-\pi\zeta/\sqrt{1-\zeta^2}}, \tag{13.3}$$

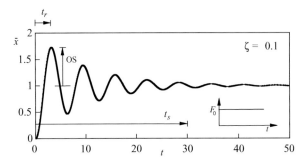

Figure 13.1. The step response of a linear system.

that is, twice the static response for an undamped ($\zeta = 0$) system, and this occurs after the rise time (t_r). Thus we may envision a load applied suddenly to be stronger in its effect than the same load applied quasi-statically (i.e., on a much slower time scale than the flexural dynamics of the system). Similarly, the time taken for the response to decay to within 5% (say) of the final value is called the settling time (t_s) and is often approximated by $3/(\zeta \omega_n)$ [6].

So much for the step response of a linear system. Things become more interesting when we study the effect of a step input on the response of a nonlinear system, and specifically one in which the nonlinear system relates to the equilibrium of an axially loaded structure. Not only may a suddenly applied load cause additional axial loading through large-amplitude effects but it may also result in collapse because of the traversing of an adjacent underlying unstable equilibrium (local potential-energy hilltop), or snap-through, for example.

13.2 Back to Link Models

A good example of where stability *in the large* may be especially important is for imperfection-sensitive structures. If we reconsider the system shown in Fig. 5.9(a) with $C = 0$ and thus $\alpha = 1$ we obtain the equilibrium condition

$$\Lambda = \frac{(\sin \theta - \sin \theta_0) \cos \theta}{\sin \theta}. \tag{13.4}$$

Note that because we are now interested in relatively large excursions from equilibrium we do not use the approximation given by Eq. (5.45). The dynamic response is governed by

$$\ddot{\theta} + \omega_n^2 \left[(\sin \theta - \sin \theta_0) \cos \theta - \Lambda \sin \theta \right] = 0, \tag{13.5}$$

where we have already established the relation between the linearized natural frequency and axial load (see Section 5.3).

We now investigate the robustness of equilibrium to relatively large disturbances as a function of the initial imperfection. Without damping we can view

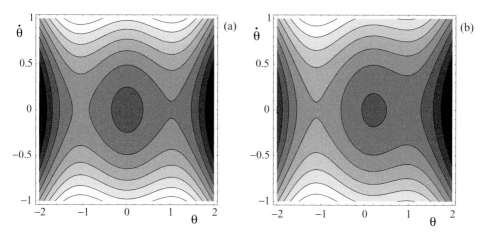

Figure 13.2. Contours of constant total energy indicating the changing catchment region as a function of initial imperfection.

free oscillations within the context of constant total energy, the level of which is determined by the initial conditions. For example, in Fig. 13.2 are shown two contour plots of total mechanical energy when the axial load has been set to $\Lambda = 0.5$, that is, at 50% of the critical load for the initially perfect geometry. In part (a), the initial imperfection is $\theta_0 = 0.01$, and in part (b) $\theta_0 = 0.1$. We see that, as expected for an imperfection-sensitive structure, the difference in potential energy between the stable and closest unstable equilibrium has diminished for the larger initial imperfection, and thus, given a certain initial velocity or step input, we would have a greater chance of an *escaping* solution. For example, when $\theta_0 = 0.01$ we have a stable equilibrium at $\theta_e = 0.020$ with adjacent saddle points at $\theta_e = -1.0538$ and 1.0404, and when $\theta_0 = 0.1$ we have a stable equilibrium at $\theta_e = 0.2055$ with the adjacent saddle points at $\theta_e = -1.1043$ and 0.9654.

Going back to Fig. 5.7, we saw some large-amplitude motion that was due to initial conditions somewhat distant from equilibrium. These were, of course, highly nonlinear, and in general recourse is made to numerical integration to solve these types of equations of motion. However, the large excursions generated by the application of a sudden load (or relatively large initial conditions) may lead to escaping solutions, which correspond in some sense to a dynamic buckling load. Consider the link model shown in Fig. 5.16. Suppose the load is held fixed at $p = 0.005$. In this case, there are three relevant equilibria: $\theta_e = 0.036789$ (stable), $\theta_e = 0.325084$ (unstable), and $\theta_e = 0.813625$ (stable).

For a system at rest (equilibrium), we may prescribe an initial velocity such that enough energy is imparted to the system that the subsequent transient traverses the hilltop. We can extract the critical velocity from the total energy and find $\dot{\theta}_0 \approx 0.0496$ if $\theta(0) = 0.036789$. This can be considered as an impulse given to the system, with an initial velocity of $\dot{\theta}(0) > 0.0497$ resulting in motion that goes beyond the distant equilibrium—undergoing oscillations that are far from sinusoidal. The trajectory passes through the unstable equilibrium position: a potential-energy hilltop.

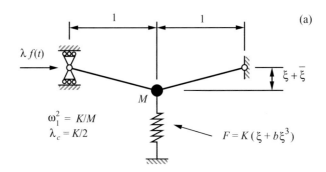

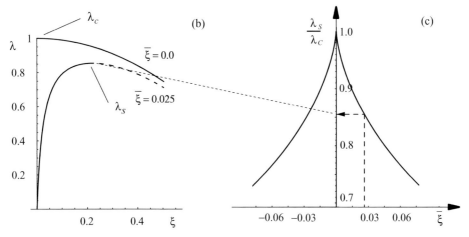

Figure 13.3. (a) A schematic of a link model, (b) equilibrium paths for the perfect and imperfect system illustrating an unstable-symmetric point of bifurcation, and (c) imperfection sensitivity [7].

An Approximate Analysis. An early study of link models subject to suddenly applied loading can be found in Budiansky [7]. In this paper, an approximate treatment was given based on perturbation theory for imperfection-sensitive structures (in which case instability is associated with some kind of severe collapse). We start by considering the case in which a step load is applied to a link model. Imperfection sensitivity was first encountered in Subsection 3.4.3 and a specific example given in Section 5.3. We again consider a simple model, shown in Fig. 13.3(a). Following the analysis in Budiansky [7] the nonlinearity in this system is solely due to the spring characteristic rather than to any kind of large-deflection effect encountered in Chapter 5. Also, the bars (of unit length) are assumed to be rigid but weightless, and all the mass is concentrated at the central hinge (to give a natural frequency of unity).

Adopting an approximate (Galerkin) approach to this system, we obtain an equation of motion,

$$\left(\frac{1}{\omega_1^2}\right)\ddot{\xi} + \left[1 - \frac{\lambda f(t)}{\lambda_c}\right]\xi + b\xi^3 = \left[\frac{\lambda f(t)}{\lambda_c}\right]\bar{\xi}, \qquad (13.6)$$

with underlying (cubic) equilibrium paths of the form [for $f(t) = 1$]

$$(1 - \lambda/\lambda_c)\xi + b\xi^3 = (\lambda/\lambda_c)\bar{\xi}. \tag{13.7}$$

Given $b = -1$, the solutions of Eq. (13.7) are plotted in Fig. 13.3(b) for both the perfect ($\bar{\xi} = 0$) and imperfect ($\bar{\xi} = 0.025$) cases. We thus confirm that in the presence of an initial imperfection (which may also be a small lateral load) the load-carrying capacity of the structure is somewhat diminished. The extent of the reduction of the maximum load is thus given by

$$[1 - (\lambda_s/\lambda_c)]^{3/2} - \frac{3\sqrt{3}}{2}(-b)^{1/2}|\bar{\xi}|(\lambda_s/\lambda_c) = 0, \tag{13.8}$$

and this expression is plotted in part (c) of Fig. 13.3. Assuming the axial load λ is applied quasi-statically (as in Chapter 5) then these results clearly resemble those shown in Fig. 5.10. For the specific case shown ($\bar{\xi} = 0.025$), the maximum load ($\lambda_s = 0.855$) corresponds to a maximum deflection of $\xi = 0.27$.

Now, suppose the load is applied suddenly as a step input, for example, by instantaneously removal of the support from a weight attached to the structure. In this case, a first integral can be performed (assuming there is no damping present) on Eq. (13.6), and from the resulting energy contours the maximum displacement (when $\dot{\xi} = 0$) is obtained from

$$[1 - (\lambda/\lambda_c)]\xi_{\text{max}}/2 + b\xi_{\text{max}}^3/4 = (\lambda/\lambda_c)\bar{\xi}. \tag{13.9}$$

This relation is shown in Fig. 13.4(a). The maximum deflection for which bounded solutions exist is defined as the dynamic buckling load λ_D and is determined from $d\lambda/d\xi_{\text{max}} = 0$, that is,

$$[1 - (\lambda_D/\lambda_c)]^{3/2} = 1.5\sqrt{6}(-b)^{1/2}|\bar{\xi}|(\lambda_D/\lambda_c) \tag{13.10}$$

now corresponds to a maximum (step) load of $\lambda_D = 0.821$, as shown in Fig. 13.4(a) with a maximum deflection of $\xi = 0.345$.

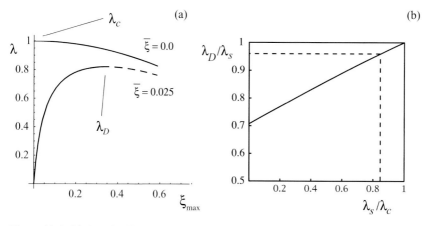

Figure 13.4. (a) Step loading vs. maximum deflection and (b) magnitude of the dynamic buckling load relative to the static case [7].

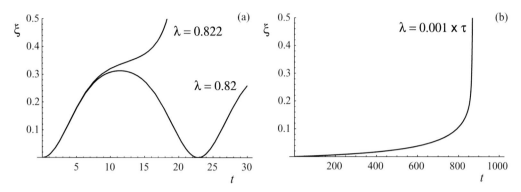

Figure 13.5. (a) Response of the structure that is due to a step load and (b) slow evolution of the load.

Equations (13.8) and (13.10) can now be combined to give the result

$$\left[\frac{1 - (\lambda_D/\lambda_s)(\lambda_s/\lambda_c)}{1 - (\lambda_s/\lambda_c)}\right]^{3/2} = \sqrt{2}\left(\frac{\lambda_D}{\lambda_s}\right), \tag{13.11}$$

which is plotted in Fig. 13.4(b), in which the same level of initial imperfection is used for both the static and dynamic cases. For example, if the structure has an initial imperfection of $\bar{\xi} = 0.025$, then there is a $0.821/0.855 = 0.96$ reduction in the maximum step load (relative to maximum statically applied load), as shown by the dashed line in Fig. 13.4(b). Thus we see, as anticipated, that when the load is suddenly applied the (imperfection-sensitive) structure is able to withstand only a reduced loading condition, and this effect is proportionately lower for a structure with a larger initial imperfection.

We finally conduct a couple of simple numerical simulations of this system. Applying the step load to Eq. (13.6) and using zero initial conditions leads to the results shown in Fig. 13.5(a). The preceding analysis gave a critical dynamic buckling load of $\lambda = 0.821$, and results are shown for the cases $\lambda = 0.82$, "stable," and $\lambda = 0.822$, "unstable." In part (b) the load parameter λ is very slowly evolved from zero (as was done in Chapter 5) until the system loses stability close to $t \approx 850$. It is worth mentioning here that given the relatively small level of initial imperfection there is not a large difference between the static and dynamic conditions. Also, the trajectory shown in Fig. 13.5(b) is not quite in equilibrium but rather the load is incremented in a large number of small steps as a function of the time step. These results do not include the effect of damping, and, as such, the results of numerical error may be an issue [8].

An alternative approach to this type of problem was conducted in Thompson [9] with the introduction of the concept of an *astatic* buckling load. This approach was also tested against some experiments on a continuous type of thin elastic arch structure, an example of which is shown in Fig. 13.6. The top-right-hand portion of this figure shows an arch with pinned ends but a rigid connection at the center and also

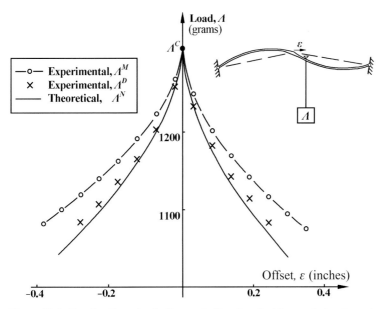

Figure 13.6. Step loading of a shallow arch. Reprinted with permission from the author [9].

shown in its deflected configuration. This structure was subject to a number of previous static tests [10] [as an example of an unstable-symmetric (subcritical) point of
bifurcation], but here the suspended weight is suddenly released. The magnitude of
the mass (Λ) and the offset from the apex (initial imperfection ϵ) are then related to
the (dimensional) dynamic buckling results of Fig. 13.6, in which Λ^M represents the
static load, Λ^D is the dynamic load, and Λ^N is the theoretical result from Thompson
[9]. Again we observe the detrimental effect of applying the load suddenly.

   In general, the applied force may have a finite duration, and thus the structure
may be subject to a pulse. For longer durations (relative to the natural dynamics of
the structure), we approach the previous results of the step load. For relatively short
pulses, we approach an impulse, which we have already shown has an equivalence
to a nonzero initial velocity. The effect of duration length was also considered in
Thompson [7] based on the earlier work described in Hutchinson and Budiansky
[11] as well as extensive studies in Simitses [12].

13.3 Dynamic Buckling of a Plate

In Subsection 10.1.4, we considered a simply supported rectangular plate subject
to a uniaxial load. Now, suppose this load is applied dynamically. Such a case was
considered in Zizicas [13] and reproduced in Bulson [14] based on small-deflection
plate theory but included the effect of an initial imperfection. In this case, it can be
shown that the governing equation of motion is

$$D\nabla^4(w - w_0) + \rho\frac{\partial^2 w}{\partial t^2} - N_x\frac{\partial^2 w}{\partial x^2} = 0, \qquad (13.12)$$

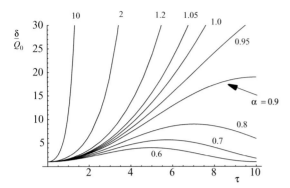

Figure 13.7. Deflection of a simply supported plate subject to a dynamic axial load [13].

in which w_0 is given by Eq. (10.49). The critical load and lowest natural frequency are given by Eqs. (10.46) and (10.45) (with $m = n = 1$), and again these provide convenient values with which to nondimensionalize the equation of motion, that is, introducing $N_x/N_{cr} = \alpha$ and $\omega t = \tau$, we can rewrite Eq. (13.12) as

$$\frac{d^2[\delta/Q_0]}{d\tau^2} + (1 - \alpha)\frac{\delta}{Q_0} - 1 = 0, \qquad (13.13)$$

in which δ is the central deflection caused by the dynamic load and Q_0 is the magnitude of the initial imperfection [see Eq. (10.49)].

Solutions to Eq. (13.13) are shown in Fig. 13.7. The axial load tends to magnify the initial imperfection according to Eq. (10.50), and it is apparent that the maximum response of the oscillation when the load is applied dynamically is larger than that for the quasi-static case. For example, when the magnitude of the axial load is 60% of the critical magnitude ($\alpha = 0.6$), the static deflection is 2.5 that of the initial value [from Eq. (10.50)], whereas the suddenly applied load at this level results in a peak-to-peak response that oscillates between $\delta/Q_0 = 1$ and $\delta/Q_0 = 4$ (and centered on $\delta/Q_0 = 2.5$). This result was anticipated from the introduction to this chapter and the overshoot of 100% for an undamped linear system subject to a step input. Clearly, we have the result that if the magnitude of the suddenly applied load is equal to, or greater than, the critical static load, then a monotonic growth of deflection occurs without bound (keeping in mind the limitations imposed when small-deflection theory is used).

13.4 A Type of Escaping Motion

In the next chapter, we shall focus attention on harmonically excited systems. In that case, most of the interest involves steady-state behavior. But even with harmonic excitation we can still expect a certain amount of transient behavior: For example, a system starting from rest will take awhile to settle down. It may very well be that the largest excursions (and hence proximity to instability) will occur during this transient stage [15, 16]. We go back to the basic form of an equation of motion in the vicinity of a transcritical bifurcation [Eq. (3.53)] and appeal to the analogy of a small ball rolling on the underlying potential-energy surface given by

$$V = \frac{x^3}{3} - \mu \frac{x^2}{2}.$$ (13.14)

For negative values of the control parameter μ, we have a stable (trivial) equilibrium state and an unstable equilibrium state (a maximum of the potential energy). As μ becomes less negative, these equilibria approach each other and interchange stability at $\mu = 0$ (see Fig. 3.6). However, it is important to realize that the domain of stability (against disturbances) changes. For an undamped system, we can construct a separatrix emanating from the hilltop (saddle) as a contour of constant total energy. Hence, any trajectory starting within this region will lead to constrained, or *bounded*, motion. But, again, we see the possibility of some trajectories escaping. This scenario is certainly complicated by the presence of damping. The area within the separatrix contains those initial conditions that do *not* lead to escape. However, this area changes (shrinks) as we approach the critical (buckling) condition. Thus we can imagine a situation in which a given step input for a relatively large negative level of μ will not cause escape (leading to infinity) whereas the same step applied to the system with a μ value less negative may very well lead to escape. Clearly, the magnitude of the step plays a crucial role as well, as seen in the previous section. Thus we have effectively described the two scenarios found in Fig. 1.1 at the start of this book. A structure may buckle because of:

- a deteriorating stiffness caused by increasing axial loading (the natural frequency characterizes this essentially static behavior) or
- an excessive disturbance applied to a structure with a given level of axial loading. This is essentially a transient, dynamic behavior.

Returning to the case of a harmonically forced, axially loaded structure, we can again consider a system described by Eq. (13.1), but now, rather than a linear spring, we shall assume a force that is quadratically related to deflection. Changing the step input to a harmonic drive, we consider

$$\ddot{x} + \zeta \dot{x} + x(1 - x) = F\Omega^2 \sin(\Omega t + \phi).$$ (13.15)

Assuming the initial conditions are zero, we have the possibility that the forcing may be sufficient to cause the response to exceed $x = 1$, thus leading to solutions escaping to infinity. However, this is a more involved issue than finding the critical initial conditions of the previous case. We can make use of the ball rolling on the potential-energy surface analogy but now the surface itself is shaken harmonically (thus resulting in a 3D phase space). It turns out that horizontally shaking the potential-energy surface is akin to transmissibility [17], a related but slightly different case from direct forcing, and thus an additional factor of Ω^2 also appears in the forcing magnitude [18]. That is, the mass is not subject to a direct force, but rather indirectly through a base displacement. Clearly, the larger F is, the greater the likelihood of escape, but we also anticipate a resonant type of effect if the forcing

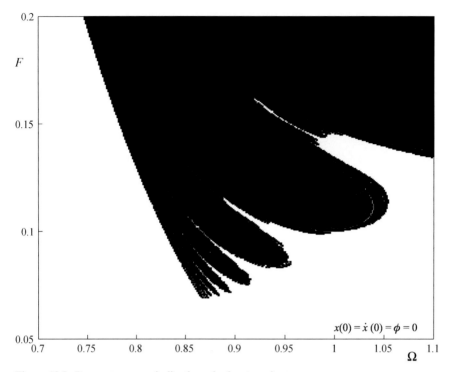

Figure 13.8. Parameter space indicating whether transients escape.

frequency Ω is somewhat close to the linear natural frequency of the system (unity). Furthermore, it would not be unreasonable to expect a cosine function to generate larger transients starting from $t = 0$ than a sine function would, and thus ϕ has an effect.

Rather than simply choose a number of forcing parameters and simulate Eq. (13.15) to determine whether steady-state oscillations persist, we can gain a more complete picture by conducting a thorough investigation of parameter space (F, Ω) by dividing it into a fine grid and labeling the outcome. An example of such a plot is shown in Fig. 13.8. What this figure describes is the result of many thousands of numerical simulations (all starting from zero initial conditions) and the areas shaded white indicate regions of parameter space that resulted in nonescaping behavior. The black-shaded regions led to escaping motion, which can be thought of as dynamic buckling. It can be seen that the boundary between escape and no-escape is not simple, and in fact exhibits certain *fractal* properties [19]. That is, on close inspection the border is nonsmooth (and self-similar at finer and finer scales) such that given any reasonable (small) uncertainty in the forcing parameters it may be difficult to tell whether the motion will escape or not, even though this is a thoroughly deterministic scenario. There is an increased likelihood of escape when the forcing frequency (Ω) is relatively close to one, as there is some associated softening resonant effect. This aspect of nonlinear behavior will be explained more throughly in later chapters.

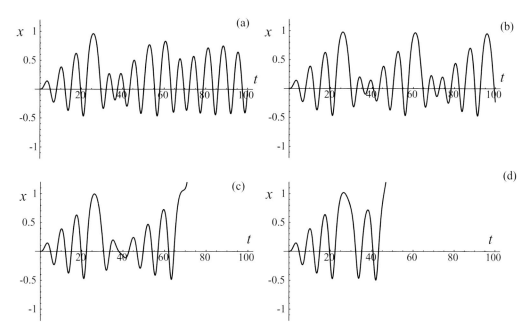

Figure 13.9. Some typical time series for parameter values spanning the escape boundary, $\Omega = 0.9$, $\zeta = 0.1$: (a) $F = 0.107$, (b) $F = 0.1075$, (c) $F = 0.108$, (d) $F = 0.1085$.

Figure 13.9 shows four typical time series generated with the same (quiescent) initial conditions and forcing frequency except for a very slightly different forcing amplitude in each case. We see that whether the trajectory escapes or not is a sensitive function of the forcing parameters (at least in certain ranges of the parameter space). In part (a), the motion passes quite close to the potential hilltop (at $x = 1$) but does *not* lead to escape. There is a small possibility of escape after the time range of the simulation. An informative view of a trajectory can be found in the phase *projection* (the phase space is 3D) shown in Fig. 13.10. This picture corresponds to the time series in part (d) of Fig. 13.9. We see that these oscillations are far from sinusoidal (which would yield elliptical trajectories), and it is interesting to see that the trajectory even passes beyond $x = 1$ (briefly) before coming back

Figure 13.10. A phase projection corresponding to the escaping trajectory when $F = 0.1085$.

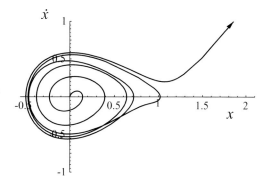

and then finally escaping [15, 19–21]. A final feature worth mentioning here is that although Fig. 13.8 divides the parameter space into two regions, within the black areas it is not unreasonable to ask the question: How long does it take for a given trajectory to escape? It turns out that trajectories corresponding to relatively high forcing magnitudes (F) tend to escape quickly (within a couple of forcing cycles or so), and trajectories corresponding to parameters very close to the boundary may take a relatively long time before finally escaping. This is also an aspect of nonlinear behavior that will be revisited in the final chapter.

13.5 Impulsive Loading

In this section, we focus attention on what happens as the duration of a suddenly applied load approaches zero. In the limit, we deal with impulsive loading. For a SDOF system, it can easily be shown that this situation is equivalent to a system subjected to a nonzero initial velocity. The situation is a little more complicated for higher-order systems [22]. Here, we consider a two-DOF (2DOF) link model that is configured as an arch, such that the loading-deflection path has a bifurcation point before a limit point is reached, and the structure buckles into an asymmetric mode.

Consider the system shown in Fig. 13.11. This system was analyzed extensively in [12, 22], and here we focus on impulsive loading and use a total energy approach following that reference. This system has two hinges of rotational stiffness β with concentrated masses m, and these are the locations at which two equal vertical loads P are applied. Deflection of the system is allowed by horizontal sliding at the right-hand support, resisted by a linear spring of stiffness k. The initial *rise* of the arch is characterized by the angle α (initially equal at each end). The deflected state of the system is described in terms of the end angles θ and ϕ [22]. Other than the method of external forcing, this structure is similar to the one shown in Fig. 5.15. Assuming small angles, the total potential energy can be written as

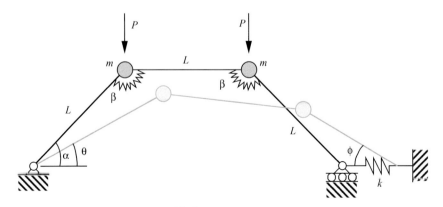

Figure 13.11. A 2DOF link model [22].

$$U_T^P = \frac{1}{2}\beta(5\theta + 5\phi^2 - 2\alpha^2 - 8\phi\theta - 2\theta\alpha - 2\phi\alpha)$$

$$+ \frac{kL^2}{2}(\alpha^2 - \theta^2 - \phi^2 + \phi\theta)^2 - PL(2\alpha - \theta\phi). \tag{13.16}$$

It is convenient to introduce new displacement variables

$$r = (\theta + \phi)/2\sqrt{\bar{\beta}},$$
$$s = (\theta - \phi)/2\sqrt{\bar{\beta}}, \tag{13.17}$$

in which $\bar{\beta} = \beta/(kL^2)$. In terms of these new variables, we can write the total potential energy as

$$\bar{U}_T^P = (r^2 + 9s^2 - 2\sqrt{\Lambda}r + \Lambda) + \frac{1}{2}(\Lambda - r^2 - 3s^2)^2 - 2p(\sqrt{\Lambda} - r), \tag{13.18}$$

in which the following nondimensional parameters have also been used

$$\bar{U}_T^P = U_T^P/(\bar{\beta}^2 kL^2), \quad p = P/(kL\bar{\beta}^{3/2}) \quad \Lambda = \alpha^2/\bar{\beta}. \tag{13.19}$$

13.5.1 Equilibrium Behavior

In the usual way, equilibrium is obtained from stationary values of the potential energy ($\partial \bar{U}/\partial r = 0$ and $\partial \bar{U}/\partial s = 0$):

$$2(r - \sqrt{\Lambda}) - (\Lambda - r^2 - 3s^2)2r + 2p = 0,$$
$$18s - (\Lambda - r^2 - 3s^2)6s = 0. \tag{13.20}$$

The solutions to these equations give the symmetric response ($s = 0$)

$$(\Lambda - 1 - r^2)r = p - \Lambda \tag{13.21}$$

and the asymmetric response ($s \neq 0$)

$$\Lambda - 3 = r^2 + 3s^2,$$
$$2r = p - \Lambda. \tag{13.22}$$

The type of resulting behavior obviously depends on the magnitude of Λ. It can be shown [12] that if $\Lambda > 4$ then the system encounters an unstable point of bifurcation, and because this is a new feature we now focus on the specific case $\Lambda = 6$.

Figure 13.12(a) shows the potential-energy contours in terms of the two coordinates r and s. Five equilibrium points exist: Points 1 and 3 are stable equilibria, with an unstable point 2 in between; points 4 and 5 are unstable saddle points.

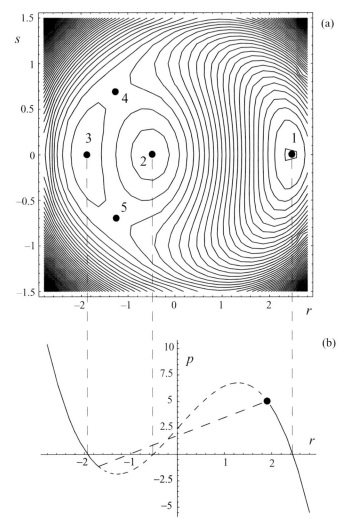

Figure 13.12. (a) Contours of potential energy indicating five extremum points, $\Lambda = 6$, (b) corresponding equilibrium curves [12].

13.5.2 Behavior under Sudden Loading

If the system is initially located at point 1 in stable equilibrium and is then subject to a disturbance (impulse), it is apparent that a sufficiently large input may cause the system to traverse the unstable equilibrium (which is a local maximum of the potential energy). Furthermore, we can imagine the situation in which sufficient kinetic energy is imparted to the system such that either of the saddle points is traversed, typically leading to very large-amplitude oscillations. The equilibrium curves are projected in Fig. 13.12(b) as a function of p.

Simitses [12] used an energy approach and the impulse-momentum theorem in which an impulse (PT_0) associated with a load P over a short time duration (T_0) is related to kinetic energy and the potential energy at the hilltop equilibrium to arrive

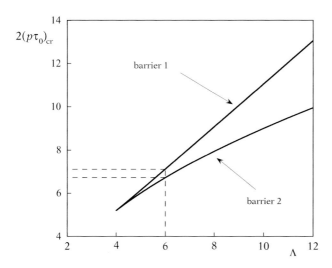

Figure 13.13. Upper and lower bounds for the critical ideal impulse [22].

at analytical expressions for the two instability situations described in the previous paragraph. Using the nondimensional time $\tau = t\sqrt{(\bar{\beta}k/2m)}$, we find that the relevant expressions are

$$2(p\tau_0)_{cr} = 3\sqrt{\Lambda - 1},$$

$$2(p\tau_0)_{cr} = [5\Lambda - 2 - 3\sqrt{\Lambda^2 - 4\Lambda} + 0.25(\Lambda + 2 + \sqrt{\Lambda^2 - 4\Lambda})^2]^{1/2}, \quad (13.23)$$

where τ_0 is the nondimensional duration of the impulse. The expressions in Eqs. (13.23) are plotted in Fig. 13.13. The barriers represent levels of the dynamic forcing required to cause instability. Thus we see that when $\Lambda = 6$, the critical impulse that causes dynamic buckling is given by $2(p\tau_0)_{cr} = 6.7$.

   These results on discrete-link models can also be extended to continuous, suddenly loaded structures liable to snap-through [23]. For example, some early studies were conducted by Hsu (e.g., [24]) on shallow elastic arches under the action of various time-dependent lateral loads, including sinusoidal, arbitrary, concentrated, etc. The effects of initial thrust and elastic foundations were investigated [25] and interaction curves developed to assess the effect of various load combinations on the snap-buckling of shallow arches [23, 26–30]. Some interesting features associated with this type of problem are described in [31, 32] in which impulsive and harmonic loading may lead to counterintuitive behavior.

13.6 Snap-Through of a Curved Panel

We considered the snap-through (saddle-node bifurcation) of a simple link model under the action of a quasi-statically increasing load in Section 5.6. For a continuous (shell-like) system under suddenly applied (step) loading, it is also possible to

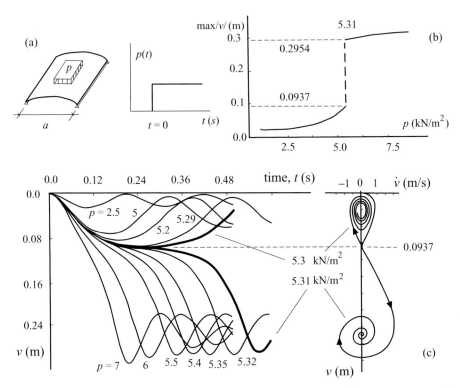

Figure 13.14. Snap-through of a shallow panel subject to step loading: (a) the geometry and load-ing, (b) maximum response as a function of loading intensity, and (c) time series and phase projec-tion. Adapted from Krätzig and Nawrotzki [35] and Krätzig [36].

encounter snap-through [33]. However, for such a system the FE technique is the method of choice.

An example of this behavior was given by Krätzig [34], in which a shallow curved panel is subject to a uniformly distributed vertical step load of magnitude p, as shown in Fig. 13.14(a). For this particular simulation, a square panel of width 10 m, thickness 0.1 m, and a radius of curvature of 100 m is considered. Young's modulus is 3.4×10^7 kN/m^2 and Poisson's ratio is 0.2. With the rest position used as the initial condition, the results of numerical simulation shown in part (b) indi-cate that provided the magnitude of the step load is less than $p = 5.31$ kN/m^2 the response is insufficient to traverse the potential-energy hilltop associated with the unstable equilibrium. Thus there is a considerable difference in the maximum re-sponse of the system depending on whether the system tends toward its inverted equilibrium or not. In both cases damped oscillations characterize the system re-sponse as energy is dissipated.

A set of superimposed times series and phase projections is shown in Fig. 13.14(c) for various values of the applied load. The phase projection shows two trajectories, one on each side of the critical value. Unlike the simple link mod-els of Chapter 5, this type of modeling requires special numerical procedures and is the subject of current research [35, 36], including similar sensitivity in spherical caps [37].

Suddenly Loaded Column. We conclude this chapter by considering an axially loaded column in which the end load is applied suddenly [1, 2, 38–41]. The following chapter will look at a pulsating end load in which case it is possible to get buckling for loads lower than the static buckling load. In this final section, we see that it is possible to have stability for loads far in excess of the static buckling load depending on the duration of the load. We focus attention on pulse loading of the type shown schematically (but with finite duration) in the inset of Fig. 13.1 [42, 43], and this is closely related to the impact loading of a bar, for example, a hammer hitting a nail [44]. Although the load is transmitted as an axial stress wave, it has been shown that typically buckling motion as the wave passes can be neglected, and the total length of the column is relatively unimportant; that is, the duration of the axial loading is relatively large compared with the period of longitudinal vibration of the bar. One consequence of this is that the buckling modes can be quite complicated (and associated with short buckling wavelengths), with divergent (hyperbolic) and bounded (trigonometric) modes both present and their separation depending on various characteristics of the system. A subtlety associated with this problem (other than the direct influence of dynamics of course) is that initial imperfections are necessary, and thus, rather than having a distinct bifurcational event, the loss of stability is most appropriately couched in terms of a dynamic growth, or amplification, of the initial geometric imperfection. We assume in the analysis that the behavior is elastic, although of course in practice very often plastic deformation is encountered [45].

We briefly discuss the simple case of a uniform, simply supported bar, subject to a suddenly applied load:

$$EI\frac{\partial^4 w}{\partial x^4} + P\frac{\partial^2}{\partial x^2}(w + w_0) + m\frac{\partial^2 w}{\partial t^2} = 0. \tag{13.24}$$

Note that in contrast to the initial imperfection encountered in Section 7.3, here, we measure the deflection $w(x, t)$ that is due to axial loading *from* the initial imperfection $w_0(x)$. In light of the earlier comments about the relative unimportance of the column length (see Lindberg and Florence [46]) we introduce a characteristic length of $1/k$ (where $k^2 = P/EI$) and then nondimensionalize Eq. (13.24) by using

$$\bar{x} = kx, \quad \bar{w} = w/r, \quad \bar{t} = trk^2 c, \tag{13.25}$$

where $c = E/m$ is the speed of wave propagation and $r = \sqrt{I/A}$ is the radius of gyration. The result is

$$\bar{w}'''' + \bar{w}'' + \ddot{\bar{w}} = -\bar{w}_0''. \tag{13.26}$$

Application of the simply supported boundary conditions ($\bar{w} = \bar{w}'' = 0$) at ($\bar{x} = 0$ and $\bar{x} = l = kL$) leads to a solution of the form

$$\bar{w}(\bar{x}, \bar{t}) = \sum_{n=1}^{\infty} g_n(\bar{t}) \sin\frac{n\pi\bar{x}}{l}, \tag{13.27}$$

and we assume a spatial distribution of initial imperfection according to

$$\bar{w}_0(\bar{x}) = \sum_{n=1}^{\infty} a_n \sin \frac{n\pi\bar{x}}{l}, \tag{13.28}$$

where

$$a_n = \frac{2}{l} \int_0^l \bar{w}_0(\bar{x}) \sin \frac{n\pi\bar{x}}{l} d\bar{x}. \tag{13.29}$$

Thus the solution for the amplitudes is obtained from

$$\ddot{g}_n + \eta^2(\eta^2 - 1)g_n = \eta^2 a_n, \tag{13.30}$$

where $\eta = n\pi/l$ is the wavenumber. We see the form of solution is quite different according to whether η is greater than, or less than, unity.

This was a distinction in the form of the solution we first encountered at the start of Chapter 7, and assuming the bar is initially at rest (i.e., $\bar{w} = \dot{w}$ when $t = 0$) we obtain the solution

$$\bar{w}(\bar{x}, \bar{t}) = \sum_{n=1}^{\infty} \frac{a_n}{1 - \eta^2} (\cos p_n \bar{t} - 1) \sin \frac{n\pi\bar{x}}{l} \tag{13.31}$$

when $\eta > 1$, and

$$\bar{w}(\bar{x}, \bar{t}) = \sum_{n=1}^{\infty} \frac{a_n}{1 - \eta^2} (\cosh p_n \bar{t} - 1) \sin \frac{n\pi\bar{x}}{l} \tag{13.32}$$

when $\eta < 1$, and where

$$p_n = \eta |1 - \eta^2|^{1/2}. \tag{13.33}$$

In terms of amplification, or lateral growth of motion, it is convenient to scale Eqs. (13.31) and (13.32) according to the underlying static amplification of an imperfect simply supported, axially loaded bar, and thus

$$G_n(\bar{t}) = \frac{g_n(\bar{t})}{a_n} = \frac{1}{1 - \eta^2} \begin{bmatrix} \cosh \\ \cos \end{bmatrix} p_n \bar{t} - 1 \end{bmatrix} \sin \frac{n\pi\bar{x}}{l}, \tag{13.34}$$

in which we take the cosine term for $\eta > 1$ and the hyperbolic cosine term for $\eta < 1$. Equation (13.34) is plotted in Fig. 13.15(a) for two values of the nondimensional time $\bar{t}$. From this we can see that greatest amplification takes place in a narrow band of wavelengths (the preferred mode of buckling). We can then make use of the derivative to find the maximum value and then plot that maximum value versus time, as shown in Fig. 13.15(b). By use of an approximate analysis [46], it can be shown that under very high compression the bar will buckle into wavelengths close to $8.88r/\sqrt{P/AE}$ with very rapid growth in motion after approximately $\bar{t} = 4$. The effect of the velocity of impact loading was studied in Holzer and Eubanks [47] and Hayashi and Sano [48, 49].

We can also approach this type of problem by considering the dropping of a weight onto the end of a long strut, which would be an easy scenario to set up in

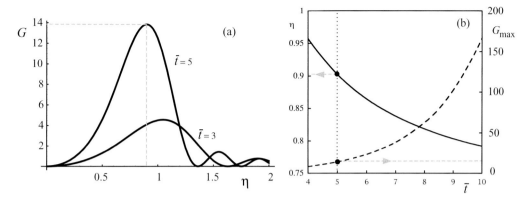

Figure 13.15. Amplification vs. wavenumber, (b) corresponding maximum amplification as a function of time [46].

the laboratory. In this case, it can be shown that the buckling wavelength scales with the inverse square root of the impact speed [50]. In a practical sense, the distribution of initial geometric imperfections might have a random element, and this has also been treated in the literature [51]. Extensive studies using similar approaches have focused on rapid application of axial loading on cylinders and shells [51–53]. Finally, the case of a dynamic application of the end load that is relatively slow is mentioned. In this type of ramp function, which is somewhat representative of what would happen in a loading machine, there is also some interesting behavior [55].

References

[1] J.H. Meier. On the dynamics of elastic buckling. *Journal of the Aeronautical Sciences*, 12:433–40, 1945.
[2] J.F. Davidson. Buckling of struts under dynamic loading. *Journal of the Mechanics and Physics of Solids*, 2:54–66, 1953.
[3] G. Herrmann. *Dynamic Stability of Structures*. Pergamon, 1967.
[4] G.J. Simitses, A.N. Kounadis, and J. Giri. Dynamic buckling of simple frames under a step load. *Journal of Engineering Mechanics*, 105:896–900, 1979.
[5] A.T. Brewer and L.A. Godoy. On interaction between static and dynamic loads in instability of symmetric or asymmetric structural systems. *Journal of Sound and Vibration*, 147:105–14, 1991.
[6] I. Cochin and H.P. Plass. *Analysis and Design of Dynamic Systems*. Harper & Row, 1990.
[7] B. Budiansky. Dynamic buckling of elastic structures: Criteria and estimates. In G. Herrmann, editor, *Dynamic Stability of Structures*. Pergamon, 1967.
[8] G.J. Simitses. Effect of static preloading on the dynamic stability of structures. *AIAA Journal*, 21:1174–80, 1983.
[9] J.M.T. Thompson. Dynamic buckling under step loading. In G. Herrmann, editor, *Dynamic Stability of Structures*. Pergamon, 1967.
[10] J. Roorda. Stability of structures with small imperfections. *Journal of the Engineering Mechanics Division, ASCE*, 91:87, 1965.

[11] J.W. Hutchinson and B. Budiansky. Dynamic buckling estimates. *AIAA Journal*, 4:525–30, 1966.

[12] G.J. Simitses. Instability of dynamically-loaded structures. *Applied Mechanics Reviews*, 40:1403–8, 1987.

[13] G.A. Zizicas. Dynamic buckling of thin elastic plates. *Transactions of the ASME*, 74:1257–68, 1952.

[14] P.S. Bulson. *The Stability of Flat Plates*. Chatto and Windus, 1970.

[15] B. Budiansky and E.S. Roth. Axisymmetric dynamic buckling of clamped shallow spherical shells. Technical Report, NASA TND-1510, 1962.

[16] M.W. Hilburger, A.M. Waas, and J.H. Starnes. Modeling the dynamic response and establishing post buckling snap-through equilibrium of discrete structures via a transient analysis. *Journal of Applied Mechanics*, 64:590–5, 1997.

[17] D.J. Inman. *Engineering Vibration*. Prentice-Hall, 2000.

[18] L.N. Virgin. *Introduction to Experimental Nonlinear Dynamics: A Case Study in Mechanical Vibration*. Cambridge University Press, 2000.

[19] J.M.T. Thompson. Chaotic phenomena triggering escape from a potential well. *Proceedings of the Royal Society of London*, A421:195–225, 1989.

[20] L.N. Virgin, R.H. Plaut, and C.-C. Cheng. Prediction of escape from a potential well under harmonic excitation. *International Journal of Non-Linear Mechanics*, 27:357–65, 1992.

[21] J.A. Gottwald, L.N. Virgin, and E.H. Dowell. Routes to escape from an energy well. *Journal of Sound and Vibration*, 187:133–44, 1995.

[22] G.J. Simitses. *Dynamic Stability of Suddenly Loaded Structures*. Springer-Verlag, 1989.

[23] D.L.C. Lo and E.F. Masur. Dynamic buckling of shallow arches. *Journal of the Engineering Mechanics Division, ASCE*, 102:901–17, 1976.

[24] C.S. Hsu. Stability of shallow arches against snap-through under timewise step loads. *Journal of Applied Mechanics*, 35:31–9, 1968.

[25] R.H. Plaut and E.R. Johnson. The effect of initial thrust and elastic foundation on the vibration frequencies of a shallow arch. *Journal of Sound and Vibration*, 78:565–71, 1981.

[26] N.J. Hoff and V.G. Bruce. Dynamic analysis of the buckling of laterally loaded flat arches. *Journal of Mathematical Physics*, 32:276–88, 1954.

[27] A.M. Liapunov. *Stability of Motion (Collected Papers)*. Academic, 1966.

[28] R.E. Fulton and F.W. Barton. Dynamic buckling of shallow arches. *Journal of Engineering Mechanics*, 97:865–77, 1971.

[29] K.-Y. Huang and R.H. Plaut. Snap-through of a shallow arch under pulsating load. In F.H. Schroeder, editor, *Stability in the Mechanics of Continua*. Springer, 1982, pp. 215–33.

[30] M.T. Donaldson and R.H. Plaut. Dynamic stability boundaries for a sinusoidal shallow arch under pulse loads. *AIAA Journal*, 21:469–71, 1983.

[31] E.R. Johnson and I.K. McIvor. The effect of spatial distribution on dynamic snap-through. *Journal of Applied Mechanics*, 45:612–18, 1978.

[32] R.H. Plaut and J.-C. Hsieh. Oscillations and instability of a shallow arch under two-frequency excitation. *Journal of Sound and Vibration*, 102:189–201, 1985.

[33] J.S. Humphreys and S.R. Bodner. Dynamic buckling of shallow shells under impulsive loading. *Journal of Engineering Mechanics*, 88:17–36, 1962.

[34] W.B. Krätzig. Nonlinear responses. In A.N. Kounadis and W.B. Krätzig, editors, *Nonlinear Stability of Structures (Theory and Computational Techniques)*. Springer-Verlag, 1995.

[35] W.B. Krätzig and P. Nawrotzki. Computational concepts for kinetic instability problems. In A.N. Kounadis and W.B. Krätzig, editors, *Nonlinear Stability of Structures (Theory and Computational Techniques)*. Springer-Verlag, 1995.

[36] W.B. Krätzig. Eine einheitliche statische und dynamische Stabilitatstheorie fur Pfadverfolgungsalgorithmen in der numerischen Festkorpermechanik. Zeitschrift für Angewandte Mathematik und Mechanik, 69:203–13, 1989.

[37] D. Dinkler and J. Pontow. A model to evaluate dynamic stability of imperfection sensitive shells. *Computational Mechanics*, 37:523–9, 2006.

[38] C. Koning and J. Taub. Impact buckling of thin bars in the elastic range hinged at both ends. *Luftfahrforschung*, 10:55–64, 1933.

[39] G. Gerard and H. Becker. Column behavior under conditions of impact. *Journal of the Aeronautical Sciences*, 19:58–60, 1952.

[40] E. Sevin. On the elastic bending of columns due to dynamic axial forces including effects of axial inertia. *Journal of Applied Mechanics*, 27:125–31, 1960.

[41] R. Grybos. Impact stability of a bar. *International Journal of Engineering Science*, 13:463–77, 1975.

[42] N.J. Huffington. Response of elastic columns to axial pulse loading. *AIAA Journal*, 1:2099–2104, 1963.

[43] I.K. McIvor and J.E. Bernard. The dynamic response of columns under short duration axial loads. *Journal of Applied Mechanics*, 40:688–92, 1973.

[44] J.M. Housner and N.F. Knight. The dynamic collapse of a column impacting a rigid surface. *AIAA Journal*, 21:1187–95, 1983.

[45] W. Abramowicz and N. Jones. Dynamic progressive buckling of circular and square tubes. *International Journal of Impact Engineering*, 4:247–70, 1986.

[46] H.E. Lindberg and A.L. Florence. *Dynamic Pulse Buckling: Theory and Experiment*. Nijhoff, 1987.

[47] S.M. Holzer and R.A. Eubanks. Stability of columns subjected to impulsive loading. *Journal of Engineering Mechanics*, 95:897–920, 1969.

[48] T. Hayashi and Y. Sano. Dynamic buckling of elastic bars (the case of low velocity impact). *Bulletin of the Japanese Society of Mechanical Engineering*, 15:1167–75, 1972.

[49] T. Hayashi and Y. Sano. Dynamic buckling of elastic bars (the case of high velocity impact). *Bulletin of the Japanese Society of Mechanical Engineering*, 15:1176–84, 1972.

[50] J.R. Gladden, N.Z. Handzy, A. Belmonte, and E. Villermaux. Dynamic buckling and fragmentation in brittle rods. *Physical Review Letters*, 94(035503), 2005.

[51] I. Elishakoff. Axial impact buckling of a column with random initial imperfections. *Journal of Applied Mechanics*, 45:361–5, 1978.

[52] H.E. Lindberg and R.E. Herbert. Dynamic buckling of a thin cylindrical shell under axial impact. *Journal of Applied Mechanics*, 33:105–12, 1966.

[53] N. Jones and C.S. Ahn. Dynamic elastic and plastic buckling of complete spherical shells. *International Journal of Solids and Structures*, 10:1357–74, 1974.

[54] R. Kao. Nonlinear dynamic buckling of spherical caps with initial imperfections. *Computers and Structures*, 12:49–63, 1980.

[55] N.J. Hoff. The dynamics of the buckling of elastic columns. *Journal of Applied Mechanics*, 17:68–74, 1953.

14 Harmonic Loading: Parametric Excitation

14.1 An Oscillating End Load

In Chapter 7 we considered the free vibrations of a simply supported beam subject to an axial load of constant magnitude, and then in Chapter 13 the load was applied suddenly. Now suppose the end load is pulsating harmonically, that is, we replace P in Fig. 7.1 with $P + S \cos \Omega t$. Because the forcing will appear in the stiffness term we are dealing with parametric excitation, and this is sometimes referred to as vibration buckling in the literature [1]. The governing equation of motion is thus

$$m\frac{\partial^2 w}{\partial t^2} + EI\frac{\partial^4 w}{\partial x^4} + (P + S\cos\Omega t)\frac{\partial^2 w}{\partial x^2} = 0. \tag{14.1}$$

With simply supported boundary conditions we take the solution in the form of a single half-sine wave,

$$w = f(t)\sin\frac{\pi x}{L}. \tag{14.2}$$

Substituting this back into Eq. (14.1), reintroducing the term from Eq. (7.20), that is,

$$\omega^2 = \frac{EI\pi^4}{mL^4}\left(1 - \frac{PL^2}{EI\pi^2}\right), \tag{14.3}$$

again normalizing the axial load and natural frequency by the Euler load ($P_{cr} = \pi^2 EI/L^2$) and frequency without load [$\omega_0^2 = \pi^4 EI/(mL^4)$], respectively, and rescaling time according to $\tau = \Omega t$, we arrive at

$$f''(\tau) + (\alpha + \epsilon\cos\tau)f(\tau) = 0, \tag{14.4}$$

where

$$\alpha = (\omega_0^2/\Omega^2)(1 - p), \quad \epsilon = -(\omega_0^2/\Omega^2)s. \tag{14.5}$$

Although Eq. (14.4), which is called *Mathieu's* equation, is linear, it is by no means easy to solve, and the character of the solutions (specifically their stability) depends in a nonsimple way on the parameters α and ϵ. This equation is also encountered in the dynamics of a pendulum with a harmonically shaken pivot [2], and it is related to the stability of forced oscillations which will be studied in more detail in the final chapter. Sinha [3] also considers the problem of pulsating axial load but applied to Timoshenko beams. To consider the stability of this type of system (which can be viewed as a natural extension of the stability of equilibria introduced in Chapter 3), we again make use of linearization before proceeding to Floquet theory [4–11].

14.2 The Variational Equation

In this chapter, we consider nonautonomous dynamical systems of the form

$$\dot{\mathbf{x}} = \mathbf{f}(\mathbf{x}, t), \tag{14.6}$$

and proceed to consider the stability of solutions based on the behavior of small perturbations. In Chapter 3 we also considered the behavior of transients in the vicinity of equilibria (point attractors) in autonomous systems, whereas now we focus on transients in the vicinity of periodic solutions in nonautonomous (especially periodically forced) systems. Again *linearization* is a key concept, and use will be made of Poincaré sampling and fixed points of maps.

Equation (14.6) is in general a nonlinear system. Following the development in Subsection 4.2.3, we obtain the linear variational equation

$$\dot{\eta} = \mathbf{DF}(t)\eta. \tag{14.7}$$

It is this equation that governs the stability of solutions in the vicinity of special solutions (which will be steady-state oscillations in this context).

Equations with Periodic Coefficients. Equation (14.4) is a specific case of the wider class of forced vibration problems, and is clearly related to Eq. (14.7) in which the coefficients of the Jacobian $\mathbf{DF}(t)$ are periodic. Consider Hill's equation [12],

$$\ddot{x} + G(t)x = 0, \tag{14.8}$$

where $G(t + T) = G(t)$. Although this is a rather benign-looking (linear) equation, it does not, in general, submit to closed-form analytical solutions, and hence a variety of approximate techniques have been developed. We also note that a simple transformation allows the related case of damped parametric oscillations to be brought into the standard form of Eq. (14.8). Hence this equation represents quite a wide class of problems and has received considerable scrutiny over the years [12–15].

Placing Eq. (14.7) in a little more general (state-variable) context, and simplifying the notation by using $A \equiv \mathbf{DF}$ and $x \equiv \eta$, we consider the solutions of

$$\dot{x} = A(t)x, \quad t \in R, \tag{14.9}$$

where x is an n-dimensional state vector and $A(t)$ is a continuous T-periodic $n \times n$ matrix, that is, $A(t + T) = A(t)$. This system has n linearly independent fundamental solutions ϕ_i, where $i = 1, 2, \ldots, n$, which can be expressed as a fundamental solution matrix $\Phi(t)$ [15]. Shifting in time by T, we see that $\Phi(t + T)$ is also a fundamental matrix solution, and because they are linearly independent there is a nonsingular $n \times n$ matrix C such that

$$\Phi(t + T) = \Phi(t)C. \tag{14.10}$$

This constant matrix C (which depends on T but not on t) is called the *monodromy* matrix of Eq. (14.9) and contains the key stability information. The eigenvalues ρ of C are called *characteristic* (or *Floquet*) *multipliers (CMs)*. They are uniquely determined and govern local divergence or convergence about a periodic orbit [16]. The matrix C can also be expressed as

$$C = e^{BT}, \tag{14.11}$$

where B is a constant matrix. The Floquet theorem then states that the fundamental matrix $\Phi(t)$ of Eq. (14.9) can be written as

$$\Phi(t) = P(t)e^{Bt}, \tag{14.12}$$

where $P(t)$ is T periodic and B is a constant $n \times n$ matrix. The eigenvalues γ of B are the *characteristic* (or *Floquet*) *exponents (CEs)* and are essentially the same as encountered earlier for equilibria, and they govern the stability of the trivial solution of Eq. (14.9). They are unique only to within an integer multiple of $2\pi i/T$.

It is helpful to consider the monodromy matrix C as a Floquet operator, which maps $\Phi(t)$ onto $\Phi(t + T)$, and taking the initial condition as the identity vector $\Phi(0) = I$, we then have from Eq. (14.10), $C = \Phi(T)$. Furthermore, if the eigenvalues ρ_i, $i = 1, 2, \ldots, n$, of C (the CMs) are distinct, then Eq. (14.9) has n linearly independent *normal solutions* of the form

$$\mathbf{x}_i = \mathbf{p}_i(t)e^{\gamma_i t}, \tag{14.13}$$

where the $\mathbf{p}_i(t)$ are periodic functions with period T. Thus we see the fundamental relationship between the CEs and the CMs:

$$\rho = e^{\gamma t}. \tag{14.14}$$

Now, again with distinct eigenvalues we can diagonalize C; that is, we define a transformed system

$$\Psi(t + T) = \Psi(t)J, \tag{14.15}$$

where $\Psi = M^{-1} CM$, M is a nonsingular $n \times n$ constant matrix (chosen to simplify J), and $\phi(t) = \Psi P^{-1}$. In this case, we can again consider the relation between the CEs and the CMs but in component form,

$$\psi_i(t + T) = \rho_i \psi(t), \tag{14.16}$$

with the *Mathieu* functions

$$\psi_i(t) = e^{\gamma_i t}\phi_i(t). \tag{14.17}$$

The stability of the periodic solutions is now emerging. Equation (14.16) can be extended to

$$\psi_i(t + NT) = \rho_i^N \psi(t), \tag{14.18}$$

where N is an integer. Therefore we see that it is the *magnitude* of ρ that determines stability:

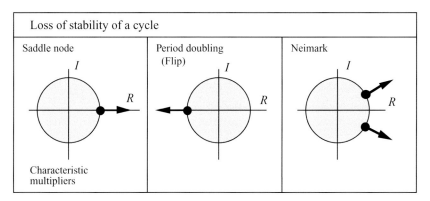

Figure 14.1. The generic routes to instability (in terms of $|\rho_i|$) for a system under the action of a single control parameter.

- For stability of a periodic orbit, we have $\psi_i(t) \to 0$ as t (and hence N) $\to \infty$ if $|\rho_i| < 1$ (i.e., the real part of γ_i is negative).
- For instability, we have $\psi_i(t) \to \infty$ as t (and hence N) $\to \infty$ if $|\rho_i| > 1$ (i.e., the real part of γ_i is positive).

In practice, the most difficult part of any analysis of this kind is determining the matrix $\Phi(t)$. We shall see that this can be achieved numerically by means of the Poincaré map or by using various approximate analytical schemes. These will be dealt with in more detail in later chapters but we will see that the three typical ways in which multipliers leave the unit circle are shown in Fig. 14.1 [15, 17], with R and I signifying real and imaginary, respectively. We note that the Neimark bifurcation is less commonly encountered in the types of structural system encountered in this book.

We can gain some stability insight by considering certain constraints, akin to the Routh–Hurwitz criterion for the stability of equilibria, placed on the eigenvalues. To do this, use is made of the Wronskian determinant of the fundamental matrix corresponding to Eq. (14.9):

$$\text{Det } \Phi(t) = \exp\left[\int_0^t \text{Tr } A(s)\, ds\right], \qquad (14.19)$$

where Tr $A(s)$ is the trace of $A(s)$. The Floquet theorem [Eq. (14.12)] tells us that

$$\text{Det } \Phi(t) = \text{Det}[P(t)e^{Bt}], \qquad (14.20)$$

which leads to

$$\text{Det}(e^{BT}) = \exp\left[\int_0^T \text{Tr } A(t)\, dt\right]. \qquad (14.21)$$

Now we are in a position to state that

$$\rho_1 \rho_2 \dots \rho_n = \exp\left[\int_0^T \text{Tr } A(t)\, dt\right] \qquad (14.22)$$

and

$$\sum_{i=1}^{n} \gamma_i = \frac{1}{T} \int_0^T \text{Tr } A(t)\, dt \left(\text{mod} \frac{2\pi i}{T}\right). \tag{14.23}$$

The procedure is quite straightforward for obtaining stability information *provided* the fundamental matrix of normal solutions, Φ, is known. This is seldom the case, and a variety of approximate analytical and numerical techniques have been developed to account for this. Fortunately, there are a number of shortcuts that can be taken to determine stability without needing the full solution.

14.3 Mathieu's Equation

Using the information concerning the stability of periodic motion, we now return to Mathieu's equation [Eq. (14.4)], where f is replaced with x:

$$x''(\tau) + (\alpha + \epsilon \cos \tau)x(\tau) = 0. \tag{14.24}$$

This can be expressed in state matrix terms (with $\dot{x} \equiv y$, replacing the primes):

$$\begin{bmatrix} \dot{x} \\ \dot{y} \end{bmatrix} = \begin{bmatrix} 0 & 1 \\ -\alpha - \epsilon \cos t & 0 \end{bmatrix} \begin{bmatrix} x \\ y \end{bmatrix}. \tag{14.25}$$

We see that the trace of the matrix in the preceding equation is zero, and using Eq. (14.22) applied to Eq. (14.25) we have $\rho_1 \rho_2 = e^0 = 1$, and thus the roots satisfy the quadratic equation

$$\rho^2 - \phi(\alpha, \epsilon)\rho + 1 = 0. \tag{14.26}$$

The solutions are given by

$$\rho_1, \rho_2 = \frac{1}{2}\left[\phi(\alpha, \epsilon) \pm \sqrt{\phi(\alpha, \epsilon)^2 - 4}\right]. \tag{14.27}$$

In general, we might be more interested in determining whether the motion is bounded or not, rather than in being able to write the specific form of the solution, and hence the *transition curves* between stable and unstable behavior are of paramount importance, and these occur when $\phi(\alpha, \epsilon) = \pm 2$. For Mathieu's equation, these transition curves correspond to the specific combinations of α and ϵ for which periodic solutions, with period 2π or 4π, occur [12].

The standard analytical approach to obtaining the transition curves involves a Hill determinant [12, 17]. However, a useful approximate technique based on the perturbation method is introduced that will also be useful when we consider large-amplitude vibration in the final chapter [2, 15].

A Perturbation Solution. For relatively small values of the parameter ϵ, the transition curves can be computed with a perturbation method [15]. The solutions to Mathieu's equation are assumed to be of the form

$$x(t) = x_0(t) + \epsilon x_1(t) + \cdots +, \tag{14.28}$$

in which $x_0, x_1, \ldots,$ have period 2π or 4π, and the transition curves are given by

$$\alpha = \alpha(\epsilon) = \alpha_0 + \epsilon\alpha_1 + \cdots + . \tag{14.29}$$

On substituting these expressions into Mathieu's equation, we obtain

$$\begin{aligned}(\ddot{x}_0 + \alpha_0 x_0) + \epsilon(\ddot{x}_1 + \alpha_1 x_0 + x_0 \cos t + \alpha_0 x_1) \\ + \epsilon^2(\ddot{x}_2 + \alpha_2 x_0 + \alpha_1 x_1 + x_1 \cos t + \alpha_0 x_2) + \cdots + = 0,\end{aligned} \tag{14.30}$$

and setting the coefficients of each power of ϵ equal to zero we obtain

$$\ddot{x}_0 + \alpha_0 x_0 = 0, \tag{14.31}$$

$$\ddot{x}_1 + \alpha_0 x_1 = -(\alpha_1 + \cos t)x_0, \tag{14.32}$$

$$\ddot{x}_2 + \alpha_0 x_2 = -x_0\alpha_2 - (\alpha_1 + \cos t)x_1, \ldots, \tag{14.33}$$

and so on. If we consider the solutions of Eq. (14.31) first, we see that we have simple harmonic motion of period 2π or 4π if $\alpha_0 = (1/4)n^2$ with $n = 0, 1, \ldots$.

With $n = 0$, we have $\alpha_0 = 0$, $x_0 = 1$ (assuming a unit displacement as the initial condition), and Eq. (14.32) becomes

$$\ddot{x}_1 = -\alpha_1 - \cos t, \tag{14.34}$$

and for periodic solutions, we require $\alpha_1 = 0$ and thus,

$$x_1(t) = \cos t + c, \tag{14.35}$$

where c is a constant. Equation (14.33) then becomes

$$\ddot{x}_2 = -\alpha_2 - 1/2 - c\cos t - 1/2 \cos 2t, \tag{14.36}$$

and again, for periodic solutions we require $\alpha_2 = -1/2$, and thus, up to terms of second order in ϵ we have

$$\alpha = -1/2\epsilon^2 + \cdots + . \tag{14.37}$$

Repeating the analysis for $n = 1$, we arrive at

$$\alpha = 1/4 \pm (1/2)\epsilon - (1/8)\epsilon^2 + \cdots +, \tag{14.38}$$

and for $n = 2$,

$$\alpha = 1 + (5/12)\epsilon^2 + \cdots +, \tag{14.39}$$

$$\alpha = 1 - (1/12)\epsilon^2 + \cdots + . \tag{14.40}$$

These transition curves are plotted in Fig. 14.2, with the shaded regions indicating regions of unbounded growth of motion. A couple of numerical simulations show the form of the stable and unstable motion. The dashed curves within the unstable zones indicate the transition curves when a small amount of damping is added. This diagram is symmetric about the α axis but is plotted only for positive ϵ here.

We are now in a position to interpret the dynamic response of the beam with a pulsating end load [18, 19]. The relation between α and ϵ and the forcing characteristics of the end load lead to the plot shown in Fig. 14.3. Here, we have focused on

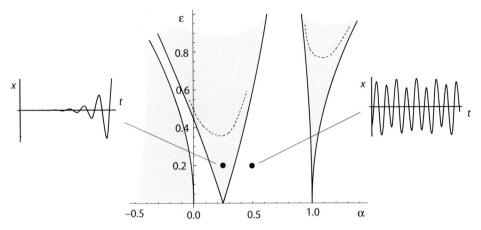

Figure 14.2. The response of Mathieu's equation in terms of the parameters α and ϵ, and based on a perturbation solution.

the instability arising from $\alpha = 0.25$. When the static component of the axial load is zero, we see that an instability occurs if the forcing frequency of the oscillating part of the load is close to *twice* the natural frequency of the system. With either additional static compression or tension the parametric instability shifts according to the fundamental frequency of the beam with an axial load—a situation described at length in earlier parts of this book. Again, damping tends to have a stabilizing effect, such that even when the forcing frequency is exactly twice the natural frequency, there needs to be a certain amount of forcing magnitude to cause instability, and this tends to make the higher-order zones of instability practically disappear. We also note that this behavior is related to the issue of quasi-perodicity and Arnold tongues, found for example in the sine map [20–22].

14.4 Pulsating Axial Loads on Shells

In much the same way that an oscillating axial load produces some interesting dynamic behavior in a column, a similar effect occurs in plates, panels, and shells [23, 24].

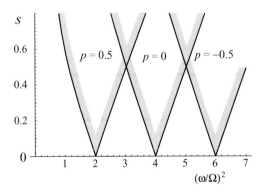

Figure 14.3. The stability of a prismatic beam with end load in which a portion of the load is oscillating. Principal parametric resonance.

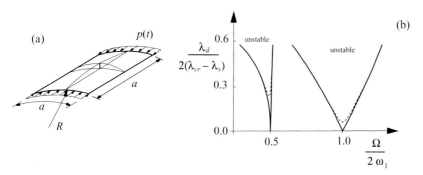

Figure 14.4. Regions of instability computed for a cylindrical panel with a periodically varying axial load. Adapted from [25].

14.4.1 A Curved Panel

We now consider a shallow panel that has some unidirectional curvature in a direction perpendicular to the loading. The analysis of such a system becomes increasingly complicated and recourse to numerical (FE) techniques is often used. Figure 14.4(a) shows an example taken from Krätzig and Nawrotzki [25] in which a shallow cylindrical panel segment is subject to a harmonically oscillating axial load. The geometry of the shell is defined by $a = 10$ m, $R = 83.33$ m, thickness $= 0.1$ m, and material properties for mild steel were used in the authors' time integration. They produced both stable and unstable time series, depending on the parameter values used, specifically the constant (static) load (λ_s), the forcing amplitude (λ_D), and the forcing frequency (Ω), with the results normalized with respect to the elastic critical buckling load (λ_{cr}) and linear natural frequency (ω_1), as shown in Fig. 14.4(b). The dashed curve within each instability zone indicates the effect of damping. Krätzig and Nawrotzki [25] also use a similar FE technique to assess the parametric instability of a truncated conical shell and compute the magnitude of the Floquet multipliers.

14.4.2 A Cylindrical Shell

When the shell is a complete cylinder, rather than a segment, we may still get parametric instability. This subsection describes the behavior reported in Popov et al. [26], based on an approximate analytical treatment. Given a cylindrical shell of the type shown in Fig. 10.14, we suppose that instead of a fixed axial load of magnitude N_x we now have a pulsating axial load of the form $p(t) = p_1 \cos \omega t$. We then make a single-mode Galerkin analysis (based on the Donnell shell theory) for the shell with the following properties: $R/h = 100$, $L/R = 2$, and a little damping added. With these parameters, it is appropriate to take an assumed form for the solution of

$$w(x, y, t) = f_1(t)h \cos \frac{\pi x}{L} \cos \frac{5y}{R} + f_2(t)h \cos \frac{2\pi x}{L}. \tag{14.41}$$

(a)   (b)

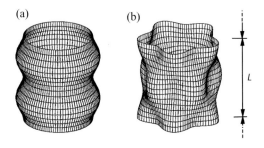

L

Figure 14.5. (a) Axisymmetric, or concertina mode of vibration, (b) asymmetric, or checkerboard, mode of vibration. Reproduced with permission from [27].

The two modes are shown in Fig. 14.5. Popov et al. [28] show that the dynamic response in mode 1 (for example) is related to Mathieu's equation in the form

$$\frac{d^2 f_1}{d\tau^2} + \left[1 - 2\mu \cos\left(\frac{\omega}{\omega_1}\tau\right)\right] f_1 = 0. \tag{14.42}$$

Furthermore, the control parameter $\mu = p_1/(2p_c)$ is introduced, in which p_c is the linear-elastic (static) buckling load.

Popov et al. [28] use a continuation technique to track the loss of stability of the trivial solution, that is, the transition from purely extensional to bending behavior. In the first part of this chapter we considered the solutions of Mathieu's equation in terms of whether the motion grew with time (or not). In a comprehensive analysis, it is possible to more fully characterize the instability phenomena, as shown in Fig. 14.6(a). Here the regions of principal parametric ($\omega/\omega_1 \approx 2$) and fundamental resonance ($\omega/\omega_1 \approx 1$) can again be observed, with the transition curves labeled S_p^1 and so on. These bifurcations indicate the nature of the instability, with the superscript 1 corresponding to a flip bifurcation and the superscript 2 indicating a pitchfork bifurcation. The former leads to the buildup of motion at twice the period of the external excitation, with the latter leading to motion at the same period as the forcing. Furthermore, we see the appearance of a couple of additional transition curves (B) that correspond to saddle-node bifurcations. All these transitions occur when a Floquet multiplier is equal to one in magnitude. For example, the flip bifurcation is characterized by a Floquet multiplier $= -1$ (see Fig. 14.1), which will be identified with the onset of period doubling in the final chapter. This diagram was based on a single-mode solution [i.e., with $f_2(t) = 0$].

Also shown in Fig. 14.6 are the bifurcation diagrams at a number of frequency ratios [and indicated by the vertical dashed lines in part (a)]. Starting at $\omega/\omega_1 = 0.9$ we see that as μ is increased the trivial solution loses stability (the dashed lines in these plots indicate unstable paths) and oscillations occur, which gradually grow as the unstable region is further penetrated. The subscript p indicates that this pitchfork bifurcation is supercritical (see Subsection 3.4.2). For $\omega/\omega_1 = 1.1$ the bifurcation is now subcritical and leads to a sudden jump in response. An interesting example is found when $\omega/\omega_1 = 1.27$ in which the system initially loses stability by means of a supercritical flip but then restabilizes and a subcritical pitchfork is encountered. Other modes can be included in the analysis [e.g., the $f_2(t)$ term in Eq. (14.41)] and although they have no effect on the initial loss of stability of the trivial state, they can

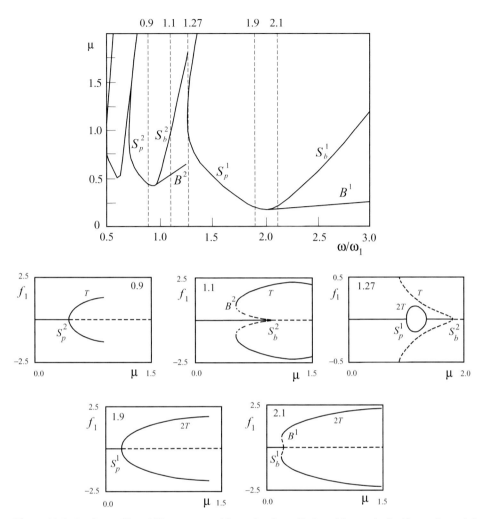

Figure 14.6. Regions of instability computed for a circular cylinder with a periodically varying axial load. Adapted from Popov et al. [28].

play an important role in subsequent postcritical behavior (for example the B curves tend to emanate from the bottom of the transition curves toward *lower* frequency ratios), and significant modal interactions can occur [28, 29].

Popov et al. [28] also consider a shallow panel by using basically the same technique, with the major difference indicating loss of stability by means of a transcritical bifurcation. Also, the proportion of static load to the magnitude of the pulsating part is an interesting parameter that relates back to the underlying buckling behavior. The behavior of these types of system may be especially complicated when parametric and direct external excitation take place simultaneously [30, 31], and combination resonances can also occur [32, 33]. At the other end of the loading regime, we have creep buckling. A comprehensive account of this phenomenon can be found in Bazant and Cedolin [18]. This chapter finishes with a mention of Meissner's problem. This is similar to the Mathieu equation but in this case the periodic variation in the stiffness is piecewise constant (a square wave) and is amenable to analytic treatment [34, 35].

References

[1] J. Singer, J. Arbocz, and T. Weller. *Buckling Experiments*, Vol. 2. Wiley, 2002.

[2] A.H. Nayfeh and D.T. Mook. *Nonlinear Oscillations*. Wiley, 1979.

[3] S.K. Sinha. Dynamic stability of a timoshenko beam subjected to an oscillating axial force. *Journal of Sound and Vibration*, 131:509–14, 1989.

[4] R.M. Evan-Iwanowski. On the parametric response of structures. *Applied Mechanics Reviews*, 18:699–702, 1965.

[5] D. Krajcinovic and G. Herrmann. Parametric resonance of straight bars subjected to repeated impulsive compression. *AIAA Journal*, 6:2025–7, 1968.

[6] A.D.S. Barr and G.T.S. Done. Parametric oscillations in aircraft structures. *The Aeronautical Journal*, 75:654–8, 1971.

[7] C.E. Hammond. An application of Floquet theory to prediction of mechanical instability. *Journal of the American Helicopter Society*, 4:14–23, 1974.

[8] R.A. Ibrahim and A.D.S. Barr. Parametric resonance, part I: Mechanics of linear problems. *Shock and Vibration Digest*, 10(1):15–29, 1978.

[9] R.A. Ibrahim and A.D.S. Barr. Parametric resonance, part II: Mechanics of nonlinear problems. *Shock and Vibration Digest*, 10(2):9–24, 1978.

[10] G.J. Simitses. *Dynamic Stability of Suddenly Loaded Structures*. Springer-Verlag, 1989.

[11] J.P. Cusumano. *Low-Dimensional, Chaotic, Nonplanar Motions of the Elastica: Experiment and Theory*. Ph.D. dissertation, Cornell University, 1990.

[12] C. Hayashi. *Nonlinear Oscillations in Physical Systems*. Princeton University Press, 1964.

[13] V.V. Bolotin. *The Dynamic Stability of Elastic Systems*. Holden-Day, 1964.

[14] M.A. Souza. Vibration of thin-walled structures with asymmetric post-buckling characteristics. *Thin-Walled Structures*, 14:45–57, 1992.

[15] D.W. Jordan and P. Smith. *Nonlinear Ordinary Differential Equations*. Oxford University Press, 1999.

[16] J. Guckenheimer and P.J. Holmes. *Nonlinear Oscillations, Dynamical Systems, and Bifurcations of Vector Fields*. Springer-Verlag, 1983.

[17] N.W. McLachlan. *Theory and Applications of Mathieu Functions*. Dover, 1964.

[18] Z.P. Bazant and L. Cedolin. *Stability of Structures*. Oxford University Press, 1991.

[19] J.F. Doyle. *Nonlinear Analysis of Thin-Walled Structures*. Springer, 2001.

[20] E. Ott. *Chaos in Dynamical Systems*. Cambridge University Press, 1993.

[21] P.R. Everall and G.W. Hunt. Arnold tongue predictions of secondary buckling in thin elastic plates. *Journal of the Mechanics and Physics of Solids*, 47:2187–2206, 1999.

[22] P.R. Everall and G.W. Hunt. Quasi-periodic buckling of an elastic structure under the influence of changing boundary conditions. *Proceedings of the Royal Society of London A*, 455:3041–51, 1999.

[23] K.K.V. Devarakonda and C.W. Bert. Flexural vibration of rectangular plates subjected to sinusoidal distributed compressive loading on two opposite sides. *Journal of Sound and Vibration*, 283:749–63, 2005.

[24] Y. Basar, C. Eller, and W.B. Krätzig. Finite element procedures for parametric phenomena of arbitrary elastic shell structures. *Computational Mechanics*, 2:89–98, 1987.

[25] W.B. Krätzig and P. Nawrotzki. Computational concepts for kinetic instability problems. In A.N. Kounadis and W.B. Krätzig, editors, *Nonlinear Stability of Structures (Theory and Computational Techniques)*. Springer-Verlag, 1995.

[26] A.A. Popov, J.M.T. Thompson, and J.G.A. Croll. Bifurcation analyses in the parametrically excited vibrations of cylindrical panels. *Nonlinear Dynamics*, 17:205–25, 1998.

[27] A.A. Popov. Parametric resonance in cylindrical shells: A case study in the nonlinear vibration of structural shells. *Engineering Structures*, 25:789–99, 2003.

[28] A.A. Popov, J.M.T. Thompson, and F.A. McRobie. Low dimensional models of shell vibrations: Parametrically excited vibrations of cylindrical shells. *Journal of Sound and Vibration*, 209:163–86, 1998.

[29] F.A. McRobie, A.A. Popov, and J.M.T. Thompson. Auto-parametric resonance in cylindrical shells using geometric averaging. *Journal of Sound and Vibration*, 227:65–84, 1999.

[30] C.S. Hsu. Impulsive parametric excitation: Theory. *Journal of Applied Mechanics*, 39:551–8, 1972.

[31] N. HaQuang, D.T. Mook, and R.H. Plaut. A non-linear analysis of the interactions between parametric and external excitations. *Journal of Sound and Vibration*, 118:425–39, 1987.

[32] T. Iwatsubo, Y. Sugiyama, and S. Ogino. Simple and combination resonances of columns under periodic axial loads. *Journal of Sound and Vibration*, 33:211–21, 1974.

[33] R.H. Plaut, N. HaQuang, and D.T. Mook. Simultaneous resonances in non-linear structural vibrations under two-frequency excitation. *Journal of Sound and Vibration*, 106:361–76, 1986.

[34] A.P. Seyranian and A.A. Mailybaev. *Multiparameter Stability Theory with Mechanical Applications*. World Scientific, 2003.

[35] C.-H. Xei. *Dynamic Stability of Structures*. Cambridge University Press, 2006.

15 Harmonic Loading: Transverse Excitation

15.1 Introduction: Resonance Effects

We have already seen many examples of how the presence of axial load tends to reduce the lateral stiffness and hence natural frequencies. In this chapter, we shall consider the effect of axial loads on the steady-state response of forced structural systems. This section will focus on an important class of forcing functions, that is, harmonic excitation. Thus, in Fig. 3.1 we might have $F(t) = F_0 \sin \omega t$, or $z(t) = z_0 \sin \omega t$, say. In the former case, we have a governing equation of motion of the form

$$M\ddot{x} + C\dot{x} + K(1 - p)x = F_0 \sin \omega t, \tag{15.1}$$

in which M, K, and C represent physical properties associated with a slender structural system. We again assume that the spring stiffness is reduced by the presence of a parameter p, later to be identified with axial load [1].

The solution of Eq. (15.1) consists of the summation of two parts. First, the homogeneous solution is obtained from the free vibration and was derived previously. For typical damping values, it consists of an exponentially decaying oscillation (assuming $p < 1$). Second, the particular solution consists of a steady-state oscillation, $X_0 e^{i\omega t}$, where the magnitude of the steady-state response (relative to the forcing magnitude) is given by

$$\frac{X_0}{F_0} = R(\omega) = \frac{1}{\sqrt{[K(1 - p) - \omega^2 M]^2 + (C\omega)^2}}, \tag{15.2}$$

and is often referred to as the receptance, amplitude-response, or frequency-response function (FRF) [2]. We observe the important resonant effect if the driving frequency ω is close to the natural frequency $\omega_n = \sqrt{K/M}$ of the system, or $\Omega = \omega/\omega_n = 1$. Figure 15.1 shows the receptance for the parameter values $K = M = 1$ and $C = 0.2$ as a function of the destabilizing parameter p. We note that increasing p tends to shift the resonant peaks toward lower frequencies [with negative p (tensile) having the opposite effect]. The receptance could also have been nondimensionalized with respect to the effective stiffness in which case the curves would have emanated from a common point on the y axis.

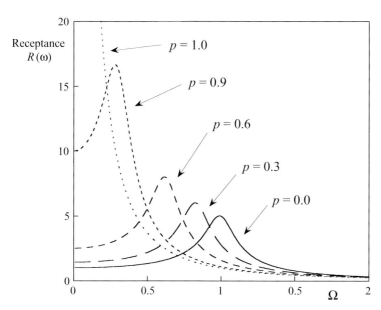

Figure 15.1. Effect of diminished stiffness on the receptance of a spring–mass–damper.

15.1.1 A Single-Mode Approximation

We have seen a number of times how it is possible to make a relatively accurate single-mode analysis of beam dynamics. Consider a thin elastic beam of length L, flexural rigidity EI, and mass m (per unit length), which is clamped at both ends and subject to a compressive axial load of magnitude P. A single-mode energy analysis of this system, assuming a mode shape of the form

$$w(x, t) = \frac{Q(t)}{2}\left[1 - \cos\frac{2\pi x}{L}\right],\tag{15.3}$$

can be conducted along the lines of Chapter 7, resulting in a natural frequency of

$$\omega^2 = \frac{1}{3m}\left[EI\left(\frac{2\pi}{L}\right)^4 - P\left(\frac{2\pi}{L}\right)^2\right].\tag{15.4}$$

From this we immediately see that buckling occurs when $P_{cr} = EI(2\pi/L)^2$ (exact), and in the absence of the axial load, we obtain a natural frequency of $\omega_0 = 22.79\sqrt{EI/(mL^4)}$ (exact coefficient = 22.37). Using these to nondimensionalize ($\bar{p} = P/P_{cr}$ and $\bar{\omega} = \omega/\omega_0$) we have

$$\bar{\omega}^2 = 1 - \bar{p}.\tag{15.5}$$

Now, subjecting the system to a transverse harmonic point force at mid-span, $F(t)$, and assuming a small amount of linear-viscous damping, C, we have the equation of motion given by Eq. (15.1) in which

$$x = Q, \quad M = 3m, \quad K = EI(2\pi/L)^4, \quad p = P/P_{cr},\tag{15.6}$$

and the corresponding receptance is still given by Eq. (15.2).

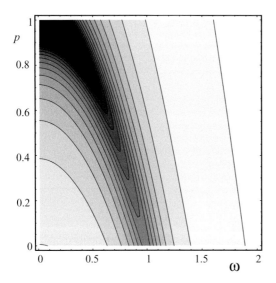

Figure 15.2. Effect of axial load on the receptance of a damped, SDOF model of a beam as a contour plot. Darker shades are higher in magnitude. $C = 0.1$.

The response can also be plotted as a contour plot in terms of axial load and forcing frequency. This is shown in Fig. 15.2. In this plot, we can observe how the resonant peaks spread as the axial load is increased. This can be viewed as an increase in the damping ratio (as this is relative to the natural frequency and hence stiffness). A closely related circumstance is what happens when the support upon which the mass is supported is excited [e.g., $z(t) = z_0 \sin \omega t$]. This is called *transmissibility*, and will be studied in detail a little later in relation to vibration isolation. We conclude that an axial load not only has the effect of shifting resonant peaks to lower frequencies but also increases the effective damping in the system.

We can again obtain a useful physical sense of the effect of the changing axial load on the forced vibration problem by evolving the axial load as a linear function of time: $p = 0.002t$. In this way, the stiffness of the system will reduce to zero when $t = 500$. Figure 15.3 shows an example based on numerical simulation of the governing equation of motion, including the diminishing stiffness. The forcing parameters are fixed at $F_0 = 1$ and $\omega = 0.5$, and hence resonance should occur when $t \approx 375$. Note that there is again a small amount of overshoot in the nonstationary (slowly evolving, or swept) response.

15.1.2 Beyond Buckling

We can extend the single-mode energy analysis of this system by including higher-order terms in the potential energy of this system (truncated after the second term):

$$V = \frac{1}{2}EI \int_0^L [w''^2 + w''^2 w'^2]\, dx - \frac{1}{2}P \int_0^L \left[w'^2 + \frac{1}{4}w'^4\right] dx, \qquad (15.7)$$

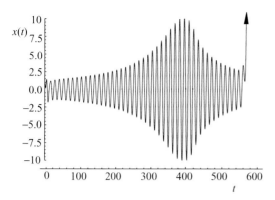

Figure 15.3. A sweep through decaying stiffness and passing through resonance.

and the kinetic energy remains the same. Evaluating the potential-energy expression now gives

$$V = \frac{1}{16}EIL\left(\frac{2\pi}{L}\right)^4 Q^2 + \frac{1}{256}EIL\left(\frac{2\pi}{L}\right)^6 Q^4$$

$$-\frac{1}{16}PL\left(\frac{2\pi}{L}\right)^2 Q^2 - \frac{1}{128}P\frac{3L}{8}\left(\frac{2\pi}{L}\right)^4 Q^4. \tag{15.8}$$

In addition to the trivial ($Q = 0$) solution, we now have a nontrivial (post-buckled) path given by

$$Q^2 = \frac{8(\bar{p} - 1)}{\left(\frac{2\pi}{L}\right)^2\left(1 - \frac{3}{4}\bar{p}\right)}, \tag{15.9}$$

where

$$\bar{p} = \frac{P}{EI\left(\frac{2\pi}{L}\right)^2}. \tag{15.10}$$

After buckling, the nondimensional natural frequency is given by

$$\bar{\omega}^2 = 1 + \frac{3(\bar{p} - 1)}{(1 - \frac{3}{4}\bar{p})} - \bar{p}\left[1 + \frac{\frac{9}{4}(\bar{p} - 1)}{(1 - \frac{3}{4}\bar{p})}\right] = \bar{\omega}^2 = 2(\bar{p} - 1), \tag{15.11}$$

that is, half the prebuckling slope of the load–frequency (-squared) relation [Eq. (15.5)]. This is a result anticipated from the normal form of the supercritical pitchfork bifurcation (Section 7.4).

15.2 The Poincaré Section

Before moving on to consider the resonance response of axially loaded continuous systems we introduce the concept of Poincaré sampling. The response of the forced nonlinear oscillator of the type of Eq. (15.1) with $p = 0$ is typically given in terms of transient (see Chapter 3) and steady-state parts. In the framework of dynamical system theory, we can view the steady state as a periodic attractor for the surrounding transients. For a linear system, the periodic attractor is unique. We shall see in the next chapter that this is not necessarily the case for nonlinear oscillators. A typical engineering approach is then to plot the maximum amplitude of response as a function of the forcing frequency (see Fig. 15.1). Often the phase difference between the forcing function and the response is also plotted and a sudden shift in phase is associated with resonance [3, 4]. However, an alternative description of the response is to reduce the 3D phase space in continuous time to a 2D phase space in discrete time by Poincaré sampling [5, 6].

The complete solution to Eq. (15.1) with $p = 0$ can be written as

$$x(t) = \frac{F_0}{K} \frac{\sin(\omega t - \phi)}{\sqrt{[1 - (\omega/\omega_n)^2]^2 + [2\zeta\omega/\omega_n]^2}} + X_1 e^{-\zeta\omega_n t} \sin(\sqrt{1 - \zeta^2}\omega_n t + \phi_1),$$

(15.12)

and focusing on the steady-state solution we ignore the second term and write Eq. (15.12) in the alternative form

$$x(t) = a \cos(\omega t) + b \sin(\omega t).$$

(15.13)

Differentiating this to get the velocity, we have

$$y(t) \equiv \dot{x} = -a\omega \sin(\omega t) + b\omega \cos(\omega t),$$

(15.14)

and setting $t = 0$ (which effectively fixes the initial forcing phase), we simply get $x = a$ and $y = b\omega$ as the fixed point location, where

$$a = \frac{(1 - \Omega^2)f}{(1 - \Omega^2)^2 + (2\zeta\Omega)^2},$$

(15.15)

$$b = \frac{2\zeta\Omega f}{(1 - \Omega^2)^2 + (2\zeta, \Omega)^2},$$

(15.16)

and $\Omega = \omega/\omega_n$. The Poincaré section can thus be considered as an alternative to the more conventional (amplitude–phase) representation of the response of an oscillator.

The complementary function (the transient solution) can also be included in the following way to give a discrete mapping: the Poincaré map P [6],

$$\begin{pmatrix} x \\ y \end{pmatrix} \rightarrow e^{\frac{-2\pi\zeta\upsilon_n}{\omega}} \begin{bmatrix} C + \frac{\zeta\omega_n}{\omega_d}S & \frac{1}{\omega_d}S \\ -\frac{\omega_n^2}{\omega_d}S & C - \frac{\zeta\omega_n}{\omega_d}S \end{bmatrix} \begin{pmatrix} x \\ y \end{pmatrix}$$

$$+ e^{\frac{-2\pi\zeta\upsilon_n}{\omega}} \begin{bmatrix} -aC + \left(-\frac{\zeta\omega_n a}{\omega_d} - \frac{b\omega}{\omega_d}\right)S \\ -b\omega C + \left(\frac{a\omega_n^2}{\omega_d} + \frac{\zeta b\omega_n\omega}{\omega_d}\right)S \end{bmatrix} + \begin{pmatrix} a \\ b\omega \end{pmatrix},$$

(15.17)

where $C \equiv \cos(2\pi\omega_d/\omega)$, $S \equiv \sin(2\pi\omega_d/\omega)$, and $\omega_d = \omega_n\sqrt{1 - \zeta^2}$. Thus, given some initial conditions, this set of *difference* equations will map out the transient at intervals of the forcing period until converging on the fixed point

$$(x, y) = (a, b\omega).\qquad(15.18)$$

This mapping is exact for a linear oscillator (and is related to the Z transform [7]) but cannot usually be easily obtained for nonlinear systems. However, importantly, this complete mapping contains the stability information regarding the fixed point and relates back to the section on Floquet theory and characteristic multipliers (see Section 14.2).

We have already introduced the concept of characteristic eigenvalues (exponents) (CEs) for determining stability of equilibria in unforced systems. Now we see that it is the eigenvalues of the map (characteristic multipliers, or CMs) that determine the stability of cycles. The eigenvalues of the Jacobian, that is, the first partial derivatives of the map given by Eq. (15.17), $DP(a, \omega b)$, are given by

$$\lambda_{1,2} = e^{-\frac{2\pi \varsigma \omega n}{\omega} \pm i \frac{2\pi \omega d}{\omega}},\qquad(15.19)$$

which confirms that the fixed point is asymptotically stable, because the damping and natural frequency are positive numbers. This is why consideration of discrete maps plays a useful role in the study of flows.

Despite the fact that this approach tends to hide the usually important engineering aspects of amplitude and phase, it does provide a very convenient means of assessing stability, and the evolution of the system responses under the slow change in a parameter. For the types of nonlinear system to be considered later, Poincaré sampling provides a powerful tool in the numerical and experimental investigation of periodically excited nonlinear oscillators.

15.3 Continuous Systems

We now turn to consideration of axially loaded, transversely forced, continuous beams. The forced string and membrane can also easily be handled by use of these techniques. A beam of length L, mass per unit length m, uniform flexural rigidity EI, viscous damping coefficient C, axial force P, and transverse load $Q_0 F(x) \cos \Omega t$, respond as $w(x, t)$. The governing equation of motion is

$$mw_{tt} + Cw_t + EIw_{xxxx} + Pw_{xx} = Q_0 F(x) \cos \Omega T,\qquad(15.20)$$

in which subscripts on w reflect partial derivatives. This equation can be put in the nondimensional form

$$\frac{\partial^4 \bar{w}}{\partial \bar{x}^4} + c\frac{\partial \bar{w}}{\partial \bar{t}} + p\frac{\partial^2 \bar{w}}{\partial \bar{x}^2} + \frac{\partial^2 \bar{w}}{\partial \bar{t}^2} = f(x) \cos \bar{\Omega}\bar{t},\qquad(15.21)$$

with

$$\bar{x} = x/L, \qquad \bar{w} = w/L, \qquad \bar{t} = t\sqrt{EI/(mL^4)},$$

$$p = PL^2/(EI), \qquad \bar{\Omega} = \Omega\sqrt{mL^4/(EI)}, \qquad c = CL^2/\sqrt{mEI},\qquad(15.22)$$

$$f(x) = Q_0 F(X)L^3/(EI), \qquad \zeta = c/(2\Omega_1),$$

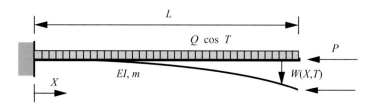

Figure 15.4. A cantilever beam subject to a constant axial load and a harmonically varying uniformly distributed lateral force.

where Ω_1 is the fundamental natural frequency and its value depends on the boundary conditions. For convenience, we now drop the overbar notation, and assuming a steady-state harmonic response of the form

$$w(x, t) = \operatorname{Re}\left[y(x)e^{i\Omega t}\right] \tag{15.23}$$

leads to

$$y''''(x) + py''(x) + (i\Omega c - \Omega^2)y(x) = f(x). \tag{15.24}$$

Equation (15.24) can then be solved for a specific set of boundary conditions and transverse forcing types. A number of examples with distributed, transverse, harmonic forcing can be found in Virgin and Plaut [1]. Here, two cases will be considered:

- an axially loaded cantilever beam with a uniformly distributed harmonic load, and
- an axially loaded, clamped–clamped beam with a harmonic central point load.

Considering the first case as shown in Fig. 15.4 we have a governing equation of motion given by Eq. (15.24) but now with unity on the right-hand side. The general solution is given by

$$y(x) = (i\Omega c - \Omega^2)^{-1} \sum_{j=1}^{2}(a_j \cosh \lambda_j x + b_j \sinh \lambda_j x), \tag{15.25}$$

where

$$\lambda_j = (v_j/\gamma_j) + i(\gamma_j/2), \quad \gamma_j = \left[2\left(-\epsilon_j + \sqrt{\epsilon_j^2 + v_j^2}\right)\right]^{1/2},$$

$$v_1 = \phi/2, \qquad\qquad v_2 = -\phi/2,$$

$$\epsilon_1 = (\Phi - p)/2, \qquad \epsilon_2 = -(\Phi + p)/2,$$

$$\Phi = -\frac{c\Omega}{2\phi}, \qquad\qquad \phi = \left\{\left[-\delta + \sqrt{\delta^2 + (c\Omega)^2}\right]/2\right\}^{1/2},$$

$$\delta = p^2 + 4\Omega^2.$$

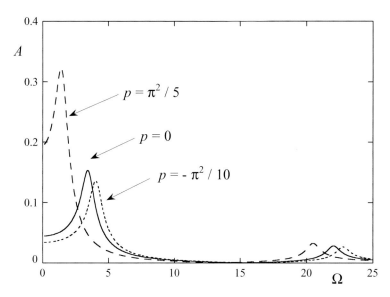

Figure 15.5. The central amplitude versus the forcing frequency for a cantilever beam with a uniformly distributed force.

The boundary conditions in this case are $y(0) = 0$, $y'(0) = 0$, $y''(1) = 0$, and $y'''(1) + py'(1) = 0$. The elastic critical load is $p_{\text{cr}} = \pi^2/4$, and the fundamental natural frequency is $\Omega_1 = 3.516$. Applying the boundary conditions and solving the resulting simultaneous equations (in a_j, b_j) leads to the results shown in Fig. 15.5 in which a damping ratio of $\zeta = 0.142$ was used. The amplitude A is the magnitude of the maximum central deflection, which depends on the static axial load and the forcing frequency, i.e., $A(\Omega, p) = |y(0.5)|$. Note the presence of the second resonant peak in the vicinity of $\Omega = 22$.

It is interesting to see how the amplitude and the corresponding resonant frequency vary with axial load. Figure 15.6 shows the relation

$$A_R(p) = \max_{\Omega \geq 0} A(\Omega, p), \qquad (15.26)$$

and the values of Ω (squared) for which this condition occurs. Three representative damping values are used. We again get a near-linear relation between frequency squared and the axial load.

Now consider the second case. Figure 15.7(a) shows the amplitude response (receptance) when $\zeta = 0.02$. We again see the anticipated increase in natural frequency for a tensile axial force, and reduction for compression.

Figure 15.7(b) shows how the receptance associated with the third (second symmetric) mode is affected by the presence of an axial load. In this case, the undamped third frequency occurs at $\Omega = 120.9$ for the unloaded case. The peaks are also shifted slightly from those of the undamped case because of the presence of a little damping.

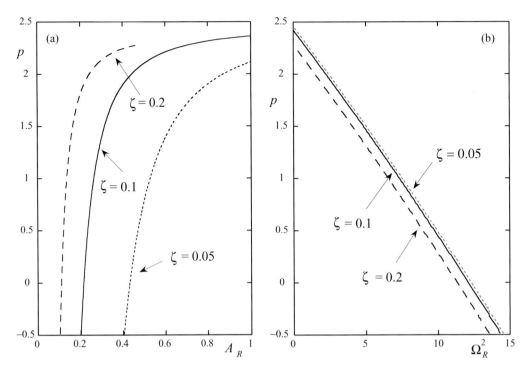

Figure 15.6. (a) The resonant amplitude, (b) the frequency squared versus the axial load for a cantilever beam with a uniformly distributed force.

In both of the preceding examples, it can be shown that the resonant amplitude and frequency are given approximately by

$$A_R(p) \approx A_R(0)/\sqrt{1 - (p/p_{cr})}, \tag{15.27}$$

$$\Omega_R^2 \approx \Omega_1^2[1 - (p/p_{cr})], \tag{15.28}$$

and we can also relate this back to the material presented in Chapter 7, for example, the $p = 0$ resonant peak for the beam in Fig. 15.7 occurs at the natural frequency coefficient of 22.4.

Experimental Verification. A thin steel strip was clamped between blocks at its ends and placed in a displacement-controlled testing machine. In this configuration, the end shortening is prescribed, and the resulting axial force is measured with a load cell. With the standard expressions based on the earlier analysis, the critical load was computed at 1235 N and the lowest natural frequency (with no axial load) at 44 Hz. The strut was struck by an impact hammer and a laser velocity vibrometer was used to measure the response [8]. The data were then acquired and analyzed with the Bruel and Kjaer PULSE system. Velocity time series were then subject to a Hann window and a fast Fourier transform algorithm to extract the frequency

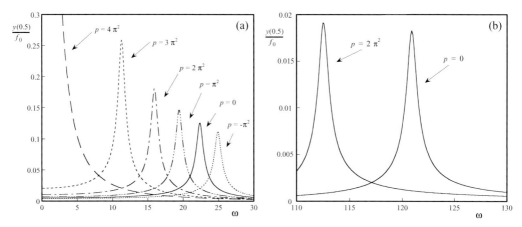

Figure 15.7. (a) Amplitude response of an axially loaded, clamped–clamped beam subject to a central point harmonic point force and (b) in the vicinity of the third mode.

content. This process was repeated 10 times at each axial-load level. The results were averaged and displayed as normalized mobility, that is, normalized with respect to the force of the impact hammer. A photograph of the experimental system is shown in Fig. 15.8(a). Figure 15.8(b) shows a summary of how the peaks shift to lower (higher) frequency as the compressive (tensile) axial load is increased. However, experimental results from this type of system need careful consideration because testing machines are sometimes referred to as *semirigid* loading devices, membrane effects in the beam may occur, and even the boundary conditions can be a function of loading [9, 10].

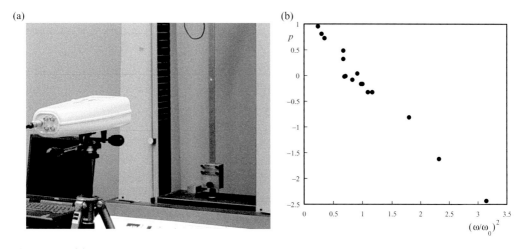

Figure 15.8. (a) A photograph of the clamped beam in the testing machine and (b) the axial load plotted as a function of the natural frequencies (squared).

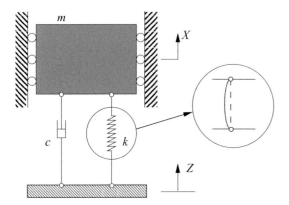

Figure 15.9. Schematic of a mass isolated from the motion of a foundation.

15.4 An Application to Vibration Isolation

In many instances, buckling is viewed as an undesirable occurrence, in particular when it precipitates a total loss of stiffness and collapse (see, for example, Section 3.4). However, there are ways in which postbuckled stiffness can be exploited, and this section introduces an example. The concept of vibration isolation, that is, how to reduce the force or motion transmitted to a device from a source of vibration, is well established [11–14]. This section describes an approach to effective vibration isolation by use of the subtle interplay of axial loads, dynamics, and stability [15].

Consider a simple mechanical system consisting of a mass (in a gravitational field) supported by a spring and damper, which are themselves supported on a base, as shown in Fig. 15.9. If the motion of the base is harmonic, for example, $Z(t) = Z_0 \sin \omega t$, then it can be shown that the steady-state displacement transmissibility, X/Z (where X is the response amplitude of the mass), is given by the expression

$$\frac{X}{Z} = \left[\frac{1 + (2\zeta\Omega)^2}{(1 - \Omega^2)^2 + (2\zeta\Omega)^2} \right]^{1/2}. \tag{15.29}$$

This is plotted in Fig. 15.10 as a function of the frequency ratio $\Omega = \omega/\omega_n$, where $\omega_n = \sqrt{k/m}$. Damping tends to severely reduce the resonant peak. This was also true for the cases examined earlier in this chapter in which the force was applied to the mass directly. In these earlier cases, damping tended to reduce the magnitude of the response at all frequencies. Here, we notice an interesting feature in which (linear-viscous) damping results in slightly larger responses at higher frequencies. We also observe that damping has no influence when Ω is exactly $1/\sqrt{2}$. In addition to direct mass and base excitation, a third type of resonance can also be found in a system with a rotating unbalance [4].

However, overall, we see that the transmissibility is small for relatively high-frequency ratios, that is, for $\Omega > \sqrt{2}$, $X/Z < 1.0$. Given a forcing frequency ω, a typical design option would be to mount the device on a *soft* spring to induce a low natural frequency, ω_n. But if the spring has a low stiffness, there is a danger that

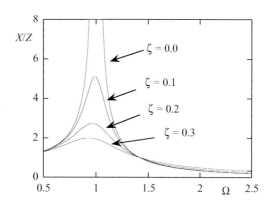

Figure 15.10. Displacement transmissibility for a SDOF oscillator for four typical damping values.

it will deflect statically too much (because $mg = kx$), and this often places practical limits on the spring stiffness.

15.4.1 Postbuckling of a Strut Revisited

In earlier chapters, we have seen how axially loaded structures typically possess nonlinear characteristics, especially close to, or beyond, initial buckling. Often this takes the form of additional postbuckled stiffness (e.g., in plates). In their postcritical state, they exhibit relatively low stiffness in the axial direction and yet they carry axial loads above their buckling load. As indicated schematically in Fig. 15.9 we see a potential opportunity in the context of vibration isolation.

Let's return to the simply supported strut as shown in Fig. 7.1. Any structure exhibiting stable postbuckled behavior can be used in this situation, but the pinned beam is easiest to analyze. Thus, we are interested in a structural system of the *supercritical* type. We have already seen [see Eq. (7.51)] that the initial postbuckled equilibrium configuration is described by

$$\frac{P}{P_e} = 1 + \frac{\pi^2}{8}\left(\frac{Q}{L}\right)^2, \tag{15.30}$$

where $P_e = EI(\pi/L)^2$ is the classical Euler critical load.

In Chapter 7 we were primarily interested in lateral stiffness effects, but now we need to consider stiffness in the axial direction as this is the direction in which the force acts. The geometric relation between the lateral deflection Q and the end shortening δ can be established as [15]

$$\left(\frac{\delta}{L}\right) = \frac{\pi^2}{4}\left(\frac{Q}{L}\right)^2 + \frac{3\pi^4}{64}\left(\frac{Q}{L}\right)^4. \tag{15.31}$$

The end shortening is approximately related to the square of the lateral deflection, and eliminating Q in Eqs. (15.30) and (15.31) leads to

$$\frac{P}{P_e} = \frac{1}{3}\left[2 + \sqrt{1 + 3\left(\frac{\delta}{L}\right)}\right] \approx 1 + \frac{1}{2}\left(\frac{\delta}{L}\right). \tag{15.32}$$

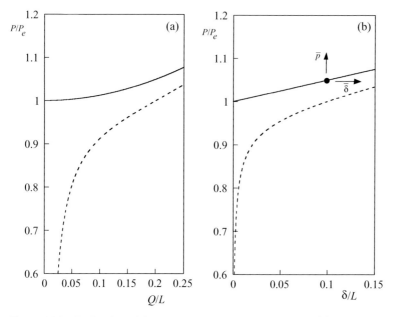

Figure 15.11. Deflection of the strut as a function of axial load: (a) central lateral deflection, (b) end shortening.

This is a result based on moderate lateral deflections, i.e., up to about 20% of the length. Recall that this is a nonlinearity induced by the curvature expression [Eq. (7.40)], rather than the membrane effect discussed in Section 7.5 for example, that is, the ends are free to move toward each other. Equations (15.30) and (15.32) are shown by the solid curves in Figs. 15.11(a) and 15.11(b), respectively.

The postbuckled *stiffness* is only mildly affected by initial imperfections (unlike in the vicinity of the critical point), and when a single mode is adopted, representative initial geometric imperfection leads to the dashed curves in Fig. 15.11. If we load the strut axially to slightly above its elastic critical load, for example, $P/P_e = 1.05$, then (for $Q_0 = 0$) we have $Q/L \approx 0.2$, $\delta/L \approx 0.1$. This specific load-deflection condition furnishes an equilibrium position from which incremental coordinates are measured: $\bar{p} = P/P_e - 1.05$, $\bar{\delta} = \delta/L - 0.1$, and indicated in Fig. 15.11(b). This strut, then, is able to support a relatively high axial load (sufficient to cause buckling) but exhibits the desirable soft spring characteristic.

15.4.2 Experimental Verification

An experimental verification is considered in this section and configured as shown in Fig. 15.12(a). Part (b) shows an alternative configuration in which four postbuckled panels provide the support. The vertical shaker was connected to a cam-shaft attached to a variable-speed motor. The forcing amplitude was fixed at 3 mm, and the shaker then imparted an almost sinusoidal motion through the isolation system (consisting of two steel struts made of spring steel) to the mass, which moved

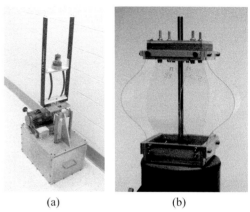

Figure 15.12. (a) Photograph of the experimental setup and (b) alternative configuration in which panels re used.

(a)   (b)

vertically, guided by low-friction linear bearings. The frequency of excitation was then varied over an appropriate range, and the vertical motion of the mass was measured [15].

Plots of axial load versus lateral deflection and end shortening are shown in Fig. 15.13, where the deflections were measured with a linear-variable displacement transformer. The two struts were 268 mm long, 19 mm wide, 0.66 mm thick (and thus $I = 4.55 \times 10^{-13}$ m^4), and taking a typical value for Young's modulus of 200 GPa, we anticipate an Euler load in the vicinity of 25 N $\equiv$ 2.55 kg. A Southwell plot can be used to recast the data from Fig. 15.13(a) to estimate a critical load of approximately 23 N. Because of initial geometric imperfections, the "critical load" is again manifest as a relatively rapid increase in the deflection that is due to additional load. The axial load versus lateral deflection, and versus end shortening, results are shown in Figs. 15.13(a) and 15.13(b), respectively, and these relations illustrate a good correlation with the corresponding theoretical curves (including an initial imperfection) given in Fig. 15.11.

By the choice of an appropriate point on the curve in Fig. 15.13(b) as the fundamental equilibrium position, for example, $P = 23.5 N \rightarrow (P/P_e \approx 1.0)$, $\delta = 15.2$ mm $\rightarrow (\delta/L = 0.057)$, the transmissibility can be assessed over a range of excitation frequencies. Locally, the stiffness is approximately 195 N/m (i.e., the slope of P/P_e versus δ/L about the chosen operating point), and because the mass is 2.4 kg, we would thus expect a natural frequency of free vibration close to 9 rad/s $\equiv$ 1.43 Hz. A free decay of this system gives a natural period of approximately 0.68 s (and hence $\omega_n = 1.47$ Hz), and thus, we anticipate effective isolation for forcing frequencies $\omega > \sqrt{2}\omega_n \approx 2.2$ Hz. If a conventional linear (helical) spring had been used instead of the buckled struts, a static deflection of approximately four times the deflection of the struts would have resulted from this load level.

15.4.3 The Forced Response

The transmissibility ratio [given by Eq. (15.29)] should be low, more specifically less than one, for $\Omega > \sqrt{2}$. Subjecting the system to a range of excitation frequencies (at

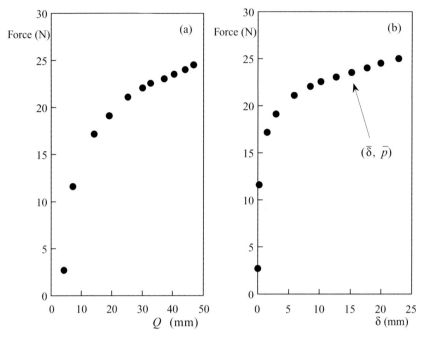

Figure 15.13. Axial load versus (a) midpoint lateral deflection and (b) end shortening for the experimental system.

constant forcing amplitude) leads to the experimental results shown in Fig. 15.14. Three typical time series are shown for the frequency ratios indicated. The response when the forcing frequency is exactly twice the natural frequency (indicated by the open circle) shows an interesting subharmonic of order two. This is a consequence of combined parametric and external forcing terms in the governing dynamic equations (see Chapter 14). In general we see a highly attenuated response for higher frequencies; that is, in this frequency range, the mass is effectively isolated from the motion of the base.

This concept can be extended in a number of ways [16], and the usual issues of avoiding stroke-out, nonlinear behavior, and fatigue still apply. Here some research is mentioned in which axially loaded structures are taken advantage of to amplify, rather than reduce, motion [17, 18], and how axial load can be used to tune resonant frequencies for the purposes of energy scavenging [19].

15.5 Forced Excitation of the Thermally Buckled Plate

In Chapter 10 we saw how axial loading (thermal) effects influenced the free vibration of thin plates. At ambient temperatures, the plate would exhibit a periodic response if subject to a periodic lateral excitation, and increasing the thermal loading would result in the now-familiar shift in resonance characteristics. For temperatures above the critical buckling level, two coexisting equilibria appear, that is,

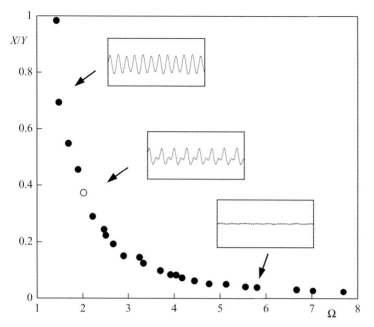

Figure 15.14. Experimental transmissibility for the displacement of the strut-supported mass.

supercritical behavior. For relatively low levels of external excitation, we still expect periodic response but now offset about some nontrivial equilibrium configuration. We shall use a panel system similar to the one described in Section 10.5 to illustrate this effect experimentally (the only difference being that now the panel is thinner). In the final chapter, we shall look at more intense excitation that can result in large-amplitude responses, including chaos and intermittent snapping between these equilibria [20].

Typical periodic responses are shown in Fig. 15.15 in which two alternative coexisting oscillations are *superimposed*, that is, motion about both static (postbuckled) equilibrium configurations within the corresponding potential energy wells. This is the situation shown schematically in Fig. 10.6. The measurand in this case is strain and is plotted against strain a quarter of a cycle later in part (b), by use of a standard embedding technique in nonlinear dynamics [21]. These results were recorded

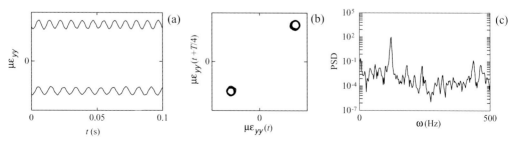

Figure 15.15. Small-amplitude periodic behavior about both postbuckled positions using microstrain and time-lag embedding: (a) time series, (b) phase projections, and (c) frequency spectrum.

at $(\zeta, \eta) = (0.583, 0.416)$ with the temperature set at $\Delta T / \Delta T^f_{cr} = 1.76$, with an excitation level[1] of 130 dB at 120 Hz (quite near the lowest natural frequency at *this*
temperature). The second attractor (which has a slightly smaller basin of attraction)
is attained by giving a variety of perturbations access to different areas of the initial condition space. Because of the inevitable initial geometric imperfections there
is a mild asymmetry, such that the postbuckled plate has a slightly tilted underlying potential-energy function [see Fig.3.8(a)]. This asymmetry is also reflected in a
slightly different location in the phase projection of the two responses as well as
a slight difference in the period in their time series. This kind of small-amplitude
periodic behavior possesses a power spectrum with a dominant spike at the forcing frequency [see Fig. 15.15(c)]. The sharpness of the resonant peak is a standard
means (the half-power method) by which the damping can be estimated [22, 23]. A
good correlation with a theoretical analysis is described in Murphy et al. [24].

References

[1]   L.N. Virgin and R.H. Plaut. Effect of axial load on forced vibrations of beams. *Journal
      of Sound and Vibration*, 168:395–405, 1993.
[2]   D.J. Ewins. *Modal Testing: Theory and Practice*. Research Studies Press, 1984.
[3]   W.T. Thomson. *Theory of Vibration with Applications*. Prentice Hall, 1981.
[4]   D.J. Inman. *Engineering Vibration*. Prentice Hall, 2000.
[5]   D.W. Jordan and P. Smith. *Nonlinear Ordinary Differential Equations*. Oxford Univer-
      sity Press, 1999.
[6]   J. Guckenheimer and P.J. Holmes. *Nonlinear Oscillations, Dynamical Systems, and Bi-
      furcations of Vector Fields*. Springer-Verlag, 1983.
[7]   K. Ogata. *System Dynamics*. Prentice Hall, 1998.
[8]   G.C. Goodwin and R.L. Payne. *Dynamics System Identification: Experiment Design and
      Data Analysis*. Academic, 1977.
[9]   A. Picard, D. Beaulieu, and B. Perusse. Rotational restraint of a simple column base
      connection. *Canadian Journal of Civil Engineering*, 14:49–57, 1987.
[10] R.H. Plaut. Column buckling when support stiffens under compression. *Journal of Ap-
      plied Mechanics*, 56:484, 1989.
[11] F.C. Nelson. Vibration isolation: A review, I. Sinusoidal and random excitations. *Shock
      and Vibration*, 1:485–93, 1994.
[12] R.H. Racca. Characteristics of vibration isolators and isolation systems. In *Shock and
      Vibration Handbook*, 4th ed. McGraw-Hill, 1996, Chapter 32.
[13] J. Winterflood, T. Barber, and D.G. Blair. High performance vibration isolation using
      spring in Euler column buckling mode. *Physics Letters A*, 19:1639–45, 2002.
[14] E.I. Rivin. *Passive Vibration Isolation*. ASME, 2003.
[15] L.N. Virgin and R.B. Davis. Vibration isolation using buckled struts. *Journal of Sound
      and Vibration*, 260:965–73, 2003.
[16] R.H. Plaut, J.E. Sidbury, and L.N. Virgin. Analysis of buckled and pre-bent fixed-end
      columns used as vibration isolators. *Journal of Sound and Vibration*, 283:1216–28, 2005.
[17] J. Jiang and E. Mockensturm. A motion amplifier using an axially driven buckling beam:
      I. Design and experiments. *Nonlinear Dynamics*, 43:391–409, 2006.

---

[1] The reference value for the decibel scale is the rms value 20 μN/m^2 for sound pressure level.

[18] J. Jiang and E. Mockensturm. A motion amplifier using an axially driven buckling beam: II. Modeling and analysis. *Nonlinear Dynamics*, 45:1–14, 2006.

[19] E.S. Leland and P.K. Wright. Resonance tuning of piezoelectric vibration energy scavenging generators using compressive axial load. *Smart Materials and Structures*, 15:1413–20, 2006.

[20] K.D. Murphy, L.N. Virgin, and S.A. Rizzi. Experimental snap-through boundaries for acoustically excited, thermally buckled plates. *Experimental Mechanics*, 36:312–17, 1996.

[21] L.N. Virgin. *Introduction to Experimental Nonlinear Dynamics: A Case Study in Mechanical Vibration*. Cambridge University Press, 2000.

[22] D.E. Newland. *An Introduction to Random Vibrations and Spectral Analysis*. Longman, 1984.

[23] J.S. Bendat and A.G. Piersol. *Random Data: Analysis and Measurement Procedures*. Wiley, 1986.

[24] K.D. Murphy, L.N. Virgin, and S.A. Rizzi. Characterizing the dynamic response of a thermally loaded, acoustically excited plate. *Journal of Sound and Vibration*, 196:635–58, 1996.

16 Nonlinear Vibration

PART I: FREE VIBRATION

16.1 Introduction

This last chapter considers the dynamic response of axially loaded structural systems in which the motion is *not* necessarily confined to the local vicinity of an underlying equilibrium position and dynamic behavior is not necessarily harmonic. In a number of places throughout this book, a statement has been made to the effect that large-amplitude behavior will be described later. We now finally consider such situations, largely in terms of revisiting examples detailed in earlier examples, but now, *not* relying on certain restrictions, for example, linear, or small-amplitude, behavior. Both free and forced vibrations will be considered [1, 2].

By way of a simple introduction, we go back to the softening cable example described in Section 3.5, and specifically consider the context of Fig. 3.12. This is a free vibration started (with initial conditions) some distance from any of the three available stable equilibrium points present at this level of loading. Because there is no damping, the total energy is conserved, and thus phase trajectories can be viewed as contours of constant total energy. Figure 16.1(a) illustrates the energy levels as a contour plot, and thus we can view the phase trajectory of Fig. 3.12 living in the second darkest shade within the contours of Fig. 16.1(a). Parts (b)–(d) give specific examples of time series generated (numerically) by different initial conditions. We see that the time series in part (b), which corresponds to the phase trajectory shown in Fig. 3.12, is far from sinusoidal. The time series shown in part (c) has relatively small amplitude with a natural frequency close to that predicted by linear theory; see Fig. 3.2 (but still slightly nonlinear; note the expanded y-axis). The time series shown in part (d) is initiated from a position very slightly removed from one of the *unstable* equilibria. After remaining in the local vicinity of the unstable point (previously calculated to be at $q = 1.92$), the trajectory moves away, undergoing a long-period motion, which again is far from sinusoidal. Parts (b) and (d) illustrate how the frequency of the response is highly dependent on the amplitude of motion and hence on the initial conditions. Other types of motion are exhibited following different initial conditions and values of the parameter (which is assumed to be maintained at the same level during the motion).

The next section will briefly describe a couple of approximate analytical techniques primarily designed to assess the effects of moderate nonlinearity on the

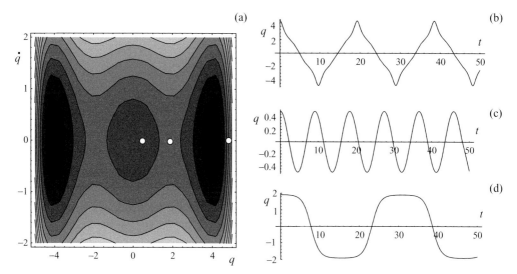

Figure 16.1. (a) A contour plot of total energy versus the state variables. $\lambda = -0.5$. Time series started from $\dot{q} = 0$, and (b) $q(0) = 4.8$; (c) $q(0) = 0.5$; (d) $q(0) = 1.91$.

system response, for example, in going from the type of motion in part (c) to that in part (d).

16.2 Abstract Models

At various points in this book, including the previous section, we have seen how the natural frequency of a system may sometimes depend on the amplitude of the motion. It is straightforward to integrate the nonlinear equation of motion numerically, but it also useful to be able to obtain an analytical relation between frequency and amplitude, for example. In free vibrations of undamped systems, it may be possible to obtain an exact relation based on elliptic integrals [3], or use can be made of the conservation of total energy to facilitate a solution [4]. A variety of approximate analytical techniques have also been developed, typically applicable to moderately large (oscillatory) behavior.

Suppose we go back to the simplest link model, first considered in Section 5.2, and set the axial load level at $p = 1.2$ corresponding to the equation of motion

$$\ddot{\theta} + \omega_n^2(\theta - 1.2\sin\theta) = 0. \tag{16.1}$$

Because there is no damping, we expect trajectories to trace phase trajectories at constant values of the total energy. Previously we expanded the sine term as a Taylor series, retaining just the first term for a standard linearization. Suppose we keep the next term in the series expansion and shift the origin to the positive equilibrium position (i.e., at the bottom of the right-hand potential-energy well with $p = 1.2$, we have an equilibrium at $\theta_e = 1.02674$). The potential energy corresponding to this situation is shown in Fig. 16.2(a) as the solid curve and corresponds to the cubic

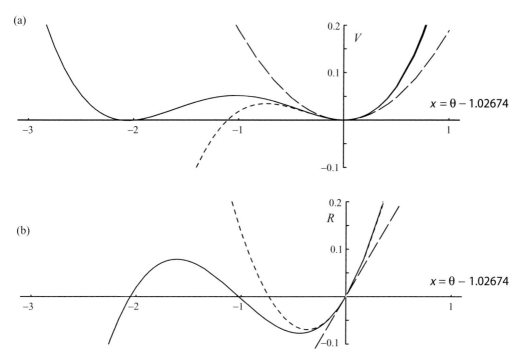

Figure 16.2. (a) Potential-energy functions for the inverted pendulum and (b) corresponding restoring force functions ($x = \theta - 1.02674$ for the specific non-truncated system).

restoring force ($R = \int V dx$) in part (b). Superimposed is the potential energy associated with linearization (long-dashed curve) together with that for a quadratic restoring force (and thus cubic potential, the short-dashed curve). There are many ways in which these curves may be fit, but the quadratic restoring force reflects the nonlinear relation between force and displacement as well as the asymmetry. We thus see how the linearization corresponds to small-amplitude motion about the equilibrium position, a distorted egg-shaped phase trajectory corresponds to motion barely contained in one of the local potential-energy wells and can be captured by the quadratic fit, and a thoroughly nonlinear cross-well behavior needs at least a cubic restoring force for global containment, or boundedness.

To consider the growth from linear to mildly nonlinear motion we can thus study an equation of the form

$$\ddot{x} + x + x^2 = 0, \tag{16.2}$$

in which x is measured from the shifted origin and the restoring force has been changed for convenience such that the unstable equilibrium position occurs at negative one. We recognize this as one of the standard forms from Chapter 3, and this system was also subject to a sudden load in Section 13.4. This system possesses a stable equilibrium at the origin and a saddle point at $x = -1$. Elliptic integral solutions are available, as well as approximate solutions based on the techniques of harmonic balance and perturbation methods. Further details of these methods can

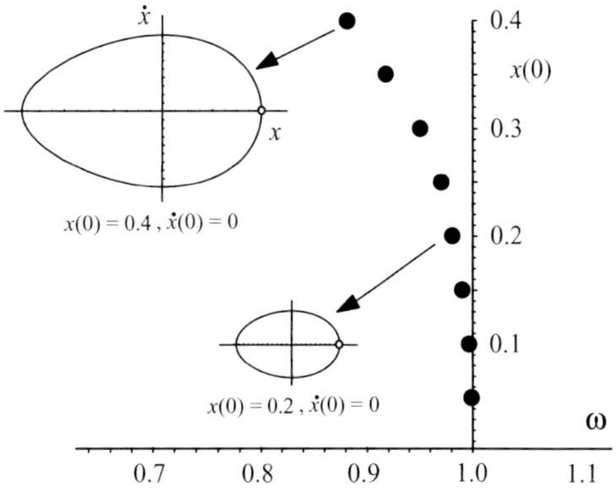

Figure 16.3. Natural frequency versus amplitude of motion.

be found in [4–8]. Numerical integration of Eq. (16.2) shows a softening nonlinearity, that is, the frequency tends to diminish with amplitude. This relation is shown in Fig. 16.3. Also shown are a couple of phase trajectories for the specific cases generated by the initial conditions $x(0) = 0.2$, $\dot{x}(0) = 0.0$, and $x(0) = 0.4$, and $\dot{x}(0) = 0.0$. The egg-shaped, asymmetric behavior of the phase trajectory started farther away from equilibrium reflects the underlying potential energy as anticipated by the form of the dashed curve in Fig. 16.2. Of course, an initial condition started *slightly* below $x(0) = -1.0$ results in a response in which the period is very long. In terms of dynamical systems theory the *separatrix* is a homoclinic orbit (i.e., an orbit that starts at the unstable equilibrium and ends there after infinite time) and separates bounded from unbounded (escaping) motion. An experimental analog of Eq. (16.2) based on the concept of a rolling point mass on a curved surface is described in Virgin [8] and Gottwald et al. [9].

The preceding section shows how the range of dynamic behavior is much broader when not confined to small amplitudes, especially when the axial load is somewhat higher than the initial buckling load. The membrane, or stretching, effect that was encountered earlier for axially constrained systems very easily leads to nonlinear vibrations.

16.3 A Mass Between Stretched Springs

Consider the analogy between a string and a point mass supported by two identical springs. One motivation for doing this is that the effects of large amplitude and of varying axial loads can be introduced without too much mathematical sophistication. Such a system is shown schematically (and in a highly deflected configuration) in Fig. 16.4. Although the springs are linear in a direction perpendicular to the SDOF, they provide a *nonlinear* restoring force in the x direction (the only allowable direction), and this increases in a disproportionate sense with displacement [10–12]. If we suppose that the natural length of the springs is less than L, that is, each was stretched by an amount d (put in tension by $T = kd$), then we have one equilibrium position at the origin. In this case, the force acting on the mass in the x

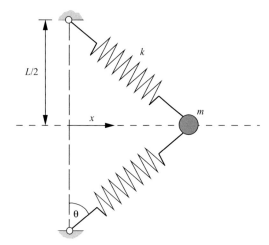

Figure 16.4. A mass supported by two stretched springs. A single-mode analog of a string.

direction (up to cubic terms in x) is

$$F(x) = -2\frac{kd}{L}x - \frac{4k(L-d)}{L^3}x^3,$$

(16.3)

and thus the equation of motion is

$$m\ddot{x} + 2\frac{kd}{L}x + \frac{4k(L-d)}{L^3}x^3 = 0.$$

(16.4)

Note that the linearization at $x = 0$ depends on d. Clearly, if the motion of the mass is small then the cubic term is negligibly small and we have a linear oscillator with natural frequency $\sqrt{(2kd)/(mL)}$, that is, the square of the natural frequency is linearly related to the tension in the springs. In fact, if the natural length of the springs is L (and therefore $d = 0$) then there is no *linear* restoring force (in the x direction). However, we also see that for moderately large oscillations (in the x direction) the natural frequency depends on the amplitude as well. This is an example of a system with a *hardening* spring stiffness in which the tension in the springs exerts a nonlinear restoring force in the direction of the motion [12].

The *effective* natural frequency can be obtained in a number of ways. We can again assume a harmonic form for the solution [5]

$$x = A\cos\omega t,$$

(16.5)

which can then be placed into Eq. (16.4), and balancing (i.e., equating the coefficients of) the cosine terms (after expanding the cubic term and ignoring the third harmonic), we find the expression

$$\bar{\omega}^2 - \bar{d} + \frac{3}{2}\bar{A}^2(1 - \bar{d}),$$

(16.6)

where

$$\bar{\omega}^2 = m\omega^2/(2k), \quad \bar{d} = d/L, \quad \bar{A} = A/L.$$

(16.7)

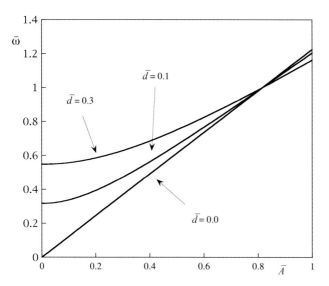

Figure 16.5. The natural frequency plotted as a function of response amplitude for a number of different pre-stress levels.

These results are plotted in Fig. 16.5. We see that if there is no initial stretching of the springs ($\bar{d} = 0$) then the natural frequency is proportional to the amplitude of the motion (within the confines of the harmonic balance approximation).

It should be borne in mind that this outcome is the result of a number of approximations. First, the restoring force was expanded as a power series up to, and including, cubic terms, and second, the harmonic balance solution procedure ignored the higher harmonic terms. Despite the fact that this might not be considered an obviously axially loaded structure, it does show how the dynamic response of a system is affected by the axial forces, and in this sense it can be compared with the large-amplitude oscillations of a stretched (continuous) string [11]. Finally, it is worth noting that in a practical context damping would tend to mitigate against a system's operating in large-amplitude motion, unless of course, the system were also subject to external forcing. This type of situation will be considered later in this chapter.

Augusti's Model Revisted. The 2DOF system considered in Section 5.7 is now revisited. We start by numerically integrating the equations of motion for the case in which there are no initial geometric imperfections and the axial load is fixed at a level of $p = 1.01134$; thus the equilibrium of the system is given by ($\theta_{1e} = 0.2598$, $\theta_{2e} = 0.0$), and the two linear natural frequencies are given by $\omega_1^2 = 0.0226$, $\omega_2^2 = 0.0535$. If initial conditions are chosen with zero initial velocity but very close to equilibrium [$\theta_1(0) = 0.25$, $\theta_2(0) = 0.001$], then the resulting motion is harmonic. This is shown in Fig. 16.6(a) as a phase projection (in terms of one of the two angles and its rate of change). The frequency spectrum in part (b) shows a spike at the natural frequency corresponding to motion in the θ_1 direction ($\omega_1^2 = 0.0226$, $f = 0.0239$ Hz) together with a small spike at twice this frequency [13]. On increasing the distance of the initial condition from equilibrium [$\theta_1(0) = 0.027$, $\theta_2(0) = 0.0001$], we

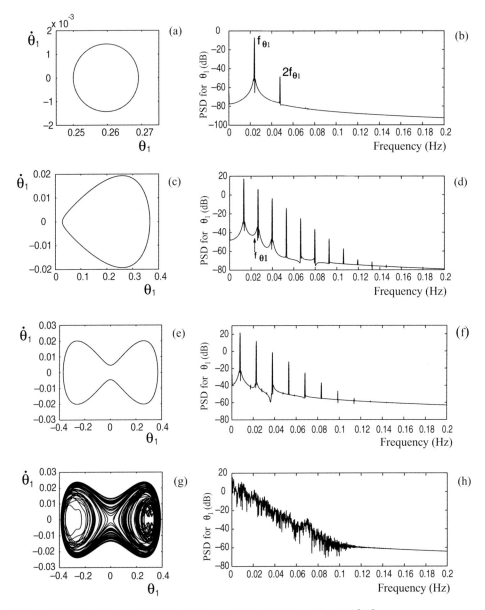

Figure 16.6. Sample numerical simulation results for the Augusti model [13].

get the results shown in parts (c) and (d). Now the asymmetry in the motion is apparent, and this is reflected in the higher harmonics in the frequency spectrum. The fundamental frequency has also shifted to a lower value, again reflecting a softening spring characteristic. Now with the initial conditions set at $\theta_1(0) = 0.37$ and $\theta_2(0) = 0.0015$, the motion has sufficient energy to traverse the potential-energy hilltop as shown in parts (e) and (f) but is still periodic as it passes around the remote equilibrium as well.

In all three of these cases, the motion in the θ_2 direction was minimal. However, now suppose the initial conditions are $\theta_1(0) = 0.027$ and $\theta_2(0) = 0.05$, that is, similar to those used in part (c) but now with a larger initial value in the θ_2 direction. This results in the behavior shown in parts (g) and (h). This motion is far from periodic as the trajectory appears to wander around and between the two stable equilibria. The frequency spectrum is now broadband with energy contributions at all frequencies. There is also now significant dynamic behavior in the θ_2 direction (not shown). This is an example of *chaotic* behavior. Despite the apparent randomlike behavior, it is characterized by some interesting underlying order. However, this behavior is occurring in a 4D phase space and is somewhat nonrepresentative because no damping is present. In fact, for this type of behavior to occur in a continuous time (as opposed to a discrete-map) system, it must be nonlinear and have at least a 3D phase space. This is the circumstance for a forced SDOF nonlinear system, and we shall come back to it a little later in this chapter.

16.4 Nonlinear Vibration of Strings

Returning now to the stretched string first considered in Section 6.2, we can restate the equation of motion but without resorting to the small-amplitude assumption (while still assuming planar motion) and thus consider

$$\frac{\partial^2 w}{\partial t^2} - c_s^2 \frac{\partial^2 w}{\partial x^2} = \frac{c_1^2}{2L} \frac{\partial^2 w}{\partial x^2} \int_0^L \left(\frac{\partial w}{\partial x}\right)^2 dx, \tag{16.8}$$

in which c_1 is the longitudinal wave speed. Given the boundary conditions $w(0, t) = w(L, t) = 0$, and the natural frequencies and mode shapes

$$\omega_n = n\pi c_s/L, \qquad \phi_n(x) = \sin(n\pi x/L) \tag{16.9}$$

from Eqs. (6.15) and (6.16), we seek a solution by using an expansion of the linear modes:

$$w(x, t) = \sum_{n=1}^{\infty} \Phi_n(t) \sin(n\pi x/L). \tag{16.10}$$

Substituting Eq. (16.10) into Eq. (16.8), we get [14]

$$\ddot{\Phi}_n + \omega_n^2 \Phi_n = -\frac{c_1^2 n^2 \pi^4}{4L^4} \Phi_n \sum_{n=1}^{\infty} m^2 \Phi_m^2. \tag{16.11}$$

Assuming a single-mode response (with appropriate initial conditions), we set $n = 1$ to get

$$\ddot{\Phi}_1 + \omega_1^2 \Phi_1 + \frac{c_1^2 \pi^4}{4L^4} \Phi_1^3 = 0, \tag{16.12}$$

which is Duffing's equation [15] with a hardening spring characteristic (because the cubic coefficient is positive).

We can solve Eq. (16.12) by using elliptic integrals, but we can easily obtain an approximate analytic solution by using the method of harmonic balance. By assuming $\Phi = A\cos\omega t$, we find an amplitude-dependent frequency

$$\omega^2 = \omega_n^2\left[1 + \frac{3}{16}\left(\frac{c_1}{c_s}\right)^2\left(\frac{\pi}{L}\right)^2 A^2\right], \tag{16.13}$$

and the anticipated stiffening effect of large-amplitude motion.

Sagging cables can also exhibit nonlinear vibrations [16], including interesting nonplanar behavior [17].

16.5 Nonlinear Vibration of Beams

In this section, we investigate the large-amplitude free oscillations of a clamped–clamped beam following the work of Yamaki [18, 19]. The beam behavior is described by the lateral deflection $w(x,t)$, and the beam has an initial axial displacement U_0, such that the governing equation of motion is given by

$$EI\frac{\partial^4 w}{\partial x^4} - \frac{EA}{L}\left[U_0 + \frac{1}{2}\int_0^L\left(\frac{\partial w}{\partial x}\right)^2 dx\right]\frac{\partial^2 w}{\partial x^2} + \rho A\frac{\partial^2 w}{\partial t^2} = 0. \tag{16.14}$$

It is relatively easy to incorporate an initial geometric imperfection into the analysis, although this is not undertaken here. Again, it is convenient to put Eq. (16.14) in nondimensional form (akin to the procedure of Section 15.3), using $\bar{w} = w/h$, $\bar{x} = x/L$, and $T = t\sqrt{EI/mL^4}$ to give

$$\bar{w}'''' - \left[u_0 + 6\int_0^1 \bar{w}'^2 d\bar{x}\right]\bar{w}'' + \ddot{\bar{w}} = 0, \tag{16.15}$$

where a prime denotes differentiation with respect to $\bar{x}$ and an overdot with respect to T, with a natural frequency $\omega = \Omega\sqrt{mL^4/EI}$ and $u_0 = (LA/I)U_0$. Note that the displacement is normalized with respect to the thickness of the beam rather than the length as used to develop Eq. (15.21), and this form of equation was also encountered in Section 7.4 before the small-deflection approximation was applied. From Eq. (16.15), we see how the effective axial load depends on the lateral deflection [20–22].

Once the boundary conditions have been specified, the solution can be assumed to take the form

$$\bar{w} = \sum_m Y_m(t)W_m(x), \quad m = 1, 2, 3, \ldots, \tag{16.16}$$

where we seek to solve for the unknown time functions $Y_m(t)$, and where $W_m(x)$ are the solutions of the underlying linear-eigenvalue problem, that is, the orthonormal modes of vibration of the beam (which satisfy the geometric boundary conditions). Use is then made of Galerkin's method to obtain a set of ordinary differential (Duffing-like) equations [23]. A few more details of this approach will be given later for the harmonically forced system.

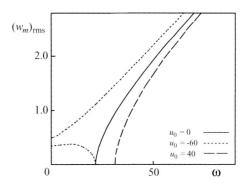

Figure 16.7. The nonlinear fundamental natural frequency as a function of response amplitude. Static component not included.

These equations can then be solved with an approximate analytical method (e.g., harmonic balance). For a single-mode analysis with small-amplitude oscillations we obtain the results discussed in Section 7.3. In Fig. 16.7 are shown some results for moderately large-amplitude oscillations. Here the response of the center of the clamped beam is plotted as an rms measure of a three-mode analysis that is due to Yamaki and Mori [18]. First consider the case with $u_0 = 0$. For linear vibrations, we have the coefficient 22.37 [24, 25] for the fundamental natural frequency, shown as the solid curve. However, this increases to a value of 60 when the response magnitude $(w_m)_{\text{rms}}$ is 2.4248, for example. When the initial end displacement is $u_0 = 40$ (stretching), we obtain a shift to higher frequencies as expected and shown as the long-dashed curve. For a typical compressive end displacement $(u_0 = -60)$, we find some interesting behavior. In this case the beam is buckled and we see the appearance of two branches (both shown as dotted curves). The first branch is close to the origin and corresponds to oscillations about the postbuckled equilibrium position with the natural frequency corresponding to vanishingly small oscillations coincidentally close to the $u_0 = 0$ case (see Fig. 7.6). However, there is also another branch and this gives finite frequencies starting from $(W_m)_{\text{rms}} \approx 0.5$. This corresponds to motion that traverses the remote equilibrium position [in the manner of Fig. 16.1(d)].

The participation of higher modes is also an important issue and will be discussed later when we deal with the nonlinear response of beams to harmonic excitation.

A Simplified Energy Approach. As we saw in earlier chapters, a useful means of estimating the fundamental frequency of vibration by Rayleigh's quotient can be based on the fact that the frequency corresponds to a stationary value in the neighborhood of a natural mode. Using an assumed displacement function, we can show that for an inexact eigenvector we get an eigenvalue that differs from the true value to the second order. This concept was introduced in Subsection 4.2.6 for discrete systems and in Subsection 4.3.2 for continuous systems in terms of the Rayleigh quotient, and we apply it here to a thin beam, including stretching effects [26].

Even without an externally applied end shortening, as we have seen, if the ends of the member are constrained in the axial direction then membrane effects can occur for deflections that are not especially large. We have just seen how strings may have this effect, and membrane effects are also important in plates, which shall be considered a little later.

We can take an approximate energy approach here for a simply supported thin beam of the type shown in Fig. 7.1 [27, 28]. The energy contributions are [29]

$$T = \frac{1}{2}m \int_0^L (\dot{w})^2 dx, \tag{16.17}$$

$$U = \frac{1}{2}EI \int_0^L (w'')^2 dx, \tag{16.18}$$

$$V_P = \frac{1}{2}P \int_0^L \frac{1}{2}(w')^2 dx, \tag{16.19}$$

in which it can be shown that the induced axial load is given by

$$P = \frac{EI}{2Lr^2} \int_0^L (w')^2 dx, \tag{16.20}$$

where, for a beam of cross-sectional area A and radius of gyration r, we have $I = Ar^2$.

Suppose the ends are fully pinned (i.e., immovable); then the lowest mode for the linear problem is simply a half-sine wave $w = Q(t) \sin(\pi x/L)$. We are primarily interested in the maximum amplitude of motion Q_m. Evaluating the energy terms and adding them, we obtain the total energy constant

$$C = \frac{\pi^4 EI}{mL^4} Q_m^2 + \frac{\pi^4 EI}{8mL^4 r^2} Q_m^4, \tag{16.21}$$

which can be used to obtain the phase trajectory as a function of the initial conditions. The equation of motion can be subjected to separation of variables and integrated (numerically) to obtain the natural period (and hence frequency) as a function of maximum amplitude. The frequency is normalized with respect to the linear natural frequency (ω_L). The result is shown in Fig. 16.8 together with a Galerkin approach (the dashed curve [30, 31]) that yields

$$\left[\frac{\omega}{\omega_L}\right]^2 = 1 + \frac{3}{16}\left(\frac{Q_m}{r}\right)^2. \tag{16.22}$$

FEA can also be used to solve this type of problem [28].

16.6 Nonlinear Vibration of a Plate

We return to consider the case in which the lateral deflections (and boundary conditions) of a plate are such that midplane stretching, or membrane response, occurs. This is somewhat similar in effect to externally applied axial loads, which will be considered subsequently. In general, these stretching effects depend on the in-plane

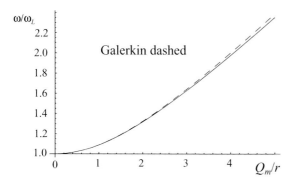

Figure 16.8. The increase in natural frequency with amplitude for a simply supported strut that is constrained from moving axially at the ends.

boundary conditions but need *not* involve amplitudes that are considered large. No exact solutions are available, but we can again make use of an approximate approach based on assumed deflection shapes (i.e., a Galerkin approach). In general we seek solutions by using the form

$$w(\xi, \eta, t) = cA(t)\Psi(x)\Phi(y),$$ (16.23)

where $\Psi(x)$ and $\Phi(y)$ are spatial mode shapes that satisfy the boundary conditions of the panel, and c is a constant. We will consider the example from Section 10.1, namely, the free-vibration behavior of a simply supported square panel, by using a single-mode approximation. Other boundary conditions will also be considered. An exact solution to the corresponding linear problem exists if two opposite sides are simply supported [32]. Importantly, we consider the case in which in-plane axial deformation is prevented along the edges.

Chu and Herrmann [33] obtained a solution for the simply supported case, and Wah [3] extended their results (using a slightly simpler approach) to include other boundary conditions. Application of the Galerkin procedure leads to an ordinary differential equation of the form (very similar to the large-deflection equation for beams)

$$\frac{d^2 A}{d\zeta^2} + \mu^4 A + 6\frac{c^2}{h^2}\lambda A^3 = 0,$$ (16.24)

where $\zeta = [t/(ab)]\sqrt{D/\rho}$, $\mu^4 = (\rho/D)a^2 b^2 p^2$, $\lambda = \eta^2/\xi^2$, where η and ξ come from the Airy stress function. Other plate characteristics include $p = \mu^2/(ab)\sqrt{D/\rho}$, the natural frequency of the linear problem, h, the plate thickness, and ρ, the mass per unit area. We recognize the equation of motion as a form of Duffing's equation with a hardening spring characteristic. A solution to the period of the response can be obtained in terms of elliptic integrals, that is,

$$\frac{T^*}{T} = \frac{2K(k)}{\pi\left[1 + 6\frac{A^2}{h^2}\frac{\lambda}{\mu^4}\right]^{1/2}},$$ (16.25)

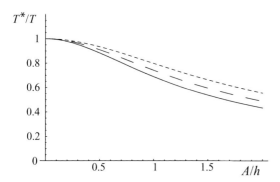

Figure 16.9. Effect of amplitude on the natural period of square plates: (a) continuous curve, SS-SS-SS-SS; (b) dashed curve, SS-C-SS-SS; (c) dotted curve, SS-C-SS-C [3].

where $K[k]$ is the complete elliptic integral of the first kind, and where

$$k = \frac{3A^2/h^2}{6A^2/h^2 + \mu^4/\lambda}, \tag{16.26}$$

and T is the period of the linear system, that is, $T = 2\pi/\mu^2$. The values of μ^4/λ depend on the boundary conditions and three types are given here (following Wah [3]): case (a), simply supported on all four sides (SS-SS-SS-SS); case (b), simply supported on three sides and clamped on the fourth (SS-C-SS-SS); and case (c) simply supported on two opposite sides and clamped on the others (SS-C-SS-C). In each case, there is also a dependency on the aspect ratio but we restrict ourselves to square plates here ($a/b = 1$). Some results are shown in Fig. 16.9. We see that with these in-plane boundary conditions the period is reduced by approximately 25% when the deflections reach the thickness of the plate (and thus the frequency increases).

16.7 Nonlinear Vibration in Cylindrical Shells

With reference to Fig. 10.14, we begin by looking at free vibrations under so-called shear diaphragm end conditions (but without end loading, i.e., $N_x = 0$). The equations of motion can be developed along similar lines to plate theory with appropriate extensions incorporating the additional geometrical effects of shells. A popular theory, mentioned in Chapter 10, is based on the Donnell–Mushtari equations [31]. Assuming a single mode of displacement in the radial direction, we have

$$w(x, y, t) = A_{mn}(t) \cos \frac{ny}{R} \sin \frac{m\pi x}{l} + \frac{n^2}{4R} A_{mn}^2(t) \sin^2 \frac{m\pi x}{l}, \tag{16.27}$$

which, on substitution into the (partial differential) equation of motion and using a Galerkin procedure, results in the nonlinear ordinary differential equation

$$\frac{d^2\zeta}{d\tau^2} + \zeta + \frac{3}{8}\epsilon\zeta\left[\zeta\frac{d^2\zeta}{d\tau^2} + \left(\frac{d\zeta}{d\tau}\right)^2\right] - \epsilon\gamma\zeta^3 + \epsilon^2\delta\zeta^5 = 0, \tag{16.28}$$

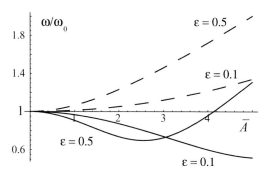

Figure 16.10. The relation between the natural frequency and amplitude of vibration for a cylindrical shell. Solid curve, $\xi = 0.5$; dashed curve, $\xi = 2.0$ [34].

where $\zeta = A_{mn}/h$, $\tau = \omega_{mn}t$, and $\epsilon = (n^2h/R)^2$. The coefficents γ and δ are complicated geometrical expressions that depend on the aspect ratio, $\xi = m\pi R/nl$, of the mode and Poisson's ratio, μ (see Leissa [31] for details).

Evensen and Fulton [34] used the method of averaging to solve Eq. (16.28). The response is given by

$$\zeta(\tau) = \bar{A}\cos\omega^*\tau, \tag{16.29}$$

in which the natural frequency depends on the amplitude in the following way:

$$\omega^{*2} = \left(\frac{\omega}{\omega_{mn}}\right) = \frac{1 - \frac{3}{4}\epsilon\gamma\bar{A}^2 + \frac{5}{8}\epsilon^2\delta\bar{A}^4}{1 + \frac{3}{16}\epsilon\bar{A}^2}. \tag{16.30}$$

The presence of cubic and quintic terms in Eq. (16.28) alerts us to the possibility of interesting behavior, and indeed, plotting Eq. (16.30) for a sample of geometries gives the results shown in Fig. 16.10. In this figure, the solid and dashed curves relate to $\xi = 0.5$ and $\xi = 2.0$, respectively, and we see that whether the nonlinear free vibrations can be classified as hardening or softening depends on ϵ and hence on geometry.

PART II: FORCED VIBRATION

16.8 Nonlinear Forced Vibration of Strings

In this section, we add external excitation (and damping) and consider the nonlinear vibrations of a number of axially loaded systems. For small-amplitude oscillations the response is typically linear. This situation was covered in Chapter 15. However, in going from free to forced vibration, we are increasing the phase space from 2D to 3D, and then with nonlinear effects we encounter a host of (possibly very complicated) behavior. Again, the steady state is of particular interest but we shall also see that the long-term dynamic behavior may no longer be independent of initial conditions. We shall start by looking at a stretched string before moving on to consider the nonlinear forced vibration of beams.

If we subject the mass-suspended model from Section 16.3, that is, a single-mode approximation of the stretched string, to harmonic excitation then the equation of motion becomes

$$m\ddot{x} + c\dot{x} + 2\frac{kd}{L}x + \frac{4k(L-d)}{L^3}x^3 = F(t),\tag{16.31}$$

in which c is a linear-viscous damping coefficient. Using the notation of Tufillaro et al. [11], we can also write Eq. (16.31) as a special case of

$$\ddot{\mathbf{r}} + \lambda\dot{\mathbf{r}} + \omega_0^2(1 + Kr^2)\mathbf{r} = \mathbf{f}(t),\tag{16.32}$$

where

$$\omega_0^2 = \frac{2kd}{mL}, \quad K = \frac{2(L-d)}{L^2 d}.\tag{16.33}$$

Here, $\mathbf{r}$ is the radial distance from the origin, and thus $r^2 = x^2 + y^2$, in which the motion in the y direction allows for the possibility of whirling motion [35]. The (unidirectional) forcing is $\mathbf{f}(t) = [A\cos(\omega t), 0]$ and in relation to the continuous string we see by analogy with Eq. (16.12) that

$$\omega_0 = \frac{c_s\pi}{L}, \quad K = \frac{\pi^2}{4\epsilon L},\tag{16.34}$$

where ϵ is the longitudinal extension of the string and $(c_s/c_1)^2 = \epsilon/L$ is assumed to be small.

Finally, we scale time and transverse deflection according to $\tau = \omega_0 t$ and $s = \mathbf{r}/(L-d)$, and, assuming planar motion [$s(t) = x(t)$], we arrive at the *forced* Duffing equation:

$$x'' + \alpha x' + (1 + \beta x^2)x = G(\gamma, \tau),\tag{16.35}$$

in which $\alpha = \lambda/\omega_0$, $\beta = 2(L-d)^3/(dL^2)$, $G = Lf/[2kd(L-d)]$, $\gamma = \omega/\omega_0$, and f is a harmonic forcing term. Again an approximate solution to this nonlinear ordinary differential equation can be obtained by a number of methods, but we shall leave consideration of specific responses until the next section when we look at the class of axially loaded structures associated with beams.

It is also worth mentioning here that for typical nonlinear string vibrations the response often involves the onset of nonplanar (or whirling) motion. In this case, the problem can be solved by use of two coupled Duffing oscillators, with the suspended-mass analogy provided by a second DOF (in a direction into the page in Fig. 16.4 [6, 11, 35]).

16.9 Nonlinear Forced Vibration of Beams

Returning to the response of a beam for which the ends are held a fixed distance apart, we now add the effect of an external (harmonic), lateral excitation, and add a small amount of linear-viscous damping. In this case, the terms $-P\cos\Omega t + C\partial w/\partial t$ are added to the left-hand side of Eq. (16.14), which corresponds to $-p\cos\omega t +$

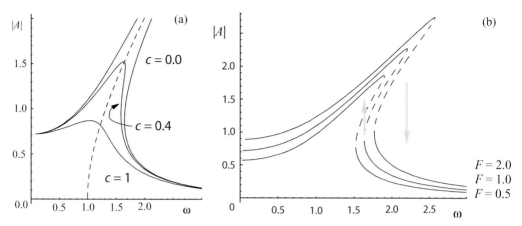

Figure 16.11. Resonance response curve based on a harmonic balance solution of Duffing's equation with $\alpha = \beta = F = 1$: (a) Various damping levels for fixed forcing magnitude ($F = 1$), (b) various forcing magnitudes for a fixed level of damping ($c = 0.2$).

$c\partial \bar{w}/\partial \bar{t}$ in nondimensional terms, in which $p = PL^4/(EIh) = 12PL^4/(Ebh^4)$ and $c = CL^2/\sqrt{mEI}$, and a rectangular beam of cross section bh has been assumed.

Solutions to the governing equation of motion are obtained in the usual way, but before looking at the detailed response we consider a simplified single-mode analysis. Using the expansion from Eq. (16.16) with $m = 1$, we again obtain the forced form of Duffing's equation,

$$\ddot{x} + c\dot{x} + \alpha x + \beta x^3 = F \sin \omega t, \tag{16.36}$$

in which α, β, and so on, depend on the various physical parameters of the beam [23]. This has the same form as Eq. (16.35), and we now consider its steady-state solution, with special attention paid to resonance phenomena. The method of harmonic balance can again be used, based on the assumed solution $x = A \cos \omega t$, to give an approximate relation between the various parameters:

$$\left[(\alpha - \omega^2)A + \frac{3}{4}\beta A^3 \right]^2 + c^2 A^2 \omega^2 = F^2. \tag{16.37}$$

In Eq. (16.37) for the unforced ($F = 0$), undamped ($c = 0$) case, we retain the hardening backbone curve (i.e., $\omega \approx 1 + (3/8)A^2 + \cdots +$), obtained earlier in this chapter for free vibration (assuming both α and β are positive unity). Furthermore, with $\beta = 0$ we get the linear resonance response curve described in Fig. 15.1.

Some typical nonlinear response curves are shown in Fig. 16.11. Here, in part (a), the dashed curve corresponds to the free- (undamped) vibration case. The solid curves correspond to different levels of damping (for a fixed level of forcing). With $c = 0$ we get the outer curve. When the damping is increased to a level of $c = 0.4$, we obtain a response that has a maximum amplitude close to $A = 1.5$ that occurs in the vicinity of $\omega = 1.6$. For the more heavily damped case ($c = 1$), the response hardly exhibits resonance at all and is similar to the linear response. This can still be considered underdamped ($\zeta = 0.5$) according to the linear description from Section 3.1.

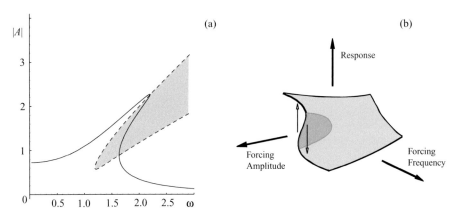

Figure 16.12. (a) Resonance response indicating stability, $\alpha = \beta = F = 1, c = 0.2$ and (b) the folded surface.

However, the lightly damped (large-amplitude) behavior is generally quite different from the linear response. Consider Fig. 16.12(b) in which $c = 0.2$, and take $F = 1$. We note that the response does not scale linearly with forcing magnitude (for a fixed level of damping). Suppose the forcing frequency were very slowly increased from an initially low value. The response would gradually grow to a maximum (close to $\omega \approx 2.2$) but then suddenly drop to the lower-amplitude branch. On subsequent slow reduction in the forcing frequency, the response would again slowly grow but with a sudden jump up (close to $\omega \approx 1.6$) before following the original path. Over the range $1.6 < \omega < 2.2$, we have two stable solutions (the curve between is unstable). Thus there is a certain dependence on initial conditions and hence path-dependent behavior. Another kind of hysteresis was encountered earlier in this book with regard to snap-through buckling. The behavior here is also characterized by sudden dynamic jumps in the response, although we now have instabilities during the evolution of oscillatory behavior rather than equilibria. However, the instability phenomena are basically the same—a saddle-node bifurcation. The link between them is in going from a time-continuous system to a time-discrete system, as discussed in Section 3.4. The hysteresis can be envisioned as passing along a folded surface [8], as shown in Fig. 16.12. The points of instability correspond to vertical tangencies in the response curve and can thus be obtained from $d\omega/dA = 0$ to give

$$\left(\alpha - \omega^2 + \frac{3}{4}\beta A^2\right)\left(\alpha - \omega^2 + \frac{9}{4}\beta A^2\right) + c^2\omega^2 = 0. \tag{16.38}$$

The roots of this equation define the region of multiple solutions [shown shaded in Fig. 16.12(b)] with one of the curves coinciding with the backbone curve when there is no damping.

In the region of hysteresis, the relative dominance of the two possible stable steady-state oscillations is reflected in their *domains of attraction*, that is, which initial conditions lead to which solution. In some ways, this is similar to a standard application of Newton's method as successive estimates of a root iterate toward a converged value that depends to an extent on the initial guess. To determine the relative dominance of the two stable solutions, it would be necessary to divide the space of initial position and initial velocity into a fine grid and then numerically solve the governing equation from these starting conditions, subsequently labeling the final outcome. Outside the region of hysteresis the final outcome does not depend on the initial conditions. We shall return to this issue a little later, and also note that there are a variety of other ways in which nonlinear dynamical systems can lose stability, especially when the nonlinearity is of the softening variety [8]. And this is where perturbations of the steady-state response lead to the variational equational (with periodic coefficients) and use of Floquet theory (see Jordan and Smith [4] for more details).

Having seen how lateral (bending) and axial (stretching) behavior, based on a single-mode analysis of the governing equation, are characterized by a bending over of the main resonance curve, consider the results of a multi-mode analysis of a clamped beam based on the work of Yamaki and Mori [18], in which they solved

$$(\bar{w} - \bar{w}_0)'''' - \left[u_0 + 6 \int_0^1 (\bar{w}'^2 - \bar{w}_0'^2) d\bar{x} x \right] \bar{w}'' + \ddot{\bar{w}} = \bar{f} \cos \omega t, \qquad (16.39)$$

where an initial imperfection $\bar{w}_0$ was included, although we just consider the perfect geometry here. Assuming a separable solution $\bar{w} = \sum_m Y_m(t) W_m(x)$ and imposing clamped boundary conditions, they arrived at a set of coupled nonlinear Duffing oscillators:

$$\ddot{Y}_n + \omega_n^2 Y_n - u_0 \sum_m \beta_{mn} Y_m + 6 \sum_k \sum_l \sum_m \beta_{kl} \beta_{mn} Y_k Y_l Y_m$$
$$= \gamma_n \bar{f} \cos \omega t, \quad k, l, m, n = 1, 2, 3, \dots, \qquad (16.40)$$

where the β terms are obtained from orthogonality. For a three-mode analysis for *symmetric* vibrations (which are most accessible experimentally), this reduces to

$$\ddot{Y}_n + \omega_n^2 Y_n - u_0(\beta_{1n} Y_1 + \beta_{3n} Y_3 + \beta_{5n} Y_5) + 6 \sum_{k=1,3,5} \sum_{l=1,3,5} \sum_{m=1,3,5} \beta_{kl} \beta_{mn} Y_k Y_l Y_m$$
$$= \gamma_n \bar{f} \cos \omega t, \quad n = 1, 3, 5. \qquad (16.41)$$

Finally, assuming the solutions are given by

$$Y_m(t) = \sum_{j=0}^3 d_m^{j\mu} \cos(j \mu \omega t), \quad \mu = 1, 1/2, 1/3, \qquad (16.42)$$

the method of harmonic balance can again be applied, and the results represented by the rms value of the response at the center of the beam:

$$(w_m)_{\text{rms}} = \frac{1}{\sqrt{2}} \left\{ \left[\sum_m d_m^1 \phi_m \left(\frac{1}{2} \right) \right]^2 \left[\sum_m d_m^2 \phi_m \left(\frac{1}{2} \right) \right]^2 \left[\sum_m d_m^3 \phi_m \left(\frac{1}{2} \right) \right]^2 \right\}^{1/2}, \quad m = 1, 3, 5.$$

$$(16.43)$$

This was the approach used to determine the free-vibration curves shown in Fig. 16.7 in which $\bar{p} = 0$. Determinig the solution involves using Newton's method to solve the 12 simultaneous cubic equations.

Figure 16.13 shows typical responses when the beam is not initially subject to any axial displacement ($u_0 = 0$) but rather axial effects are induced because of midplane stretching. The forcing magnitude is set at $\bar{f} = 200$. We see the complicated nature of the response, including various subharmonic and superharmonic waveforms. Again, regions of unstable behavior, resonant jumps, and hysteresis are apparent. In Yamaki and Mori [18], equivalent results are also shown for the buckled beam (e.g., $u_0 = -60$), which brings into play the frequencies associated with postbuckled equilibria (see Fig. 16.7) as well as snap-through (which will be considered in the next section with regard to a panel). A companion paper [19] shows remarkable experimental confirmation of this type of behavior based on the excitation of a small duralumin test specimen in which the test frame was subjected to a constant peak acceleration with a measurement of relative displacement. Some intriguing applications of this theory at very small scales are also mentioned [36].

16.10 Persistent Snap-Through Behavior in a Plate

Next, we return to the thermally loaded plate first considered in Section 10.5 in terms of free vibration and also subject to relatively mild harmonic (narrowband acoustic) excitation in Section 15.5. Now suppose that the magnitude of excitation is increased (from 130 to 155 dB). In this case sufficient energy is applied to the system so that the plate can exhibit snap-through behavior, i.e., both stable equilibria are traversed (see Fig. 10.6).

Figure 16.14 shows a typical response (in terms of a strain measurement), with the temperature change held fixed at 32 °F above ambient. At this post-critical level of thermal loading the lateral deflection at the center of the panel is of the order of the panel thickness. The excitation magnitude is 155 dB and is narrowly focused in the vicinity of 115 Hz. Figure 16.14(a) shows an experimental time series of strain-gauge data taken from a point on the panel (see the photo in Fig. 10.9) at $(\xi, \eta) = (0.583, 0.416)$, where $\xi = x/a$, $\eta = y/b$. Note the contrast with Fig. 15.15. Part (b) again shows a time-lag embedded phase projection that also contrasts with the closed orbit corresponding to periodic motion, and the frequency spectrum in part (c) is decidedly broadband. This type of intermittent snap-through behavior can also be observed in the vibrations of a buckled beam and has clear implications for fatigue [37]. Figure 16.15 shows a time series in terms of lateral deflection but based on a nine- (cosine) mode Galerkin analysis for nominally the same parameter values as those of the experimental data. More details of this analysis can be found in Murphy [12].

Another way of encountering snap-through is through a slow sweep of a system parameter. This was a technique used earlier in this book to allow the evolution of a buckling instability. Figure 16.16 shows two examples of nonstationary snap-through

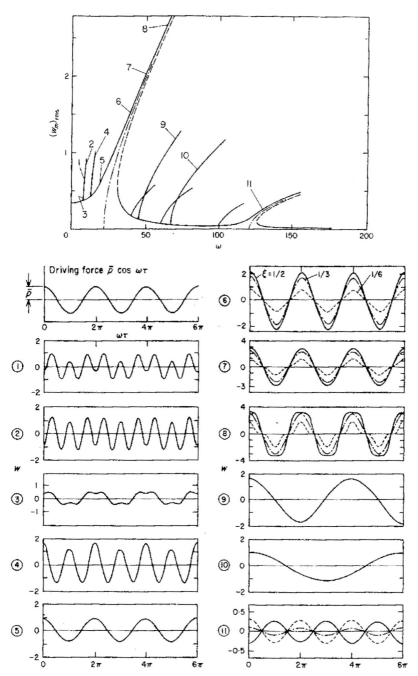

Figure 16.13. Frequency-response characteristics for a clamped beam including typical waveforms. Reproduced with permission from Elsevier [18].

behavior where the sound pressure level (SPL) is gradually ramped up and then down to produce a nonstationary transition through large-amplitude behavior. The parameters used to generate these experimental results correspond to $\Delta T/\Delta_{cr}^{f} = 1.76$ with a *baseline* forcing of 130 dB at 120 Hz. In both parts, the SPL is ramped

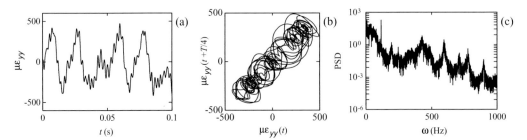

Figure 16.14. Large-amplitude snap-through behavior: (a) strain time series, (b) time-lag embedded phase projection, and (c) frequency spectrum.

from 130 dB up to 150 dB and then back down to 130 dB, at roughly the same rate. In part (a), the motion is initiated as a small-amplitude periodic oscillation that grows very gradually until it is sufficiently large that the motion escapes the confines of its local potential-energy well and snap-through occurs, that is, an erratic snapping organized around the two stable equilibria. The motion then reduces in amplitude as the SPL is ramped down and settles to small-amplitude motion about the other equilibrium configuration (in the adjacent potential-energy well). In part (b) the system is initiated with nominally the same conditions. After the burst of cross-well motion, the system, on subsequent reduction in the applied SPL, settles back to small-amplitude motion about the *original* equilibrium.

In Section 13.4, and specifically in Fig. 13.8, it was shown how the strength (in terms of forcing amplitude and frequency) of harmonic excitation could lead to a transient resonantlike condition in which the motion might *escape* the confines of its local potential-energy well. In essence, this is the situation with the snap-through behavior shown in Fig. 16.16.

Fixing the (postcritical) temperature at $\Delta T/\Delta T_{cr}$ thus determines the fundamental natural frequency. For a given frequency of excitation, the SPL is then quasi-statically increased until the first occurrence of snap-through behavior occurs, that is, the system goes beyond the potential-energy hilltop associated with the unstable (almost flat) equilibrium configuration. The frequency is then incremented, and the procedure is repeated. Figure 16.17 shows a summary of 30 such runs plotted against nondimensional frequency, clearly separating regions of parameter space into *snap-through* and *no-snap-through* regions [38].

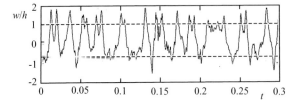

Figure 16.15. A numerical simulation based on using nine (cosine) modes [12].

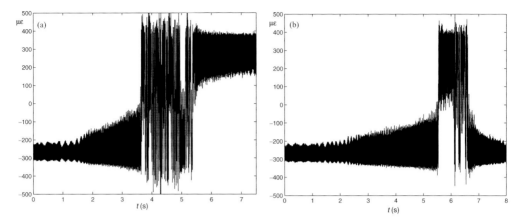

Figure 16.16. Evolving time series showing transient snap-through caused by sweeping the SPL from 130 dB → 150 dB → 130 dB. (a) The motion was initiated around the secondary equilibrium but then settled around the primary equilibrium. (b) The motion was initiated around the secondary equilibrium and returned there after a burst of transient snap-through.

The similarity between Figs. 16.17 and 13.8, at least in terms of the boundary between escape (snap-through) and no escape (no snap-through) is noted. Although there are many details associated with this behavior, we can envision this situation as the growth of a softening resonance effect that encounters instability and subsequent transition beyond the locally bounded area of phase space.

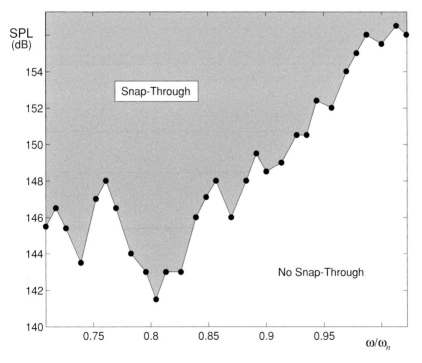

Figure 16.17. Snap-through boundary plotted in parameter space using tonal inputs $\Delta T / \Delta T_{cr} = 1.95$, $\omega_n \approx 111$ Hz at this temperature.

16.11 A Panel in Supersonic Flow

Some axially loaded slender structures are also subject to nonconservative forces, e.g., in fluid–structure interaction, or Beck's problem (see Section 7.8), and some of this behavior can lead to nonlinear oscillations [39, 40]. We shall look at a typical situation in which axial load and fluid loading conspire to produce some complex dynamic behavior. This will also provide an example of chaotic oscillations that are *not* generated by periodic forcing [41]. Only a simplified analysis will be conducted in which a number of assumptions are made in order to ease the analysis [42, 43].

Consider the thin elastic simply supported panel shown in Fig. 16.18. It has a length L, is subject to a constant axial load N_x, and has flow velocity U that produces a dynamic pressure p. We assume that there is no structural damping (although some energy dissipation will be introduced by the aerodynamic modeling). If we assume that the width of this panel (into the page) is infinite then we can use a 2D form of Von Karman's plate theory [39] to describe lateral deflections $w(x, t)$ by means of solutions of

$$\rho_m h \frac{\partial^2 w}{\partial t^2} - \left[N_x + \frac{E\alpha h}{2a} \int_0^a \left(\frac{\partial w}{\partial x} \right)^2 dx \right] \frac{\partial^2 w}{\partial x^2} + D \frac{\partial^4 w}{\partial x^4} + \hat{p} = 0. \qquad (16.44)$$

The parameters used are panel thickness h, Young's modulus E, material density ρ_m, in-plane stiffness α, D is the same as in Chapter 10, that is, $D = Eh^3/(12(1 - v^2))$, and $\hat{p}$ is the aerodynamic pressure loading.

The spatial parameters and time can be nondimensionalized following the procedure of Section 7.2, that is,

$$\bar{w} = w/h, \quad \bar{x} = x/L, \quad \tau = t\sqrt{\frac{D}{\rho_m h L^4}}, \qquad (16.45)$$

such that Eq. (16.44) can be written as

$$\frac{\partial^2 \bar{w}}{\partial \tau^2} - \left[R_x + 6\alpha(1 - v^2) \int_0^1 \left(\frac{\partial \bar{w}}{\partial \bar{x}} \right)^2 d\bar{x} \right] \frac{\partial^2 \bar{w}}{\partial \bar{x}^2} + \frac{\partial^4 \bar{w}}{\partial \bar{x}^4} + \hat{P} = 0, \qquad (16.46)$$

where we now have scaled axial load $R_x = N_x a^2/D$ and scaled dynamic pressure that can be obtained from piston theory for supersonic flow speeds, that is, Mach number $M \gg 1$. This is a linear aerodynamic theory [39, 42], and in nondimensional terms,

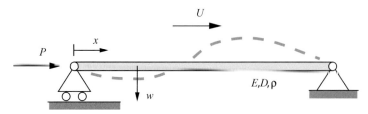

Figure 16.18. A slender panel subject to both axial loading and supersonic flow over one face.

causes a pressure on the panel of the form

$$\hat{P} = \lambda \left[\frac{\partial \bar{w}}{\partial \bar{x}} + \sqrt{\frac{\mu}{M\lambda}} \frac{\partial \bar{w}}{\partial \tau} \right], \tag{16.47}$$

where μ and λ are nondimensional flow density and flow velocity, respectively.

We next conduct a standard Galerkin procedure and assume a series expansion for the motion

$$\bar{w}(\tau, \bar{x}) = \sum_{m}^{N} q_m(\tau) \sin m\pi\bar{x}. \tag{16.48}$$

Placing this expression back in Eq. (16.46), we obtain a set of N coupled ordinary differential equations:

$$\frac{\partial^2 q_m}{\partial \tau^2} + (m\pi)^2 q_m \left[R_x + (m\pi)^2 + 3\alpha(1-v^2) \sum_i^N (i\pi)^2 q_i^2 \right] + 2\hat{P}_m = 0, \tag{16.49}$$

in which

$$\hat{P}_m = \int_0^1 \hat{P} \sin m\pi\bar{x}\, d\bar{x} \tag{16.50}$$

$$= \lambda \sum_i^N q_i \left(\frac{mi}{m^2 - i^2} \right) [1 - (-1)^{m+i}] + \sqrt{\frac{\mu\lambda}{M}} \frac{dq_m}{d\tau}. \tag{16.51}$$

Previous studies on this system have indicated that assuming two terms in the solution [Eq. (16.48)] gives the correct qualitative results, although six or eight terms should be retained for accurate quantitative results [39]. It is interesting to note that these equations may be numerically "stiff," which requires special care in their solution [44]. In relating this analysis to a more practical setting it may be necessary to incorporate some in-plane stiffness at the boundaries [45]. Taking the first two harmonics leads to the coupled equations

$$\ddot{q}_1 + U\dot{q}_1 - U^2 q_2 + q_1(1-P) + 4q_1(q_1^2 + 4q_2^2) = 0,$$
$$\ddot{q}_2 + U\dot{q}_2 + U^2 q_1 + 4q_2(4-P) + 16q_2(q_1^2 + 4q_2^2) = 0, \tag{16.52}$$

where some additional nondimensionalization has been undertaken, and we focus on the role of two nondimensional parameters: a flow velocity U and an axial load P. In general these equations can be solved only numerically, and we might anticipate quite complicated solutions depending on the parameters.

Before conducting some numerical experiments, it is instructive to consider the loss of stability from the trivial (flat) equilibrium configuration. In this case, if the axial loading is gradually introduced we encounter our familiar transition to buckling. Given the scaling in Eqs. (16.52) we see the Euler load of $P=1$ that corresponds to $R_x = -\pi^2$. But a loss of stability can also be induced by the fluid loading through increasing the parameter U. Suppose we have no axial loading, and we restrict ourselves to relatively small-amplitude oscillations. In this (linear) case we can assume that the solution consists of exponential terms of the form $r = Ae^{\lambda t}$, which leads directly to the state matrix

$$[\dot{r}] = \begin{bmatrix} -U & -1 & 0 & U^2 \\ 1 & 0 & 0 & 0 \\ 0 & -U^2 & -U & -16 \\ 0 & 0 & 1 & 0 \end{bmatrix} [r], \tag{16.53}$$

in which $r = (dq_1/dt, q_1, dq_2/dt, q_2)^T$, the eigenvalues of which determine the form of the motion. Specifically, we have the possibility that positive eigenvalues lead to a local growth of motion and hence instability. In general, in this book, the monitoring of an eigenvalue has typically been associated with real eigenvalues (which is proportional to the square of the frequency) and their sign changing from negative to positive.

However, we now consider another route to instability for the case in which the system eigenvalues are complex, and the imaginary part is associated with frequency. Setting $U = 0$ in Eq. (16.53) leads to purely imaginary eigenvalues in this state-variable format. With Euler identities this means oscillatory motion (which is what we would expect for a plate without external forcing or damping—recall that with our simplified modeling the aerodynamics is the only source of energy dissipation). However, as the flow rate (described by U) increases, the system eigenvalues change and we can track their movement in the complex plane as shown in Fig. 16.19(a). Part (b) shows the variation of the real part of the system eigenvalues with changing U, and instability occurs when the critical value $U_c = 3.59$ is reached. This is a Hopf bifurcation (and quite different from Beck's problem) and is the other way in which a system under the action of a single control parameter might generically lose its stability (along with the saddle-node). In aeroelasticity, this phenomenon is known as *flutter* (and the static instability that is due to the axial load is often called *divergence*). Both of these instabilities can also be traced back to the root structure of the linear oscillator (Fig. 3.4), and thus we identify the Hopf bifurcation with the onset of negative damping, in addition to the loss of stiffness encountered throughout this book.

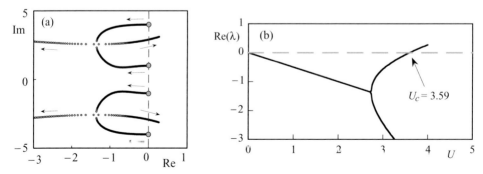

Figure 16.19. (a) A root locus of the state eigenvalues, (b) the real part of the eigenvalues plotted as a function of the control parameter U.

When the two control parameters operate together on this moderately high-order system we can expect a myriad of types of response. The following presents a summary of typical behavior. In Fig. 16.20 a variety of time-series responses are shown for the panel with no axial load (on the left) together with a similar set when a constant axial load of twice the buckling load is present, that is, $P = 2$. In all these cases, the inital conditions consist of a half unit in each of the two modal amplitudes but with zero initial velocity. As the flow velocity is increased, we have increased damping and the initial condition leads to a stable equilibrium for the flat panel. The damping increases with flow rate up to a certain extent and then starts to decrease. When $U = 4.2$, the flow speed is greater than the critical value corresponding to Fig. 16.19, and we observe the appearance of a limit cycle oscillation (LCO). The amplitude of this periodic motion grows with flow rate and follows a classic (nonlinear) Hopf bifurcation scenario [46]. The right-hand column indicates roughly similar behavior but now the stable equilibrium has shifted to a nontrival (buckled) value. However, for a certain range of flow rates (including $U = 2.2$), the panel returns to a stable equilibrium at the origin, that is, the flat panel. Subsequently, flutter occurs again, but this time at a lower flow rate than was the case for the unloaded panel.

Figure 16.21 shows a more complicated periodic response when the axial load is $P = 5$ and the flow rate is $U = 1.85$. Here the periodic, but non-simple, trajectory seems to dwell in the vicinity of the positive and negative underlying equilibria. The phase projection in part (b) shows an alternative view. This is a relatively high-order (phase-space) system (depending on how many modes we retain in the expansion), and thus we should not be surprised to learn that chaos is not uncommon in this particular system.

It is instructive to summarize these responses in term of the control parameters P and U. Figure 16.22 shows how the response of the panel depends on the combination of control parameters. We can thus locate the specific responses illustrated from Figs. 16.20 and 16.21 as the black data points, as well as the critical flow rate when no axial load is acting. It is interesting to note that it is possible for the panel to be flat and stable for certain flow rates even though the panel is buckled and stable for lower flow rates. To the left of the wide-dashed gray curve in the LCO region is where nonsimple oscillations may typically occur. The figure is the result of many numerical simulations but with the same initial conditions, and different responses are possible depending on the choice of initial conditions. Previous studies have included the effect of a static pressure differential across the panel (e.g., the top surface of an airplane wing), and a distributed stiffness in the form of a supporting elastic foundation [39].

16.12 Chaotic Behavior

In a number of situations in this final chapter, we have observed behavior that appears to be erratic in nature, for example, Figs. 16.6(g) and 16.14. Both of these randomlike responses are characterized by broadband power spectra, that is,

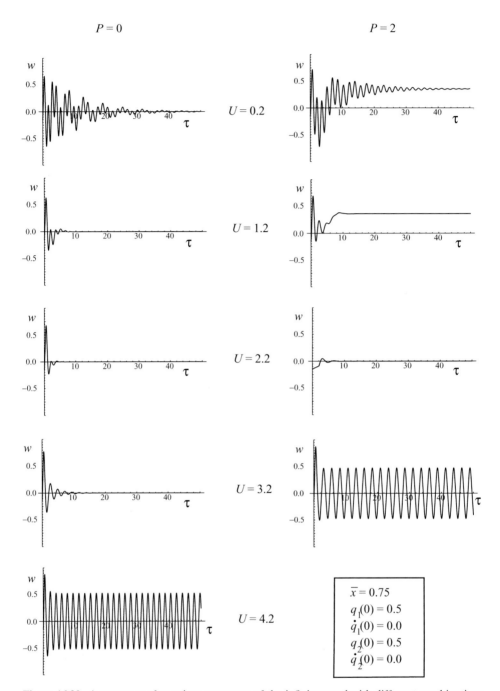

Figure 16.20. A summary of transient responses of the infinite panel with different combinations of the control parameters U and P.

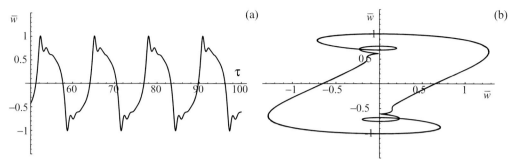

Figure 16.21. A nonsimple periodic response when $P = 5$ and $U = 1.85$.

motion in which a multitude of frequencies actively participate. However, although a similar response might be obtained from a very high-order system or one in which noise were present, these types of responses are examples of low-order deterministic *chaos*. This is a feature of nonlinear dynamical systems that has received considerable attention over recent times, and we finish this final chapter by examining an abstract model of an axially loaded structure that exhibits chaotic behavior.

We start by recalling the forced form of Duffing's equation:

$$\ddot{x} + c\dot{x} + \alpha x + \beta x^3 = F \sin \omega t. \tag{16.54}$$

Depending on the parameters, this nonlinear oscillator with its 3D phase space, that is, requiring the position, velocity, and forcing phase to uniquely determine the solution, is capable of exhibiting a vast array of behavior. Earlier in this chapter we saw how a region of hysteresis was possible, including a dependence on initial conditions. We now look at some typical responses of Eq. (16.54) in which the forcing is such that chaos occurs.

Suppose we set the parameters as $\alpha = \beta = 1, c = 0.3, \omega = 1.2$, and $F = 0.5$. Numerically integrating Eq. (16.54) leads to the results shown in Fig. 16.23. The time

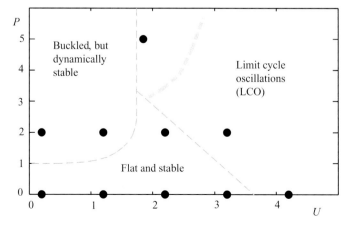

Figure 16.22. Dependence of the panel response as a function of axial load P and flow rate U.

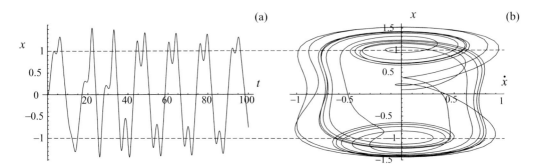

Figure 16.23. A typical chaotic response, $c = 0.3$, $\omega = 1.2$, $F = 0.5$: (a) time series; (b) phase projection.

series in part (a) shows a randomlike hopping around and between the equilibria (at ± 1). Clearly, this behavior bears a strong similarity to the intermittent (post-buckled) snap-through motion shown in Fig. 16.15. Again, a convenient alternative form for displaying this response is the phase projection (velocity versus position), and this is shown for the same data in part (b).

However, despite the apparent randomness of these responses, they are de-terministic and there is a good degree of *order* underlying this behavior. A useful technique (especially for SDOF oscillators) is the Poincaré section. This was intro-duced in Section 15.2 in terms of an analytical expression for a linear forced oscil-lator. This can still be obtained for nonlinear dynamical systems but not typically in closed form. If a time series is sampled at intervals of the forcing period, as shown in Fig. 16.24, then a periodic orbit would penetrate this section at the same loca-tion [47]. In this illustration, the periodic orbit repeats itself very two forcing cycles. However, if the response is chaotic then an interesting sequence of points is mapped because of the folding and stretching evolution of the *chaotic attractor*. This is per-sistent behavior and not associated with any initial transients that may be present. It is also in stark contrast to a damped unforced oscillator in which equilibrium repre-sents a *point attractor*, or a damped forced oscillator in which a steady-state motion is represented by a *periodic attractor*.

An example of a chaotic attractor is shown in Fig. 16.25 for the same motion as in Fig. 16.23. This fine structure shows some fractal characteristics and displays an extreme sensitivity to initial conditions (about 10,000 points after transients have been allowed to die out are plotted). Many numerical tools have been developed to shed light on chaos. We have already noted the broadband nature of the frequency spectrum [48, 49], but the complex geometry of the attractor can be described in terms of dimension [50] (and this is where certain fractal features are apparent). Initial conditions do not affect the qualitative nature of behavior in the vicinity of isolated point and periodic attractors, but we saw earlier how hysteresis allowed for some dependence of the final outcome on initial conditions. For a chaotic re-sponse, this dependence is *extreme*. There is a local exponential divergence of ad-jacent points on a trajectory and this feature is described by a positive Lyapunov

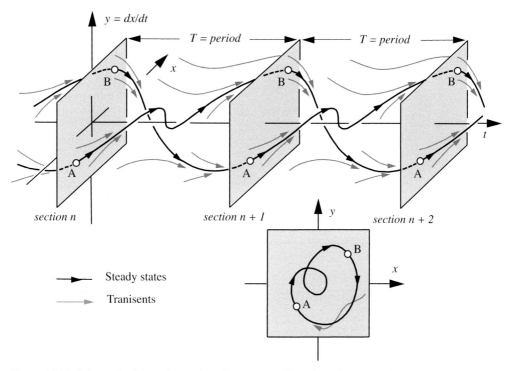

Figure 16.24. Schematic of a stroboscopic trajectory sampling: the Poincaré section. Reproduced with permission from Thompson and Stewart [47].

exponent (LE). The LEs are related to the characteristic eigenvalues from Chapter 3 [51]. There are other measures, including the autocorrelation function [8]. The reader is referred to specialized texts for more details [47, 52].

The sensitivity to initial conditions is illustrated in Fig. 16.26. Here, Duffing's equation, with the parameters set for periodic motion, is numerically integrated by a fine grid of initial conditions, and the black and white regions correspond to those initial conditions (basins of attraction) that lead to periodic motion about the $+1$

Figure 16.25. A chaotic attractor based on a numerical simulation of Duffing's equation. $c = 0.3, \alpha = -1, \beta = 1, \omega = 1.2, F = 0.5$.

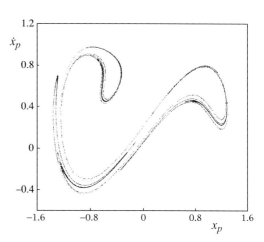

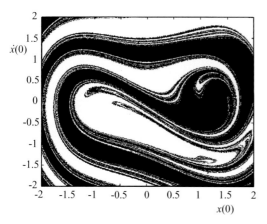

Figure 16.26. Fractal basin boundaries based on the numerical simulation of Duffing's equation from a fine grid of initial conditions. $c = 0.168$, $\alpha = -0.5$, $\beta = 0.5$, $\omega = 1$, $F = 0.15$.

and -1 equilibrium positions, respectively [53–55]. Hence there is a degree of un-certainty about the exact location of an initial condition; it may be very difficult to say which of the possible steady states the transient will be attracted to. The fractal nature of these basin boundaries remains no matter how fine the grid, and of course, in an experimental context there is always a degree of imprecision. Thus we see that sensitivity to initial conditions in terms of basin boundaries may occur even when steady-state chaos is not present, as only periodic solutions are present in Fig. 16.26.

There are some other universal features of chaos that have made its study fas-cinating. Many nonlinear structures can exhibit chaos, including, for example, the shallow arch [56, 57], although it should be mentioned that most practical designs would not typically encounter this type of thoroughly nonlinear behavior. Often the broad characteristics of chaotic attractors are quite similar. For example, Fig. 16.27 shows a typical Poincaré section taken from the model of the forced suspended mass from Section 16.3 (with $d = 0$), and thus we have a purely cubic oscillator that can be thought of as analogous to a laterally excited strut in which an axial load is maintained at its critical buckling value. It should be pointed out that to induce chaos in this particular case the forcing needs to be relatively large. Also, chaos can often occur after a sequence of period-doubling bifurcations [58] or can be mani-fest in other standard sequences including intermittency and quasi-periodicity [59]. Clearly, the study of chaos relies heavily on numerical simulation (and graphics), but some progress has been made analytically, for example, in the development of Melnikov theory to predict the onset of *strange attractors*, in which use is made of perturbation theory [60–62].

Finally, some experimental evidence for chaos in axially loaded structures is described. Figure 16.28 is a Poincaré section based on the response of a magnetoe-lastic thin strip (a buckled beam) [63], which can be modeled by Duffing's equation [64, 65]. The double-well shape of the underlying potential energy function is a form we have come to know well in this book.

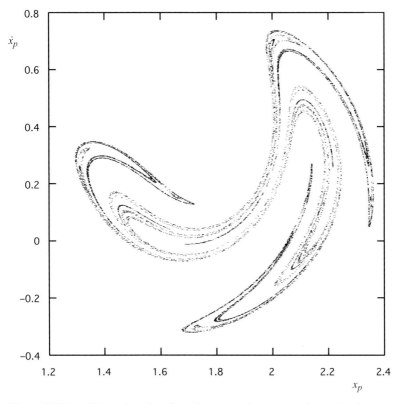

Figure 16.27. A Poincaré section from the suspended mass problem, $d = 0$.

Epilogue

We have now covered all the scenarios laid out in the point of departure at the start of this book. Vibration and buckling play out in a variety of interesting ways, ranging from linear free vibrations of axially loaded rigid-link models all the way to large-amplitude forced vibrations of axially loaded continuous systems. On this journey, we have come to rely on approximate techniques or numerical simulation as access to exact solutions has become limited. However, the dynamic behavior of structures in which there is a degree of axial loading is important and has widespread

Figure 16.28. A Poincaré section taken from an experimental nonlinear beam. Reproduced with the permission from Elsevier [63].

application. The quest for lighter aerospace structures is just one example of how the types of problem discussed in this book have an increasingly key role to play.

References

[1]   P. Hagedorn. *Nonlinear Oscillations*. Clarendon, 1981.
[2]   N. Krylov and N. Bogoliubov. *Introduction to Non-Linear Mechanics*. Princeton University Press, 1949.
[3]   T. Wah. Large amplitude flexural vibration of rectangular plates. *International Journal of Mechanical Sciences*, 5:425–38, 1963.
[4]   D.W. Jordan and P. Smith. *Nonlinear Ordinary Differential Equations*. Oxford University Press, 1999.
[5]   C. Hayashi. *Nonlinear Oscillations in Physical Systems*. Princeton University Press, 1964.
[6]   A.H. Nayfeh and D.T. Mook. *Nonlinear Oscillations*. Wiley, 1979.
[7]   J. Kevorkian and J.D. Cole. *Perturbation Methods in Applied Mathematics*. Springer-Verlag, 1981.
[8]   L.N. Virgin. *Introduction to Experimental Nonlinear Dynamics: A Case Study in Mechanical Vibration*. Cambridge University Press, 2000.
[9]   J.A. Gottwald, L.N. Virgin, and E.H. Dowell. Experimental mimicry of Duffing's equation. *Journal of Sound and Vibration*, 158:447–67, 1992.
[10] G.H. Argyris and H.-P. Mlejnek. *Dynamics of Structures*. North-Holland, 1991.
[11] N.B. Tufillaro, T. Abbott, and J. Reilly. *An Experimental Approach to Nonlinear Dynamics and Chaos*. Addison-Wesley, 1992.
[12] K.D. Murphy. *Theoretical and experimental studies in nonlinear dynamics and stability of elastic structures*. Ph.D. dissertation, Duke University, 1994.
[13] H. Chen. *Nonlinear analysis of post-buckling dynamics and higher order instabilities of flexible structures*. Ph.D. dissertation, Duke University, 2004.
[14] J.W. Miles. Resonant, nonplanar motion of a stretched string. *Journal of the Acoustical Society of America*, 75:1505–10, 1984.
[15] G. Duffing. *Erzwungene Schwingungen bei veranderlicher Eigenfrequenz*. F. Vieweg u. Sohn, 1918.
[16] H.M. Irvine. *Cable Structures*. MIT Press, 1981.
[17] O.M. O'Reilly and P.J. Holmes. Non-linear, non-planar and non-periodic vibrations of a string. *Journal of Sound and Vibration*, 153:413–35, 1992.
[18] N. Yamaki and A. Mori. Non-linear vibrations of a clamped beam with initial deflection and initial axial displacement, part I: Theory. *Journal of Sound and Vibration*, 71:333–46, 1980.
[19] N. Yamaki, K. Otomo, and A. Mori. Non-linear vibrations of a clamped beam with initial deflection and initial axial displacement, part II: Experiment. *Journal of Sound and Vibration*, 71:347–60, 1980.
[20] J.G. Eisley. Large amplitude vibration of buckled beams and rectangular plates. *AIAA Journal*, 2:2207–9, 1964.
[21] G.-B. Min and J.G. Eisley. Nonlinear vibration of buckled beams. *ASME Journal of Engineering for Industry*, 94:637–46, 1972.
[22] A.H. Nayfeh, W. Kreider, and T.J. Anderson. Investigation of natural frequencies and mode shapes of buckled beams. *AIAA Journal*, 33:1121–6, 1995.
[23] W.Y. Tseng and J. Dugundji. Nonlinear vibrations of a buckled beam under harmonic excitation. *Journal of Applied Mechanics*, 38:467–76, 1971.

[24] W.T. Thomson. *Theory of Vibration with Applications*. Prentice Hall, 1981.

[25] D.J. Inman. *Engineering Vibration*. Prentice Hall, 2000.

[26] G.V. Rao and K.K. Raju. Large amplitude free vibration of beams – an energy approach, *Zeitschrift für Angewandte Mathematik und Mechanik* 83:493–8, 2003.

[27] A.V. Srinivasan. Large amplitude free oscillations of beams and plates. *AIAA Journal*, 3:1951–3, 1965.

[28] C. Mei. Nonlinear vibration of beams by matrix displacement method. *AIAA Journal*, 10:355–7, 1972.

[29] J.M.T. Thompson and G.W. Hunt. *Elastic Instability Phenomena*. Wiley, 1984.

[30] H. Wagner. Large-amplitude free vibrations of a beam. *Journal of Applied Mechanics*, 82:887–90, 1965.

[31] A.W. Leissa. Vibration of plates. Technical Report SP–160, NASA, 1969.

[32] S.P. Timoshenko and S. Woinowsky-Krieger. *Theory of Plates and Shells,* 2nd ed. McGraw-Hill, 1968.

[33] H.-N. Chu and G. Herrmann. Influence of large amplitudes on free flexural vibrations of rectangular elastic plates. *Journal of Applied Mechanics*, 23:532–40, 1956.

[34] D.A. Evensen and R.E. Fulton. Some studies on the nonlinear dynamic response of shell-type structures. Technical Report, NASA TMX 56843, 1965.

[35] J.M. Johnson and A.K. Bajaj. Amplitude modulated and chaotic dynamics in resonant motion of strings. *Journal of Sound and Vibration*, 128:87–107, 1989.

[36] L. Nicu and C. Bergaud. Experimental and theoretical investigations on nonlinear resonances of composite buckled microbridges. *Journal of Applied Physics*, 86:5835–40, 1999.

[37] B.L. Clarkson. Review of sonic fatigue technology. Technical Report, NASA Contract Report 4587, 1994.

[38] K.D. Murphy, L.N. Virgin, and S.A. Rizzi. Experimental snap-through boundaries for acoustically excited, thermally buckled plates. *Experimental Mechanics*, 36:312–17, 1996.

[39] E.H. Dowell. *Aeroelasticity of Plates and Shells*. Noordhoff, 1975.

[40] P.J. Holmes. Bifurcations to divergence and flutter in flow-induced oscillations. *Journal of Sound and Vibration*, 53:471–503, 1977.

[41] B. van der Pol. The nonlinear theory of electric oscillations. *Proceedings of the Institute of Radio Engineers*, 22:1051–86, 1934.

[42] R.L. Bisplinghoff, H. Ashley, and R.L. Halfman. *Aeroelasticity*. Addision-Wesley, 1955.

[43] J.M.T. Thompson. *Instabilities and Catastrophes in Science and Engineering*. Wiley, 1982.

[44] C.W. Gear. *Numerical Initial Value Problems in Ordinary Differential Equations*. Prentice Hall, 1971.

[45] C.S. Ventress and E.H. Dowell. Comparison of theory and experiment for nonlinear flutter of loaded plates. *AIAA Journal*, 8:2022–30, 1970.

[46] P.J. Holmes. Nonlinear dynamics, chaos, and mechanics. *Applied Mechanics Reviews*, 43:23–39, 1990.

[47] J.M.T. Thompson and H.B. Stewart. *Nonlinear Dynamics and Chaos*, 2nd ed. Wiley, 2002.

[48] D.E. Newland. *An Introduction to Random Vibrations and Spectral Analysis*. Longman, 1984.

[49] V. Brunsden, J. Cortell, and P.J. Holmes. Power spectra of chaotic vibrations of a buckled beam. *Journal of Sound and Vibration*, 130:1–25, 1989.

[50] P. Grassberger and I. Procaccia. Measuring the strangeness of strange attractors. *Physica D*, 9:189–208, 1983.

[51] A. Wolf, J.B. Swift, H.L. Swinney, and J.A. Vastano. Determining Lyapunov exponents from a time series. *Physica D*, 16:285–317, 1985.

[52] C. Grebogi, E. Ott, and J.A. Yorke. Chaos, strange attractors, and fractal basin boundaries in nonlinear dynamics. *Science*, 238:632–8, 1987.

[53] C.S. Hsu. *Cell-to-Cell Mapping: A Method of Global Analysis for Nonlinear for Nonlinear Systems*. Springer-Verlag, 1987.

[54] E. Eschenazi, H.G. Solari, and R. Gilmore. Basins of attraction in driven dynamical systems. *Physical Review A*, 39:2609–27, 1989.

[55] H.E. Nusse and J.A. Yorke. Basins of attraction. *Science*, 271:1376–80, 1996.

[56] N. Sri Namachchivaya and M.M. Doyle. Chaotic motion of a shallow arch. In *Proceedings of the 29th AIAA/ASME/ASCE/AHS/ASC Structures, Structural Dynamics, and Materials Conference*. AIAA, New York, 1988, pp. 198–209.

[57] J.J. Thomsen. Chaotic vibrations of non-shallow arches. *Journal of Sound and Vibration*, 153:239–58, 1992.

[58] M.J. Feigenbaum. Quantitative universality for a class of nonlinear transformations. *Journal of Statistical Physics*, 19:25–32, 1978.

[59] S.H. Strogatz. *Nonlinear Dynamics and Chaos*. Addison-Wesley, 1994.

[60] V.K. Melnikov. On the stability of the center for time periodic solutions. *Transactions of the Moscow Mathematics Society*, 12:1–57, 1963.

[61] J. Guckenheimer and P.J. Holmes. *Nonlinear Oscillations, Dynamical Systems, and Bifurcations of Vector Fields*. Springer-Verlag, 1983.

[62] S. Wiggins. *An Introduction to Applied Dynamical Systems Theory and Chaos*. Springer-Verlag, 1990.

[63] F.C. Moon and P.J. Holmes. A magneto elastic strange attractor. *Journal of Sound and Vibration*, 65:275–96, 1979.

[64] P.J. Holmes and F.C. Moon. Strange attractors and chaos in nonlinear mechanics. *Journal of Applied Mechanics*, 50:1021–32, 1983.

[65] F.C. Moon. *Chaotic and Fractal Dynamics, An Introduction for Applied Scientists and Engineers*. Wiley, 1992.

Index